Robotics Goes MOOC

Bruno Siciliano

Editor

Robotics Goes MOOC

Design

 Springer

Editor
Bruno Siciliano
Department of Electrical Engineering
and Information Technology
University of Naples Federico II
Naples, Italy

ISBN 978-3-319-75822-0   ISBN 978-3-319-75823-7   (eBook)
https://doi.org/10.1007/978-3-319-75823-7

Preface

In effetti l'uomo si dimostra essere cosa divina perché dove la natura finisce di produrre
le sue spetie l'uomo quivi comincia colle cose naturali a fare coll'aiutorio d'essa natura
infinite spetie. (Leonardo da Vinci)

Technique is the essence of the human being. With the technique, human beings
can obtain for themselves what they once asked of the Gods. A push to overcome
one's own limits that is projected into the ambition to create artifacts in one's own
image and likeness: this is how the automaton was born, a machine that moved by
itself. Through the course of centuries, there were mechanical inventions of various
types: self-propelled fountains, clocks which moved by the force of water, heat,
or a particular mechanism, emulating human beings up to the recent zoomorphic or
anthropomorphic creatures. They marked a long history from the Christian and Arab
Middle Ages to the Italian Renaissance, with the ingenious designs of Leonardo da
Vinci, up to the flourishing in Europe and Asia of creations such as the Jaquet-
Droz family of androids and the karakuri-ningyo mechanical dolls of the eighteenth
century. The playful dimension of the first automatons was rooted in the frontier
technical knowledge of the time. In the industrial period, with the prevalence of the
concept of machine utility over any other function, the wonders of these automata
are to be considered as the parents of modern robots.

The term robot, of Slav origin and synonym of subordinate work, comes from
the pages of the Czech writer Karel Čapek in the drama Rossum's Universal Robots
(R.U.R.) from 1920 to indicate an anthropomorphic machine designed and built
with organic material to relieve the fatigue of humans. Twenty years later, in 1940,
the image of the robot changed becoming a mechanical artifact with the Russian
writer Isaac Asimov, yet at the level of science fiction.

In parallel to the development of robotics as a new science, we observed the
coexistence of myths and legends with the anthropomorphization of machines which
often interferes with what robotic technologies really are. To get to understand the

---

The original version of the book has been revised. A correction to this book can be found at https://
doi.org/10.1007/978-3-319-75823-7_9

technical meaning of the term robot, we can refer to the definition of robotics of the 1980s as the science that studies the intelligent connection between perception and action. The action is offered by a mechanical system equipped with a locomotion apparatus to move (wheels, tracks, limbs, crawlers) and a manipulation apparatus to act on the objects in the surrounding environment (mechanical arms, artificial hands, end-effectors, tools). Perception is entrusted to a sensory system capable of acquiring information on the mechanical system and on the environment (position and speed sensors, cameras, distance and proximity sensors, force and tactile sensors). The intelligent connection is entrusted to a control system that governs motion in relation to what happens in the environment, according to the same principle of feedback that regulates the functions of the human body. It is therefore understood how robots cannot ignore physical reality—not only minds and sensors as in artificial intelligence (AI), but also mechanical bodies.

In the 1990s, research was stimulated by the need to use robots to address the problem of human safety in hazardous environments, to increase the skills of operators, and to reduce their fatigue or by the desire to develop products for potentially large markets and designed to improve the quality of life. A common denominator of these application scenarios is the challenge of operating in poorly structured contexts, whose geometric or physical characteristics not completely known a priori require increased skills and a higher level of autonomy.

At the turn of the new millennium, robotics unveiled a wide spectrum of applications involving different scientific disciplines such as biomechanics, haptic perception, neuroscience, virtual simulation, animation, machine learning, and sensor networks. In perspective, its further development can be declined by themes and visions on the basis of four paradigms: **knowledge, design, interaction, and impact**.

The **Robotics Goes MOOC** project is organized into **four volumes**, each devoted to one of the above four paradigms. The contents go beyond those to be found in a standard textbook in robotics by enlightening the common threads on the advantages of resorting to robotic systems and their inherent potential of innovation. On the other hand, the topics are dealt with a lighter level of deepening on the research challenges, such as in the widely known Springer *Handbook of Robotics*, with the clear intent of an outreach toward new users and new communities of robotics science and technology in disparate applications. In return, the challenges of the new emerging areas are proving an abundant source of stimulation and insights for the field of robotics. It is indeed at the intersection of disciplines that the most striking advances are expected to happen.

The eight chapters of each book contain useful lab cases, applications, and examples. The main novelty of the project over the previous literature is the associated **MOOC course "Robotics & Robots"** available at ▶ sn.pub/NNHqHS. This unique complement features video lectures by some of the chapters' authors and supporting slides containing exercises, schemes, and tables summarizing the most important concepts explained in the online videos, with the goal of facilitating the comprehension of the various concepts.

In the following, the four paradigms are explained and contextualized, and the contents of each book are presented.

Knowledge

The robot "concept" was clearly established by those many creative historical realizations, such as those recalled above. Nonetheless, the emergence of the "physical" robot had to await the advent of its underlying technologies of mechanics, controls, computers, electronics, and sensors—in one word, mechatronics—during the course of the twentieth century. As always, new designs motivate new research and discoveries which, in turn, lead to enhanced solutions and thus to novel concepts. This virtuous circle over time produced that **knowledge** and understanding which gave birth to the field of robotics, properly referred to as the science and technology of robots.

To make robots and intelligent machines useful to humans, it is necessary to have a broad and tight intersection between robotics and AI. Sophisticated mathematical models are needed that enable the robot from a physical point of view, as well as intelligent algorithms capable of correlating all the information coming from the use of technologically advanced sensors with the data available from experience. It is expected that the synergy of model-based techniques with data-driven approaches will contribute to increasing the level of autonomy of robots and intelligent machines in the near future.

The first book of the Robotics Goes MOOC project starts with the journey of robotics in the introductory chapter by Khatib, who has pioneered our field of robotics and has ferried it to the third millennium. Sensing is crucial for the development of intelligent and autonomous robots, as covered in the second chapter by Nüchter et al. Model-based control is dealt with in the third chapter by Kröeger et al. along with motion planning, as well as in the fourth chapter by Villani and the fifth chapter by Chaumette to handle force and visual feedback, respectively, when interacting with the environment. Resorting to AI techniques is the focus of the last part of the book, namely, the sixth chapter by Peters et al. on learning, the seventh chapter by Beetz et al. on knowledge representation and reasoning, and the eighth chapter by Burgard et al. on graph-based SLAM.

Design

A robot's appearance and its way of interacting with humans are of fundamental importance. Until a few years ago there was a clear asymmetry between the typically excellent performance of industrial robots and their ugly and disharmonious bodies, with crude ways and potentially very dangerous movements for the human environment. A modern artifact can be as harmonious and beautiful as a complex biological machine or a work of plastic art, and thus, it should be clear how **design** plays a key role in robot technology to become a part of our everyday life and change it essentially in a responsible and beneficial manner. It is designers who shape the interface between humans and machines and, as such, they will contribute to make robots customizable and intuitively useful to inexperienced users according to a plug-and-play mode.

The new concept of robotronics as the mechatronics approach to designing advanced robots is the focus of the first chapter of the second book of the Robotics Goes MOOC project by Asfour et al. The main issues for robot manipulator design are covered in the subsequent material, namely redundant robots in the second chapter by Maciejewsky et al. and parallel robots in the third chapter by Müller, where widely adopted kinematic solutions are presented. Then, the adoption of flexibility, as opposed to the rigid mechanics paradigm, is discussed in the fourth chapter by Malzahn et al. with reference to elastic robots and in the fifth chapter by Laschi focused on soft robotics. Somewhat speculating on the previous two design solutions comes the sixth chapter by Cutkosky dealing with bioinspired robots. The last part of the book is devoted to robot locomotion, namely, the seventh chapter by Vendittelli on wheeled robots and the eighth chapter by Harada on biped humanoids.

Interaction

With the massive and pervasive diffusion of robotics technology in our society, we are heading toward a new type of AI, which we call "physical AI" at the intersection of robotics with AI, which is the science of robots and intelligent machines performing a physical action to help humans in their jobs of daily lives. Physical assistance to disabled or elderly people; reduction of risks and fatigue at work; improvement of production processes of material goods and their sustainability; safety, efficiency, and reduction of environmental impact in transportation of people and goods; and progress of diagnostic and surgical techniques are all examples of scenarios where the new InterAction Technology (IAT) is indispensable.

The **interaction** between robots and humans must be managed in a safe and reliable manner. The robot becomes an ideal assistant, like the tool used by a surgeon, a craftsman, or a skilled worker. The new generation of robots will co-exist—the cobots—with humans not only in the workplace but, gradually, in homes and communities, providing support in services, entertainment, education, health, manufacturing, and care.

As discussed above, interaction plays a crucial role in the development of modern robotic systems. Grasping, manipulation, and cooperative manipulators are covered in the first part of the third book of the Robotics Goes MOOC project, respectively, in the first chapter by Prattichizzo et al., the second chapter by Kao et al., and the third chapter by Caccavale. Specific interaction issues along with the development of digital and physical interfaces are dealt with in the fourth chapter by Marchal et al. and in the fifth chapter by Croft et al., respectively. Interaction between robot and human also means that a robot can be worn by a human as presented in the sixth chapter by Vitiello et al. A different type of interaction at a cognitive and planning level is the focus of the seventh chapter by Lima devoted to multi-robot systems and the eighth chapter by Song et al. on networked, cloud, and fog robotics, respectively.

Impact

It is often read in the media that AI and robotics are the primary cause of technology unemployment. AI and machine learning techniques are expected to take over lower-level tasks, while humans can spend more time with higher-level tasks. In perspective, it can be said that jobs requiring boring cognitive tasks or repeatable and dangerous physical tasks will be considerably shredded by automation thanks to the wide adoption of AI and robotics technology to replace humans, while jobs requiring challenging cognitive tasks or unstructured physical tasks will be suitably re-engineered with the progressive introduction of AI and robotics technology to assist humans.

From the discussion above, it should be clear that in a world populated by humans and robots, issues arise that go beyond engineering and technology due to the **impact** resulting from the use of robots in various application scenarios. The anthropization of robots cannot ignore the resolution of those ethical, legal, social, and economic (ELSE) problems that have so far slowed their spread in our society.

The final book of the Robotics Goes MOOC project enlightens the impact of using robotic technology in the main fields of application, namely, industrial robots as in the first chapter by Bischoff et al., medical robotics as in the second chapter by Dario et al., aerial robots as in the third chapter by Ollero et al., orbital robotics as in the fourth chapter by Lampariello, underwater robots in the fifth chapter by Antonelli, and rescue robots as in the sixth chapter by Murphy. The last part is devoted to the open dilemma of using and accepting robots in human co-habited environments which is addressed in the seventh chapter on social robotics by Pandey and the very final chapter by Tamburrini on the important issues raised by roboethics.

Some 100 years after the entry of the word robot into our lexicon, the challenge and at the same time the opportunity that the world of research will have to represent is related to future scenarios in which robotics will become an interactive means to help improve conditions of life. In this vision, the robot revolution can help us reaffirm the least artificial feature of our world: our humanity. Will human-friendly robots become commonplace, as so often imagined by science fiction? Only time will tell.

Acknowledgments

I am deeply thankful to the authors for their intellectual contributions and careful cross-reviews of the material contained in these four books, as well as for the video lectures and annexed material they produced for the MOOC course. I am indebted to Francesca Bonadei, Executive Editor at Springer Science+Business Media for MOOCs, and her staff in Milan. I wish to acknowledge the continuous support of Ilaria Merciai, PR and Media Manager at federica.eu, who has been committed to the project with her staff to produce the MOOC course at the University of Naples Federico II. Mario Selvaggio and Daniela Passariello from my team have also helped with the technical and communications contents of the MOOC course, respectively. Finally, a warm note of thanks goes to Fabrizio Bosco, the artist who has created the images on the covers to metaphorically illustrate the four paradigms of robotics; namely, hand trying to catch apple (knowledge), hand firmly grasping apple (design), hand dexterously manipulating apple (interaction), and hand holding bitten apple (impact).

Naples, Italy   Bruno Siciliano
January 2025

About "Robotics Goes MOOC" Books

Each chapter of the "**Robotics Goes MOOC**" books [1–4] contains citations and cross-references to content in other chapters belonging to the four books. The reader can find the cross-references as listed in the scheme below:

Book title	Chapter title	Cross-references
KNOWLEDGE [1] Siciliano B. (Ed) Robotics Goes MOOC – Knowledge, ISBN 978-33-197-4094-2, Springer, 2025		
KNO1	The Journey of Robotics	
KNO2	Sensing and Estimation	KNO5, KNO8, DES7
KNO3	Control and Motion Planning	KNO4, KNO5, KNO6, DES2, INT5, IMP1, IMP2, IMP3, IMP4, IMP5, IMP6
KNO4	Force Control	KNO2, KNO3, DES4, INT1, INT2, INT3, INT6, IMP2
KNO5	Visual Control	KNO2, KNO3
KNO6	Learning	KNO2, KNO3, KNO7
KNO7	Knowledge Representation and Reasoning	KNO6, INT5, INT7, INT8, IMP7
KNO8	Graph-Based SLAM	KNO2, KNO3
DESIGN [2] Siciliano B. (Ed) Robotics Goes MOOC – Design, ISBN 978-33-197-5822-0, Springer, 2025		
DES1	Robotronics: Robot Mechatronics	KNO2, KNO3, DES5, DES8, INT1, INT2

(continued)

Book title	Chapter title	Cross-references
DES2	Redundant Robots	KNO3, DES3, INT1
DES3	Parallel Robots	KNO3, DES2, INT2, IMP1, IMP2
DES4	Elastic Robots	KNO3, KNO4, KNO6, DES5, DES6, DES8, INT2, INT3, INT5, INT6
DES5	Soft Robotics	KNO6, DES6, INT6, IMP2, IMP5
DES6	Bioinspired Robots	DES5, DES8, INT1, IMP3, IMP5, IMP7
DES7	Wheeled Robots	KNO2, KNO3, KNO8
DES8	Humanoids	DES1, DES2, DES3, DES6
INTERACTION [3] Siciliano B. (Ed.) Robotics Goes MOOC – Interaction, ISBN 978-33-197-7269-1, Springer, 2025		
INT1	Grasping	KNO2, KNO4, DES5, INT2, INT3
INT2	Manipulation	KNO3, KNO4, DES6, INT1, INT5
INT3	Cooperative Manipulators	KNO3, KNO4, INT1, INT2
INT4	Virtual Reality and Haptics	DES1, DES5, INT1, INT5, INT6
INT5	Human–Robot Interaction	DES5, DES8, INT1, INT3, INT4, INT6, IMP6, IMP7, IMP8
INT6	Wearable Robotics	KNO4, DES1, DES4, INT4, INT5, IMP2
INT7	Multi-robot Systems	DES6, INT7, INT8
INT8	Networked, Cloud and Fog Robotics	KNO6, INT5, INT7
IMPACT [4] Siciliano B. (Ed.) Robotics Goes MOOC – Impact, ISBN 978-33-197-7266-0, Springer, 2025		
IMP1	Robotics Applications	KNO2, KNO3, KNO4, KNO5, KNO6, KNO7, KNO8, DES1, DES2, DES3, DES4, INT1, INT2, INT3, INT5, INT8, IMP2, IMP8
IMP2	Medical Robotics	KNO2, KNO3, KNO4, KNO5, KNO6, DES3, DES4, INT1, INT4, INT5, INT6, INT8, IMP7, IMP8

(continued)

Book title	Chapter title	Cross-references
IMP3	Aerial Robots	KNO2, KNO3, DES6, INT2, INT5, IMP5
IMP4	Orbital Robotics	KNO2, KNO3, KNO5, KNO6
IMP5	Underwater Robots	KNO2, KNO3, IMP3
IMP6	Rescue Robots	KNO2, KNO3, KNO8, DES5, DES7, INT5, INT7, INT8, IMP2, IMP3, IMP5, IMP8
IMP7	Social Robotics	KNO2, KNO3, KNO6, KNO7, DES6, DES8, INT1, INT2, INT3, INT4, INT5
IMP8	Roboethics	KNO6, KNO7, DES8, INT5, IMP1, IMP2, IMP5

Naples, Italy

Bruno Siciliano

References

1. B. Siciliano (ed.), Robotics Goes MOOC – Knowledge (Springer, 2025), ISBN 978-33-197-4094-2
2. B. Siciliano (ed.), Robotics Goes MOOC – Design (Springer, 2025), ISBN 978-33-197-5822-0
3. B. Siciliano (ed.), Robotics Goes MOOC – Interaction (Springer, 2025), ISBN 978-33-197-7269-1
4. B. Siciliano (ed.), Robotics Goes MOOC – Impact (Springer, 2025), ISBN 978-33-197-7266-0

About MOOC&BOOK

The MOOC&BOOK offers a unique opportunity to access a book complementary to an online course and thereby combine the quality of an academic/scholarly/research essay with the communicative power of an online educational product.

As a matter of fact, in addition to the classic book formats and other texts, we all have started to learn with many different multimedia formats—this applies to students and lecturers as well as to all those who are undergoing professional training. In order to meet this need for learning, we have integrated multimedia formats into our books and also developed an app to access the digital content in print books.

The Springer Nature More Media app (iOS, Android) offers the possibility to play videos and soon also other digital content in addition to the printed book content.

The content published here is linked to a series of MOOCs on Robotics specifically created and hosted by Federica Web Learning. You can access the related content via our app: download the SN More Media app for free, scan the link, and access directly to the online courses on your smartphone or tablet: ▶ sn.pub/ 9je5v2.

Contents

Robotronics: Robot Mechatronics 1
Tamim Asfour, Samuel Rader, Pascal Weiner, Felix Hundhausen, Julia
Starke, Cornelius Klas, Stefan Reither, Christian R. G. Dreher, Fabian
Reister, Miha Dežman, and Charlotte Marquardt
1   Introduction 1
2   Design Approach 3
3   Hardware Design 5
    3.1   Humanoid Arm Design 5
    3.2   Humanoid Hand Design 12
    3.3   Humanoid Head Design 23
4   Software Design 26
    4.1   Embedded Software Design 27
    4.2   ArmarX—Hardware Abstraction Layer 30
    4.3   ArmarX—High-Level Framework 32
5   Conclusions 33
References 34

Redundant Robots 39
Anthony A. Maciejewski and Biyun Xie
1   Introduction 40
2   Task-Oriented Kinematics 40
    2.1   Position Level Forward Kinematics 40
    2.2   Jacobian-Based Kinematics 43
3   Singular Value Decomposition 45
    3.1   Definition in Terms of the Jacobian 45
    3.2   Singularities and Measures of Dexterity 48
    3.3   Computing the Singular Value Decomposition 50
4   Differential Inverse Kinematics 53
    4.1   Pseudoinverse Solution 53
    4.2   Damped Least Squares 55
    4.3   Truncated SVD 62

4.4   A Simple Illustrative Example 63
5   Redundancy Resolution via Optimization 66
     5.1   Gradient Projection 66
     5.2   Task Priority 67
     5.3   Augmented Jacobian 70
     5.4   Extended Jacobian and Algorithmic Singularities 71
     5.5   Self-Motion Manifolds 72
6   Fault Tolerance 73
     6.1   Local Fault Tolerance 73
     6.2   Global Fault Tolerance 79
7   Summary and Conclusions 83
8   Exercises 83
References 86

Parallel Robots 89
Andreas Müller
1   Overview 89
2   Origin and History 91
     2.1   Spatial and Lower-Mobility PKM 91
     2.2   Redundantly Actuated PKM 94
     2.3   Kinematically Redundant PKM 97
3   Definitions 97
4   Notation 99
5   Topology of PKM 101
6   Kinematics of Parallel Manipulators with Simple Limbs 102
     6.1   Inverse Kinematics of the Manipulator 103
     6.2   Forward Kinematics of the Manipulator 105
     6.3   Limb Kinematics 106
     6.4   Kinematics of the Mechanism for Kinematically
             Non-redundant PKM 107
     6.5   Inverse Kinematics of the Mechanism for Kinematically
             Redundant PKM 112
7   Kinematics of PKM with Complex Limbs 116
     7.1   Modeling Complex Limbs by Functionally Equivalent Joints 116
     7.2   Modeling Complex Limbs by Resolution of Velocity
             Loop Constraints 116
     7.3   Inverse Kinematics 118
8   Kinematics of Manipulators with General Topology 120
9   Dynamics Modeling of Kinematically Non-redundant PKM
     with Simple Limbs 120
     9.1   Kinematics of Limbs Without Platform 121
     9.2   Motion Equations of a Single Limb 122
     9.3   Equations of Motion of the Platform 122
     9.4   Task Space Formulation of the Equations of Motion 123
     9.5   Equations of Motion in Terms of Actuator Coordinates 124

9.6   Forward Dynamics 126
10   Dynamics Modeling of Kinematically Redundant PKM with
     Simple Limbs 127
     10.1   Kinematics of Limbs Without Platform 127
     10.2   Task Space Formulation of the Equations of Motion 128
     10.3   Equations of Motion in Terms of Actuator Coordinates........... 129
     10.4   Forward Dynamics 130
11   Dynamics Modeling of Kinematically Non-redundant PKM
     with Complex Limbs 130
     11.1   Kinematics of Complex Limbs Without Platform 130
     11.2   Motion Equations of a Single Complex Limb 131
     11.3   Task Space Formulation of the Equations of Motion 132
     11.4   Equations of Motion in Terms of Actuator Coordinates........... 132
     11.5   Forward Dynamics 132
12   Dynamics of Manipulators with General Topology 132
13   EOM Linear in Dynamic Parameters 133
14   Model-Bases Control Schemes... 134
     14.1   Inverse Dynamics 134
     14.2   PID Control with Non-linear Feedforward......................... 137
     14.3   Computed Torque Control (CTC) 139
     14.4   Particular Issues of Redundantly Actuated PKM 141
15   Singularities 144
     15.1   Singularities of the Input-Output Formulation 144
     15.2   Singularity Identification Using Line Geometry 144
     15.3   Singularity-Free Assembly Mode Changing 145
     15.4   Kinematic Singularities of the Mechanism 145
16   Dexterity and Elastic Stiffness 147
17   Synthesis 149
     17.1   Specifics of PKM... 149
     17.2   Type Synthesis... 149
     17.3   Dimensional Synthesis 151
18   Exercises 152
19   Conclusion and Further Reading 156
References 157

Elastic Robots 167
Jörn Malzahn, Freia I. Muster, and Torsten Bertram
1   Introduction 167
     1.1   Is This Robot Elastic?... 168
     1.2   A (Not So) Formal Definition of Elastic Robots.................... 169
     1.3   Chapter Outline... 170
2   Elasticity of Objects 170
     2.1   Stress and Strain.. 171
     2.2   Elastic Moduli 172
     2.3   Material Strength and Fatigue 172

	2.4	Component Geometry	174
3		Dynamics of Elastic Elements	176
	3.1	The Single Mass-Spring-Damper System	176
	3.2	Multi-Degree-of-Freedom Systems	182
4		Linear-Elastic Joints	186
	4.1	Joint Shaft Torsion	187
	4.2	Torsion Wave Propagation	189
	4.3	Equivalent Spring Element	189
	4.4	Non-uniform Joint Shafts	190
5		Linear-Elastic Links	191
	5.1	Link Bending Equation	191
	5.2	Exact Solutions for Elastic Link Bending	194
	5.3	Approximate Solutions for Elastic Link Bending	200
6		Multi-Degree-of-Freedom Robots	202
	6.1	General Equations of Motion	203
	6.2	Elastic Joint Robots	204
	6.3	Elastic Link Robots	206
7		Design Considerations for Elastic Elements	210
	7.1	Elastic Elements as Force Sensors	210
	7.2	Elastic Elements for Impact Mitigation	211
	7.3	Bandwidth and Delays	214
	7.4	Efficiency	216
8		Variable Impedance Actuation	219
	8.1	Variable Stiffness Principles	220
	8.2	Variable Damping Principles	222
	8.3	Variable Inertia Principles	222
References			241

Soft Robotics 245

Cecilia Laschi

1		A Rationale for Soft Robotics: Embodied Intelligence and Morphological Computation	245
	1.1	Soft Robotics Definitions	246
2		How to Build Soft Robots: Soft Robotics Actuation Technologies	247
	2.1	Electro-active Polymers—EAPs	247
	2.2	Shape-Memory Materials	249
	2.3	Fluidic Actuation	250
	2.4	Tendons	252
	2.5	Stiffening	253
3		How to Make Soft Robots Move: Soft Robot Control	257
	3.1	Model-Based Approaches to Soft Robot Control	257
	3.2	Model-Free Approaches to Soft Robot Control	259
4		What Soft Robots Can Do: Soft Robot Abilities	260
	4.1	Basic Soft Robot Abilities	260

5   How to Use Soft Robots: Soft Robotics Applications 262
    5.1   Soft Robots for Personal Assistance: The Case
          of the I-SUPPORT Arm for Personal Hygiene 263
    5.2   Soft Robotics Technologies for Minimally Invasive
          Surgery: The Case of the STIFF-FLOP Endoscope 265
    5.3   Soft Robotics Technologies for Organ Simulators: The
          Case of the Biorobotic Larynx.................................... 266
6   Conclusions 267
Self-Assessment 268
References 269

Bioinspired Robot Design 273
Mark R. Cutkosky
1   Introduction 273
    1.1   Motivation and Themes 275
    1.2   Open Questions for Discussion 278
2   Design Case Studies 279
    2.1   Enhancing Robustness and Stability in Running 279
    2.2   Exploiting Interaction with Surfaces in the Environment 282
3   Fabrication Challenges and Opportunities 286
    3.1   Remaining Fabrication Challenges 288
4   Bioinspired Design Process 289
    4.1   Design Exercises 291
References 294

Wheeled Robots 299
Marilena Vendittelli
1   Chapter Overview... 299
2   Wheeled Mobility... 300
    2.1   Standard Wheels... 300
    2.2   Mecanum Wheels 301
    2.3   Spherical Wheels 302
3   From Wheels to Vehicles 303
    3.1   Configuration Space of Wheeled Robots........................... 303
    3.2   From Wheels to Vehicles Constraints 304
4   From Constraints to Feasible Motion 306
    4.1   Unicycle 307
    4.2   Car-Like 309
    4.3   Multi-Trailers.. 310
    4.4   Omnidirectional Vehicles 314
5   Structural Properties 319
    5.1   Controllability, Linearizability, Stabilizability 319
    5.2   Differential Flatness and Chained Form Transformability 322

6   Motion Planning and Control of Wheeled Robots 327
      6.1   Noholonomic Motion Planning 328
      6.2   Steering Nonholonomic Vehicles 329
      6.3   Feedback Control of Nonholonomic Vehicles 333
7   Problems 335
      7.1   Problem 1 335
      7.2   Problem 2 335
      7.3   Problem 3 335
      7.4   Problem 4 335
      7.5   Problem 5 335
      7.6   Problem 6 336
References 336

Humanoids 341
Kensuke Harada
1   Introduction 341
2   Foot/Leg Mechanism 343
      2.1   High-Stiffness Leg Design 343
      2.2   Compliant Leg-Joint Mechanism 344
      2.3   Foot Mechanism 345
3   Hand 347
      3.1   Electric Motor-Actuated Link Mechanism 348
      3.2   Tendon-Driven Under-Actuation 349
      3.3   Link Mechanism-Based Under-Actuation 352
4   Face and Head Mechanism 354
5   Full-Body Design 355
6   Conclusions 358
References 359

Correction to: Robotics Goes MOOC C1
Bruno Siciliano

About the Editor

Bruno Siciliano is professor of Control and Robotics at the University of Naples Federico II, Chair of the Scientific Council of the ICAROS Center, and Director of the PRISMA Lab at the Department of Electrical Engineering and Information Technology. He is also Honorary Professor at the University of Óbuda where he holds the Rudolf Kálmán chair. His research interests include robot manipulation and control, human–robot cooperation, and service robotics. He has co-authored/co-edited 24 books, more than 140 journal papers, and more than 310 conference papers/book chapters; his book *Robotics: Modelling, Planning and Control* is one of the most widely adopted textbooks worldwide and has been translated into Chinese, Greek, and Italian. He has delivered more than 30 keynotes and more than 150 invited lectures and seminars at institutions worldwide, and he has been the recipient of several awards, including the recent IEEE/RAS Pioneer Award in Robotics and Automation (2024). He is a Fellow of IEEE, ASME, IFAC, AAIA, and AIIA. He is Co-Editor of the Springer Tracts in Advanced Robotics (STAR) series and the Springer Proceedings in Advanced Robotics (SPAR) series, and he has served on the Editorial Boards of several journals as well as Chair or Co-Chair for numerous international conferences.

He co-edited the Springer *Handbook of Robotics*, which received the AAP PROSE Award for Excellence in Physical Sciences & Mathematics and was also the winner in the category Engineering & Technology (2009). His group has been granted more than 25 European projects, including an Advanced Grant and a Synergy Grant from the European Research Council. He has served the IEEE Robotics and Automation Society as President, as Vice-President for Technical Activities and Vice-President for Publications, as a member of the AdCom, and as Distinguished Lecturer. He has been Board Director of the European Robotics Association. Professor Siciliano is currently an IFAC Pavel J. Nowacki Distinguished Lecturer, a member of the International Foundation of Robotics Research Board, a member of the Nature Italy Board of Trustees, a member of the Advisory Board of Rovial Space, and chair of the Science and Innovation Committee of Enchanted Tools.

Robotronics: Robot Mechatronics

Tamim Asfour, Samuel Rader, Pascal Weiner, Felix Hundhausen,
Julia Starke, Cornelius Klas, Stefan Reither, Christian R. G. Dreher,
Fabian Reister, Miha Dežman, and Charlotte Marquardt

Abstract Building robotic systems requires a seamless integration of mechanics, electronics, embedded systems and software. The chapter provides insights into *the robotronics*, the robot mechatronics, for the design of integrated humanoid robot systems developed to perform tasks in real world environments. Selected use cases are used as examples for mechatronics design and control.

1 Introduction

Mechatronics unites many different fields of science and engineering in an interdisciplinary way [44], including electrical engineering, computer science, mechanical engineering and information technology. The term itself is a hybrid term, taken from mechanics, electronics and computer science and originated in Yaskawa, Japan in 1971 [11]. It extends mechanical systems with sensors and microcomputers to build components of intelligent systems. This includes design of actuators, embedded control, design and integration of sensor and communication [11]. Robot systems such as humanoid robots as shown in Fig. 1 feature complex mechatronics and require close cooperation of many engineering and scientific disciplines to design and build such systems. Among others, mechanical design specialists need to design body parts such as arms, hands, legs and heads and optimize the components to increase strength and decrease weight. In addition, electronic design specialists develop custom-designed and embedded electronic circuits for sensor data processing and control, which must be integrated in a very limited space. Further, specialized low-level software requires expertise in embedded programming to develop programs that run directly on custom electronic devices and connect different components through appropriate communication protocols. Finally, high-

T. Asfour (✉) · S. Rader · P. Weiner · F. Hundhausen · J. Starke · C. Klas · S. Reither ·
C. R. G. Dreher · F. Reister · M. Dežman · C. Marquardt
Institute for Anthropomatics and Robotics, Karlsruhe Institute of Technology, Karlsruhe,
Germany
e-mail: asfour@kit.edu

© Springer Nature Switzerland AG 2025

B. Siciliano (ed.), *Robotics Goes MOOC*,
https://doi.org/10.1007/978-3-319-75823-7_1

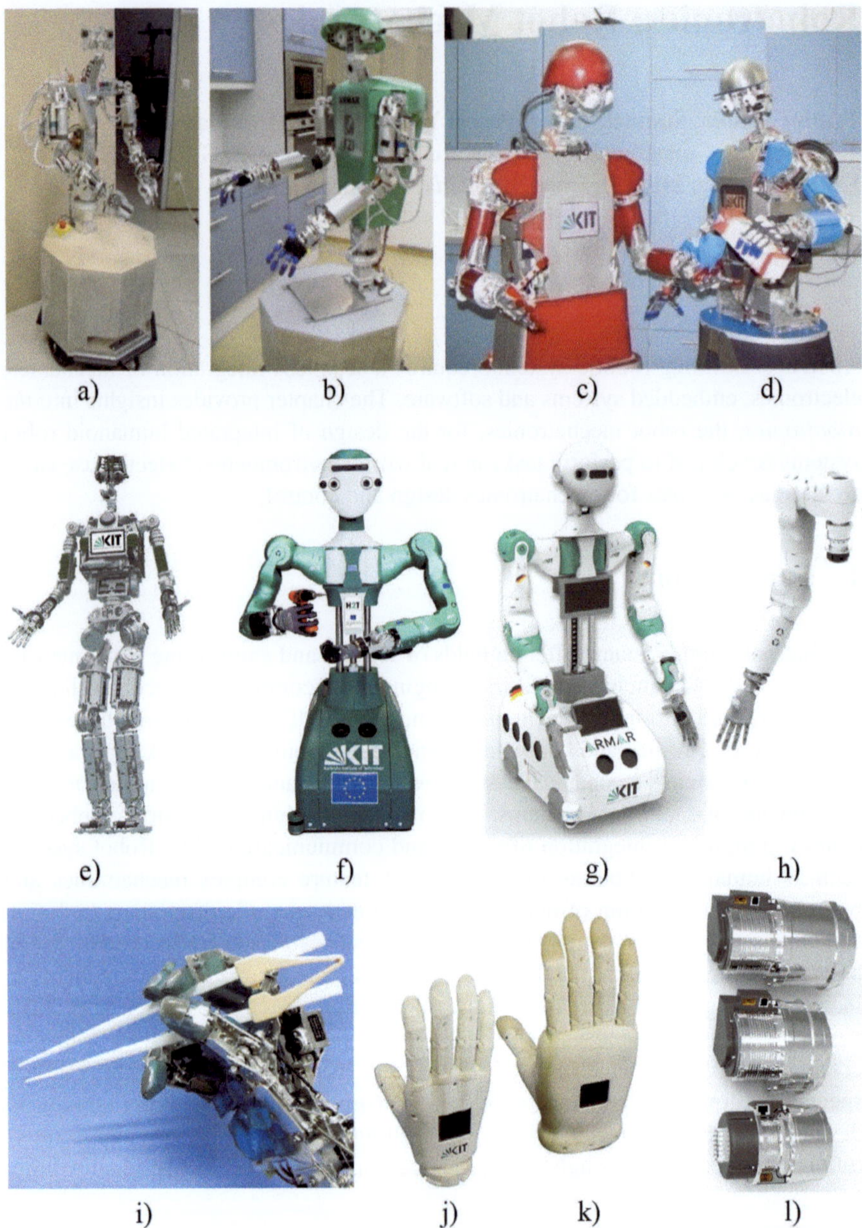

Fig. 1 Robotronics of humanoid robots and components. (**a**) ARMAR-I, (**b**) ARMAR-II, (**c**) ARMAR-IIIa, (**d**) ARMAR-IIIb, (**e**) ARMAR-4 [4], (**f**) ARMAR-6 [6], (**g**) ARMAR-DE, (**h**) ARMAR-7 arm, (**i**) TUAT/Karlsruhe Hand [19], (**j**) Female KIT Hand, (**k**) Male KIT Hand, (**l**) Sensor-actuator-controller units

level control and communication design skills are required to develop robot software architectures software architectures that integrate symbolic planning and reasoning and sensorimotor control.

As such the mechatronics of robots, the robotronics, provide the embodiment for physical, embodied intelligence. This chapter starts with a short description of methodologies of mechatronic design and continues with several examples, taken from our research on humanoid robots at the Institute for Anthropomatics and Robotics in Karlsruhe Institute of Technology.

2 Design Approach

The large design space of mechatronic systems has led to emergence of many system design methods. A design method can help the engineers from different disciplines to facilitate smooth collaboration in the development of the mechantronic systems. Many such methods emerged at the end of the twentieth century, like the waterfall model [7], the spiral model [9], and the V-model [17]. The V-model presents a general flow for the product development process, which starts with the identification of user's requirements and their decomposition into software and hardware. It continues with the fulfillment of functionality, the system integration of components and concludes with testing and user validation. The VDI guideline 2206 is a functional modeling methodology based on the V-model [22], which was developed and standardized by the VDI committee, the German engineers association. This standard provides guidance for managing the complexity of mechatronic systems.

It was developed in 2004 and builds on the good practices of previous software-based guides and represents the logical relationships of facts and interactions that form the foundation for the successful development and deployment of complex technical systems. The guide divides the design of mechatronic systems into four main phases, namely "System Design", "Domain Specific Design", "System Integration", and "Performance Check". The development process starts with the definition of requirements. Figure 2 shows our research areas at H^2T, which guide the design of our humanoid robot development.

The design of our humanoid robots is driven by the requirements related to the abilities of perception, grasping and manipulating objects as well as to skills from humans and experience. We take inspiration from the anatomy and motion behavior of the human body and design integrated components, subsystems and complete humanoid robot systems. We consider continuous integration and testing crucial to expose design flaws and weaknesses, to improve the systems performance, and to make use of gained new knowledge for the design of new robot generations. Once the requirements are identified, the "System Design" defines a cross-discipline solution concept for the system. In this phase, the overall function of the system is divided into sub-functions. Several guiding tools or design principles are employed here to define clean interfaces between software and hardware components, to improve their robustness and ease the final system integration, see Fig. 3.

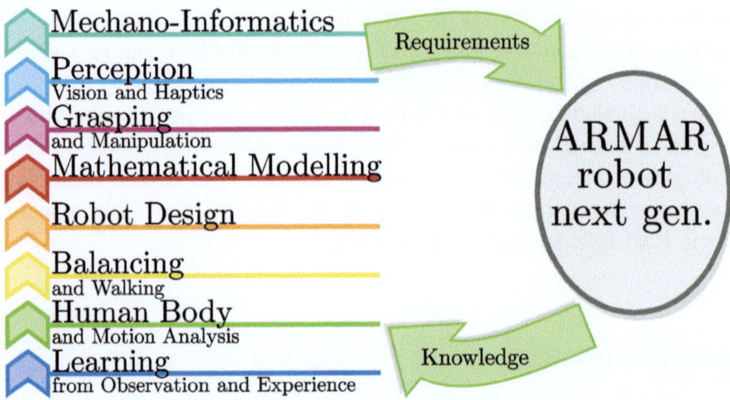

Fig. 2 Co-development of requirements, knowledge and real prototypes in the design of ARMAR robots

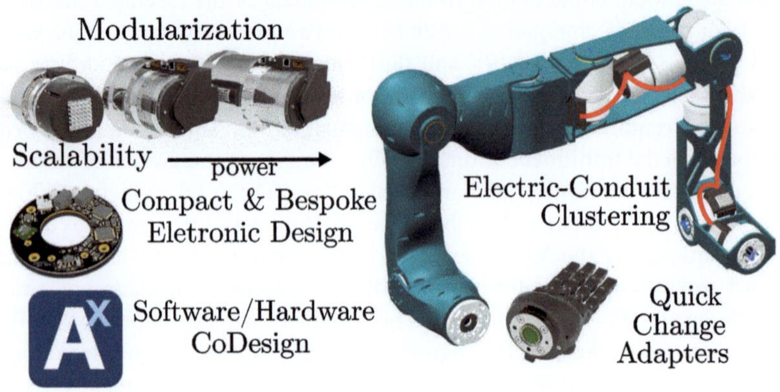

Fig. 3 Common design principles in development of ARMAR robots

An important aspect in our design approach is *modularity*, where the mechatronic system is decomposed into modules to reduce the complexity of the robot and to make developed components reusable. Such modularity also allows easier fault identification, since the underlined modules may be isolated and tested outside of the system. A good example are the sensor-actuator-controller units (SAC units), which are used for the design of the arms of ARMAR-6. These units integrate the whole mechanical drive train with sensors and electronics for control and communication [42]. These SAC units can be *scaled* to three different power levels, but they retain a similar sub-assembly structure. Careful design and calculations are necessary to ensure the performance of the SAC unit function, especially the critical functions, like emergency shut down. The development is supported by modeling and analysis of system characteristics with the help of models and computer-aided tools for simulation. *The custom electronics circuits* connect power electronics, control and communication lines into a compact PCB design to minimize the number of electronic conduits and increase their robustness. These PCB circuits have a custom shape that fits into the actuation unit resulting in a highly integrated system. The SAC units are placed throughout the robot and connected to the power and communication electrical conduits. The number of electrical conduits is minimized and *clustered* together so that only one power and one communication line connects the neighboring SAC units. The design of modular components also includes design of specialized interfaces, both on the software level and hardware level. A good example are the special *quick-change adapters* that allow quick replacement of robot hands to increase the robot's versatility. The adapter ensures a rigid connection between the hand and the arm, and connects the power and communication signal lines. The adapter also allows easy isolation of the component and testing.

The described domain specific phases are flanked by the "Modelling and analysis" of the system characteristics with the aid of models and computer-aided tools for simulation [58]. Computer Aided X (CAx) applications are used to support the product development process, generating data that is typically stored in a Mechanical Product Data Management (M-PDM) system, while electrical and electronic designers use Electrical/Electronic Engineering Solutions (EESs) to create data that is stored in Electrical PDM (E-PDM). Software designers use development solutions to create source code, which is managed through Software Configuration Management (SCM) or the Concurrent Versioning System (CVS). To reach the final product, each phase of product definition should be tested in system integration and concluded with system validation. In the specific case, the robot components are assembled together, and the joint design of software and hardware ensures that information about the status of components is readily available at all times. More information on system integration and validation is given through specific examples presented throughout the chapter.

3 Hardware Design

In order to describe the development of the hardware of mechatronic systems in detail, selected subsystems from our humanoid robots are presented in the following: Humanoid arms, heads and hands. After a general introduction to the design of these robot systems, the humanoid robots of the ARMAR family (Fig. 1) are used as examples to describe how such systems can be implemented.

3.1 Humanoid Arm Design

In contrast to industrial robot arms, humanoid robot arms are designed to be used for a wide variety of tasks, often in cooperation with humans. This results in higher demands on the robots kinematic structure as well as on the sensor setup, which must also ensure the safety of humans. This section describes a possible development procedure using the ARMAR robots as an example: After defining the workspace, the appropriate actuators and sensors are selected. Then the mechanical arm structure is described. Finally, the development of a human-like wrist follows, which is particularly challenging.

3.1.1 Workspace Consideration

Joint Configuration The most important function of a robot arm used to physically manipulate its environment is to bring an end-effector into a desired pose in the workspace. A pose describes position and orientation in R^3. To bring the end-effector into an arbitrary pose in the workspace, at least 6 degrees of freedom (DoF) are required, 3 of which are for the position and 3 for the orientation. Since industrial robots are typically used for the same task all the time, they can be designed with 6 or fewer number of DoF. However, in order to use humanoid robots for a wide variety of mobile manipulation tasks similar to humans, they are often designed with 7 or more DoF. Such redundant robot arms fulfill additional constraints given by the task such as collision avoidance or the generation of human-like movements.

ARMAR-III therefore has arms with 7 DoF each. Each arm consists of one spherical joint (3 DoF) in the shoulder and one in the wrist, and a rotational joint in between to allow flexion/extension of the elbow. This results in a large workspace that allows the execution of versatile manipulation tasks in a kitchen environments. To further increase the workspace, the arms of ARMAR-4 and ARMAR-6 each have an additional inner shoulder joint, resulting in 8 DoF arms. The advantage of these additional joints was investigated early in the design phase through simulations of reachability and manipulability [51]. Thus, with ARMAR-6, the bimanual workspace increases from 1.8 to 4.9 m^3 due to an additional inner shoulder joint on the left and right side.

When defining a joint configuration, kinematic singularities should also be avoided. It is possible to avoid singularities by adding displacements, for example between the upper arm axis and the extension/flexion joint of the elbow. In ARMAR-6, the elbow joint was moved forward by 55 mm for this purpose. In addition to avoiding singularities, this also increases the maximum angle for elbow flexion rotation.

Arm Proportions Apart from the joint configuration and the joint limits, the lengths of the individual segments of the kinematic chain influence the working space. The first step should be to determine the required reach of the arm. It will ultimately determine how strong the first drives have to be and how heavy the entire arm will be. While the arms of ARMAR-III and ARMAR-4 have human proportions, the proportions of ARMAR-6 are much larger. To ensure that the robot is able to manipulate objects at a height of 2.3 m and at the same time pick up objects from the ground, an arm with a length of about 1000 mm from the center of the shoulder to the Tool Center Point (TCP) was developed. The Tool Center Point of the arms is located in the middle of the palm. With an inner shoulder joint, it even has a reach of 1300 mm. In combination with a prismatic joint in the torso that can be moved by 400 mm, the robot is thus able to fulfill both conditions. The arm length of ARMAR-6 significantly exceeds the arm length of ARMAR-4, which measures 610 mm from shoulder to TCP and 810 mm including the inner shoulder joint. The segment lengths of the upper arm and forearm and the distance between the wrist and TCP for ARMAR-6 are based on data at the 95th percentile male. They are scaled up by a factor of 1.3.

3.1.2 Electromechanical Key Components

Actuation and Power Transmission Apart from the kinematics and the resulting workspace, the maximum payload is one of the most important requirements for a robot arm. Based on this minimum load specification, the estimated masses and inertia of the arm segments, and the desired angular accelerations, the necessary torques in the individual joints can be calculated. Since motors with the desired joint torque are usually too large and heavy, they are usually translated by gear mechanisms. A wide variety of gears can be used in robot arms. Planetary gears and harmonic drive gears can usually be attached directly to the motor. The motor axis corresponds to the output axis of the gear unit. Other gear mechanisms such as spur gears, worm gears, belt drives, wire rope drives, linear drives and parallel kinematics allow the motor axis not to correspond to the joint axis. The motor can then be spatially separated from the arm joint. This principle has been used several times in the ARMAR-III arms.

The power transmission in the shoulder joints of ARMAR-III (Fig. 4) is realized by a combination of belt drives and worm gears. The elbow is cable-driven and equipped with a complex cable roll mechanism for the transmission of mechanical

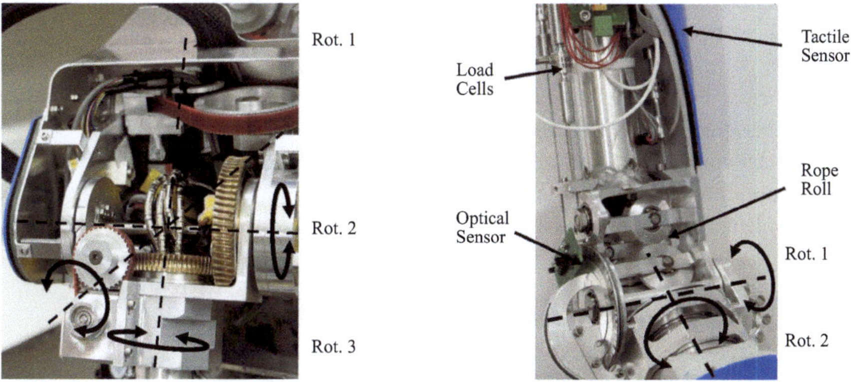

Fig. 4 Shoulder (left) and elbow (right) of ARMAR-III [1]

energy to the following joint. The elbow motor is located in the torso of the robot. Both gear mechanisms allow the motors to be placed more proximal, i.e., closer to the body. This makes the arms significantly lighter. In addition, the arm proportions are not dependent on the size of the motors.

In contrast to ARMAR-III, the arms of ARMAR-4 and ARMAR-6 use slim brushless direct current motors (BLDC motors) and harmonic drive gears. This makes it possible to accommodate the entire drive train in the joint axis, which was used consistently in both robots up to the forearm. By concentrating most electromechatronic components in the joints, the structure between the joints as well as the cabling can be significantly simplified. This can ultimately lead to better robustness.

Sensors Besides actuators, sensors are also required to control the robot arm. Position encoders are required as standard. In addition to an incremental or hall encoder on the motor shaft, modern robot arms also have an absolute encoder on the output of the gear unit. While incremental encoders only measure the relative rotation of the motor to its initial position in the form of ticks, the absolute encoder can determine the current joint position at any time.

For robots used for physical human-robot interaction, torque control is critical to ensure humans safety. To achieve this, torque sensing is needed. Torque can be measured in a variety of ways. A very accurate sensor is obtained by applying shafts or spoke wheels with strain gauges. Alternatively, it is also possible to determine the torque using two high-precision absolute encoders, placed at the beginning and end of the output shaft. The difference between the two measured angles corresponds to the twist of the shaft, from which the torque can be calculated. Finally, the motor current can also be used to roughly determine the approximate torque. Since sensor data on forces and torques are particularly helpful when gripping objects, force-torque sensors in the wrists of robots are often used in addition to torque sensors in the joints. ARMAR-III, ARMAR-4 and ARMAR-6 all have a 6D force-torque sensor in the wrist, which allows forces and torques to be measured in all spatial directions and orientations. Other sensors frequently used in robotic arms are temperature sensors, which are used to monitor the heat in the arm and to initiate measures at critical threshold values to protect heat-sensitive components. Furthermore, temperature measurements can be used to compensate for temperature drift of other sensors. In robot arms, all conceivable other sensors can be accommodated. Another example are inertial measurement units (IMU), which combine several inertial sensors such as accelerometers and angular rate sensors. But also cameras, distance sensors and pressure sensors are an option.

3.1.3 Sensor-Actuator-Controller Units

As shown by the arm of ARMAR-III (Fig. 4), actuators, sensors and motor controllers were often placed separately and scattered throughout the robot arm. While this has the advantage of utilizing every space in the arm, it makes design, assembly

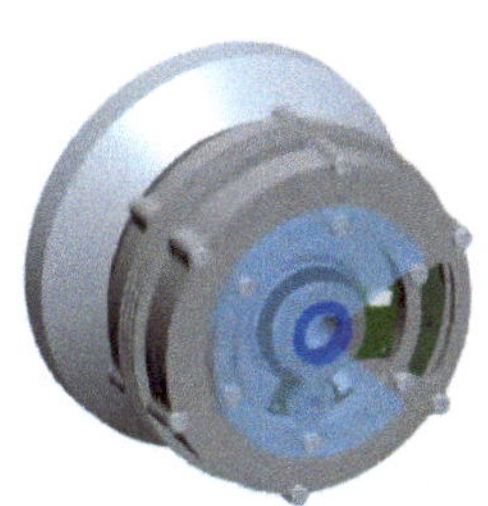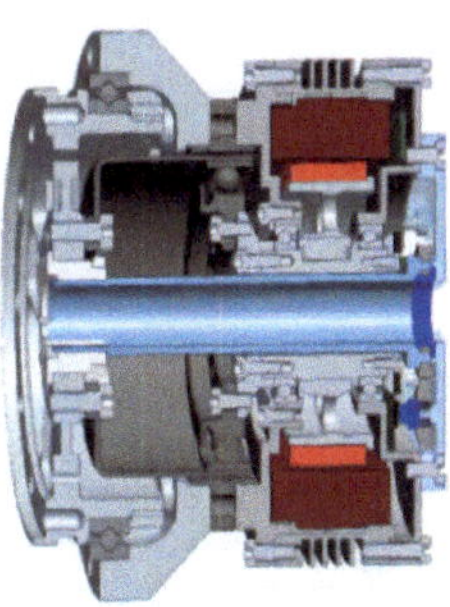

Fig. 5 Sensor-actuator unit of ARMAR-4 integrating motor, gearbox, incremental and absolute position sensors, torque and temperature sensors [4]

and maintenance extremely complex. The many distributed interfaces make it difficult to build a robust system, especially with regard to cabling. Furthermore, this rarely allows arm designs to be quickly reusable.

As a result, sensor-actuator units (SA unit) that integrate the drive train and sensors in one compact module with a well-defined interfaces have become increasingly popular in recent years. Other names for such modules are drive unit or joint unit. Figure 5 shows a sensor-actuator unit as used in the ARMAR-4 arm and leg joints. They integrate a high performance BLDC motor, a high-ratio harmonic drive gear, an incremental encoder, an absolute encode, a torque sensor and temperature sensors. However, the motor controllers are placed outside the units, so there are several cables to be routed between the SA unit and the motor controller.

Therefore, the joint units for the arms of ARMAR-6 go even further. They integrate all important mechatronic components for actuation and control of individual joints into one module, the Sensor-actuator-controller units (SAC units), which is shown in Fig. 6. In addition to the drive train and the sensors of the ARMAR-4 joint units, they also contain a motor controller as well as electronics for control and communication via EtherCAT, a high-speed Ethernet-based fieldbus system that allows for periodically sampling of all sensors at 1 kHz. Other additional components include an IMU on the sensor PCB of the SAC unit and a slip ring at its center. Slip rings are cable connections that can be rotated infinitely by sliding contacts consisting of gold rings and brushes. Slip rings allow all joints with SAC units to be rotated continuously without breaking the cables. Therefore, the upper or lower arm of ARMAR-6 can also be rotated infinitely and have no joint angle limits. In addition, slip rings make the cabling much more robust.

The slip ring is a good example of the fact that in the development of highly integrated mechatronic components, electrical and mechanical design cannot be done separately. As shown in Fig. 7, electronic requirements define which slip rings are to be considered. How many cables are needed? How much current flows through the cables? Do shielded data cables have to be used? Based on these cabling requirements the size of the slip ring is different. Hence, the construction space and the fixation of the slip ring influence mechanical parts. The slip ring is of course only one component of the SAC unit. Since there are also many other dependencies between the mechanical and electronic components, it is rather a

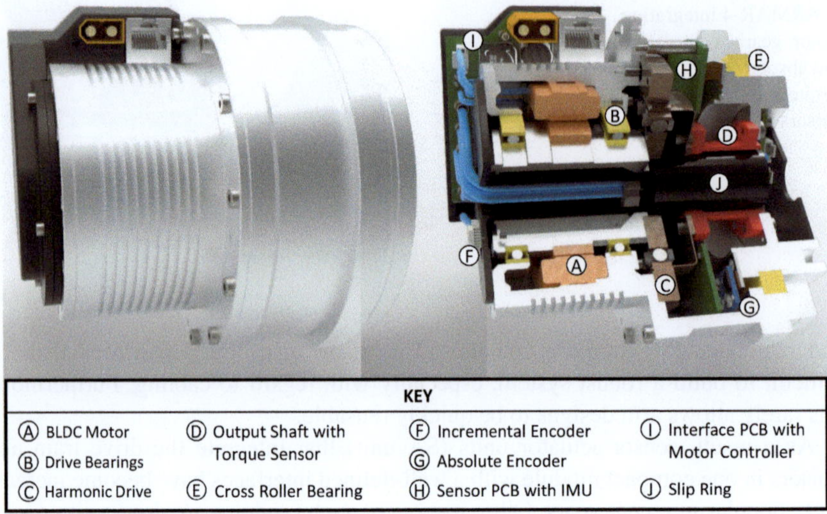

Fig. 6 Sensor-actuator-controller unit (SAC unit) of ARMAR-6

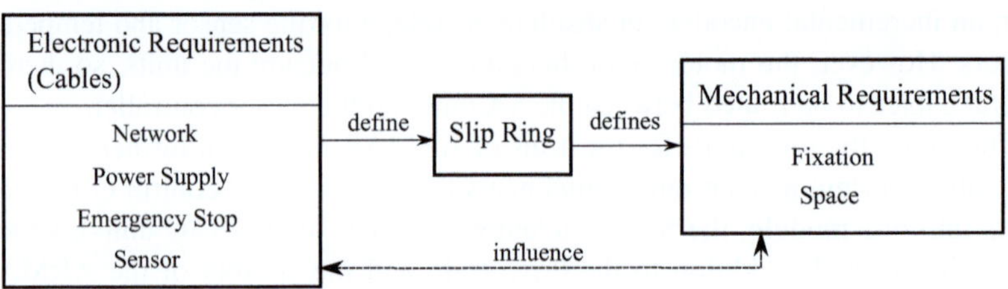

Fig. 7 Dependencies between different mechatronic parts of a SAC unit with the slip ring as an example [42]

network of dependencies. In order to nevertheless create a robust, highly integrated mechatronic system, care was taken early to consider the electronics in the CAD design and vice versa. All cables with their bending radii were modeled in CAD and it was checked at an early stage whether there was sufficient space for the electronic components on the PCB.

In parallel to the CAD design, the electronic setup of the SAC units was developed (Fig. 8). In total, there are 6 circuit boards in the SAC unit, 5 of which were self-designed to improve the integration of the electronics through an optimal form factor. The connection between two SAC units electronically consists of only 3 cable connections: Supply voltage, a network line for EtherCAT and optionally an emergency line. Multiple SAC units can be easily daisy chained together. This focus on a minimum of well-defined interfaces was also realized mechanically: In addition to two screw flanges, the units can also be mounted by a clam ring connection. To meet the different, installation space, torque and speed requirements of the

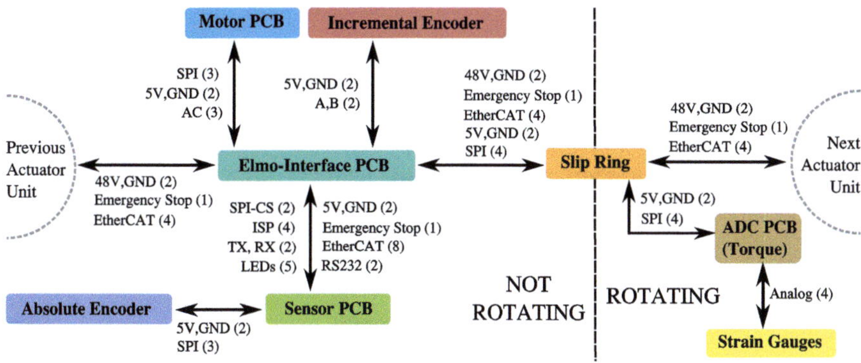

Fig. 8 Electronic setup of a SAC unit: PCB and cabling overview [42]

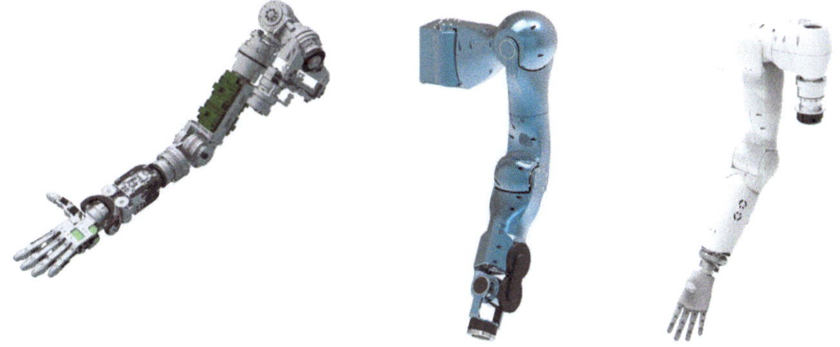

Fig. 9 Arm designs of humanoid robots: In contrast to the arm of ARMAR-4 [4] (left) that uses a frame construction, the arms of ARMAR-6 (middle) and ARMAR-7 (right) use an exoskeleton design

individual arm joints, SAC units were developed in 3 sizes providing a maximum torque of 64 N m (wrist), 123 N m (elbow) and 176 N m (shoulder). With these units, each arm of ARMAR-6 has a payload of 11 kg at long range.

3.1.4 Arm Structure

The next step is to design the arm structure. There exist two basic concepts for arm structures: Frame construction and exoskeleton design (Fig. 9). In a frame construction, very simple flat parts or profiles are screwed together (Fig. 9, left). They are easy to manufacture and therefore low-cost. Many parts can be reused in the sense of a modular system and changes can be made quickly. Thus, additional sensors and other components can also be easily attached later. To cover and protect cables and electronics, additional cover parts can be attached to the frame structure.

In contrast, exoskeleton designs use complex shell parts that serve both as load bearing structure and as covering parts. This makes the parts difficult to manufacture and expensive, but in addition to saving on covering parts, this offers several advantages. Overall, there is more freedom in the design, which makes the structure close to the joint units less bulky and simplifies higher joint limits. Furthermore, the use of a hollow structure and the resulting moment of inertia is optimal for the different load cases, which a robot arm has to withstand. Hence, lightweight design is supported. And finally, such design makes assembly and maintenance easier and lead to increased robustness, in particular with respect to cabling. For the dual arm systems of ARMAR-6 and ARMAR-7 in Fig. 9, the exoskeleton design is realized consequently thanks to the modularity of the SAC units. However, using complex power transmission mechanisms as they are used in ARMAR-III would not allow such implementation.

3.2   Humanoid Hand Design

The human hand is a complex and versatile system that allows us to intuitively grasp and manipulate a wide range of objects with different shapes and functionalities. From a kinematic point of view, the hand is a very complex system with 27 bones, 29 muscles and 21 DoF. Such complex kinematics endows our hands with versatility and dexterity in both powerful actions as well a fine-granular manipulation tasks. When building humanoid robotic hands however, the kinematic complexity poses great challenges for mechatronic integration and control. The actuation of the many degrees of freedom of humanoid hands requires high integration and smart actuator design within the tight space constraints of the human palm. In addition, the closing motion of the individual fingers and joints requires a well coordinated control to accomplish a stable grasp tailored to the object's shape and the desired manipulation goal.

Findings from neuroscience indicate that humans do not control every joint of the hand individually. Instead, correlations can be seen within the hand's DoF. These correlations can be partially explained by the mechanical structure of the forearm muscles, which are connected to several joints by the hand's tendons [35], and are partially caused by correlated muscle activation originated in the human brain [2]. Thereby, the posture of the human hand can be described by a significantly lower number of parameters—the hand synergies. Overall, more than 80% of the information transmitted by the joint angle configuration of a grasp posture can be described by two to three synergies [45].

An implementation of the hand synergies in hardware allows to accommodate both the high control complexity as well as the restricted integration space. The concept of underactuation reduces the number of *degrees of actuation* (DoA) by driving several DoF with a single motor. Thereby, an underactuation mechanism allows the implementation of hand synergies into the mechatronics of a humanoid robotic hand. In the following, a wide range of hand actuation strategies will be discussed including hydraulic actuators, underactuation based on electric motors as well as entirely soft kinematic structures. In addition, multimodal sensing and on-board computation for hand control will be focused.

3.2.1 Actuation Principles and Mechanisms

Most robotic hands are driven by electric motors or fluid actuators. In the following, both technologies are presented and it is explained how their power is converted and transmitted to actuate the various hand joints. Furthermore, softness in hands is discussed in more detail, through which grasping objects can be enhanced.

Fluid Hands Originating in biomimetics, *flexible fluidic actuators* (FFAs) provide a suitable operating principle for compliant actuation. They are also often referred to as artificial (fluidic) muscles and provide a high power-to-weight ratio. The operating principle is based on the transmission of potential energy from the pressurized fluid within a flexible shell into mechanical force. While pneumatic actuators already come with inherent compliance due to the compressibility of gases, hydraulic actuators achieve compliance only with additionally integrated compliant structures [21]. In humanoid hands, the compliance of FFAs allows for adaptive grasping of several different objects [56].

In FFAs, torque can be directly generated from compression without the need for additional transmission elements. However, to be able to control the direction of the generate torque, as well as increase its magnitude and the range of motion, a guiding structure is required. This guiding element is essential for overall performance of the actuation [21]. In fluid driven hands the (mechanical) guiding structure generally comes with anthropomorphic features. Existing prototypes consist of a palm frame with attached artificial fingers. Single mechanical phalanges connected by artificial actuated soft joints form the fingers [31, 56].

Miniaturized hydraulic systems have been developed to be included into the palm of a humanoid hand without the need for external components except a power supply. Such a system generally consists of a hydraulic pump, a fluid reservoir, electric valves and an electronic unit which can be integrated into the palm of the hand. Choosing a suitable pump is mainly based on the size limitations. Gear pumps offer a feasible compromise regarding size, weight, efficiency and complexity. The FFAs, reinforced flexible bellows attached via solid fittings to the lever of the joint, are integrated directly into the joints. To ensure a stable grasp, a correct choice of hydraulic valves is significant. As the valves control the fluid supply of the FFAs, only a complete closures and thus interruption of the fluid supply allow for specific positioning of the fingers during grasping. The modular structure of all components allows for independent actuation and positioning of the joints within the fingers [31].

Figure 10 shows an exemplary five-fingered pneumatic hand. It is a hybrid concept combining an anthropomorphic humanoid hand with a robotic gripper. While the rigid structure displays a more robotic appearance, the soft, inflatable joints in palm and fingers give it a humanoid look. To ensure precise grasping,

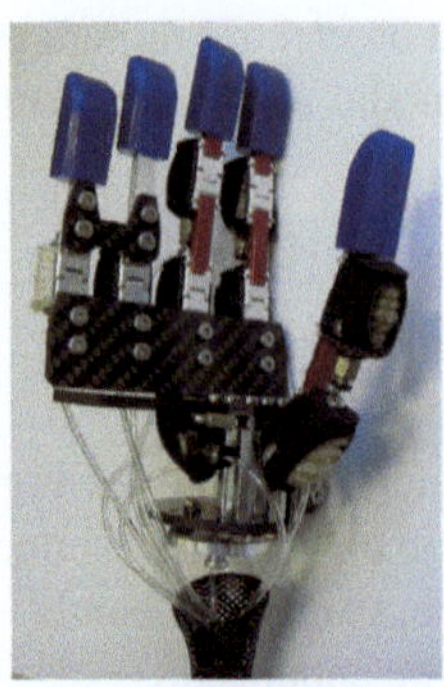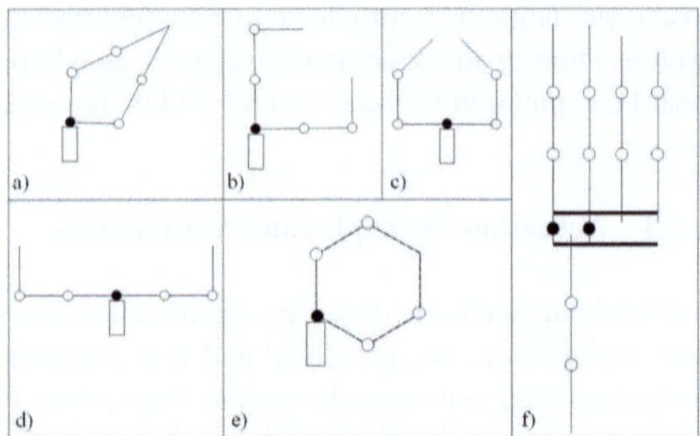

Fig. 10 The ARMAR-III pneumatic hand and its achievable grasp patterns [20]

a symmetric arrangement of all components was chosen. Mechanical joints are enhanced by reinforced flexible bellows such that a torque is applied by inflation. The compliant and lightweight system requires only an air supply and five wired cables for operation. Its modular design integrates all necessary parts into the hand, therefore no space is required proximal to the wrist and it can be attached to any robotic arm [20].

The compliant nature of pneumatic and hydraulic actuation, as well as the flexibility in actuator placement and pressure transmission facilitate the implementation of adaptive underactuation inspired by synergistic human grasping behavior. By the arrangement of fluidic chambers, actuation force can be distributed within several DoA. Communicating vessels connected to a single actuator can drive several fingers of a hand and automatically provide a balance in pressure and actuation force independent of the individual finger postures [48]. Further, pneumatically driven artificial muscles [47] allow for bio-inspired, human-like hand design. A robotic hand, that mimics human muscle actuation allows the direct transfer of human motion demonstrations into the robot control and provides insights on the biological functionality of the human hand [37].

Overall, fluidic hand actuation provides an alternative to electrically driven motors, that is more flexible in power transmission and actuator placement. The compressibility of the actuation fluid allows a humanoid hand to adapt to object shape and environmental constraints when complemented with compliant joint mechanisms or soft finger structures. By these means, grasp control can be simplified especially in highly restricted spaces. In such scenarios, the mechanical adaptation of fluid hands can directly cope with environmental constraints without the need to address them specifically in grasp control.

Tendon-Driven Hands Inspired by the musculotendon system of the human hand, tendon-transmission of actuation forces is commonly used in humanoid robotic hands. The tendon's tractability allows for flexible force transmission from distantly placed actuation motors, usually positioned in the palm or the lower arm. However, tendons only transmit traction forces, thereby limiting the transmitted actuation to a single direction and two tendons or one stiff bar are required for the full actuation of one DoF [25, 26]. In the case of humanoid hands, high forces are required only in finger flexion to grasp objects, while finger extension merely needs to support the fingers own weight. Therefore, a widely-used strategy combines a tendon-driven actuation for finger flexion with springs locally integrated into the hand's joints to drive the finger's extension [16]. A hand with such a design is thereby actively closed, since the actuation of the finger flexion needs to counteract the extension force of the springs in the joints. Similar to the human model, the succeeding joints of one finger are commonly actuated together by a single tendon.

This coupled force transmission results in an underactuation of the robotic finger [34]. The flexible coupling by a tendon-transmission additionally allows the different finger joints within the kinematic chain to adopt different angles influenced by external forces applied to the finger phalanges. This adaptive underactuation causes a mechanical adaptation of the finger to the shape of a surface it closes onto. Similarly, adaptive underactuation can be applied to actuate two or more fingers by the same motor while still allowing their individual adaptation to an object's shape. This can be done either by routing a single tendon through several fingers [12] or with the use of an underactuation mechanism distributing the motor torque onto several finger tendons [8, 10, 18].

Figure 11 shows an underactuation mechanism based on a rigid bar transmission [18, 19]. An electrical motor is attached to rod J and the actuation force is distributed equally to the thumb and the fingers respectively by the rocker H. The subsequent rockers G, E and F further divide the force equally between the rods A, that are actuating the finger flexion. All rockers, except for the transmission rocker D, are freely floating. While the hand is unobstructed, the entire mechanism is therefore pulled towards the motor, causing all fingers to close equally. If one finger is blocked by the object, the respective rocker tilts and the other finger continue closing.

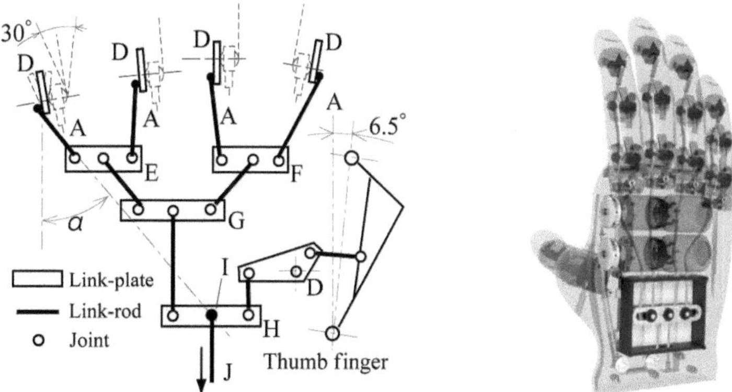

Fig. 11 Underactuation mechanism driving all five fingers of the hand with a single motor [18] (left) and the tendon routing of the male KIT Prosthetic Hand implementing the underactuation mechanism to drive the four fingers (adapted from [53]) (right)

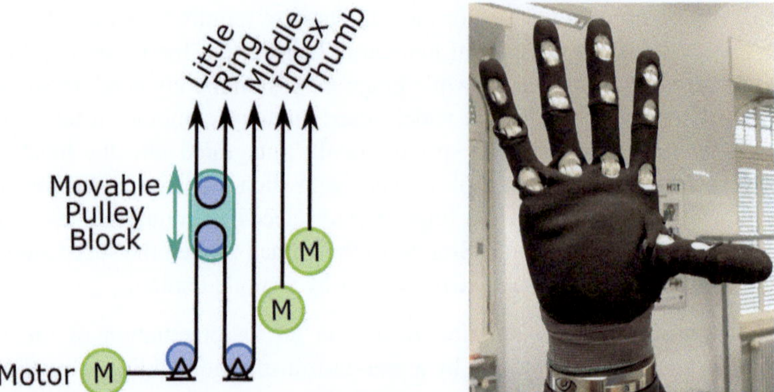

Fig. 12 Underactuation mechanism for three fingers actuated by one tendon (adapted from [28]) (left) and KIT Hand for the humanoid robot ARMAR-6 [5] (right)

A tendon-driven implementation of an adaptive underactuation mechanism is used into the KIT Prosthetic Hand [53]. The tendon routing throughout the mechanism as well as the entire hand is presented in Fig. 12. The hand is driven by two motors actuating the thumb and the fingers respectively. The thumb's tendon is directly attached to one motor and guided over two deflection pulleys in the thumb metacarpophalangeal and interphalangeal joints until it is attached within the fingertip. The second motor drives a tendon routed over the central pulley of the rocker within the underactuation mechanism and fixed at the housing thereafter. The rocker is sliding freely within the mechanism box and the motor tendon pulls the entire rocker bar down towards the palm, thereby distributing the motor torque to all four fingers. Two tendons connect index and middle finger as well as ring and little finger respectively. These tendons are routed over the deflection pulleys on either side of the mechanisms rocker. If one finger is blocked, the tendon rotates around the rocker pulley, allowing the other finger to continue closing. If both fingers connected by one tendon are blocked, the rocker tilts and thereby enables further motion of the other two fingers.

Along the fingers, the actuating tendon is routed over deflection pulleys in each joint and fixed at the fingertip, similar to the thumb. The flexion of the finger metacarpophalangeal and proximal interphalangeal joints is driven by this tendon, while the distal interphalangeal joint is fixed at an angle of 20°. Torsion springs in both joints provide a passive extension of the fingers. By varying the spring pretension for both joints, the correlation between the joint closing speed can be adjusted within the adaptive underactuation realized by the tendon. The difference in joint spring pretension thereby allows to mimic the human finger closing trajectory with a single DoA and a tendon transmission.

The underactuation mechanism has been adapted to actuate a different number of DoF as shown in [28]. By replacing the main rocker with a slider, a mechanism configuration to drive three fingers can be designed as shown in Fig. 12. Two fingers are still actuated by a single tendon, which is attached to both fingertips and routed over a pulley within the mechanism slider. This slider connects the finger tendon pulley with another pulley entangled by the actuating tendon from the motor. The third finger is directly actuated by the motor tendon. The combination of fixed and freely floating pulleys of this mechanism, which is similar to a gun tackle, ensures an equal force distribution between the uneven number of actuated fingers.

The KIT hands are applied both on humanoid robots and for prosthetics. They include a scalable finger design taking into account standardized components as well as electrical boards that can only be scaled stepwise. By these means, hands can be customized in different sizes corresponding to human hands for assistive robots or prostheses as well as larger than the human model for robots performing taxing industrial work. The hands are able to grasp a wide range of objects found in daily household and workshop tasks, with more than 80% grasp success rate. A cylindrical grasp force of more than 20 N allows the hands to lift and handle also heavy objects [53].

Soft Hands Inclusion of soft elements into robots has become a very active research topic in the last two decades [41]. Soft surfaces, actuators and structures improve both safety and compliance of robots since collisions with the environment do not result in immediate damage. Especially for robotic hands this is desirable as these are specifically build to interact with the environment. Yet, the design of soft hands presents additional challenges since both state estimation as well as dexterous control are more difficult to archive for soft and deformable joints or fingers [13]. Since the structure of the fingers is soft, they do not only allow actuation in one direction, but bend and twist in multiple, potentially undesired directions. This can be mitigated by including rigid elements such as springs into the internal structure of the finger that limit deformation in undesired directions. Additionally sensors can be embedded into the fingers that are able to sense such deformation so that a suitable control mechanism can compensate for such movements.

The amount of softness introduced into robotic grippers varies greatly among different hands based on requirements and applications. In some applications soft materials are used on just a single joint to in order to increase compliance of contacts and increase mechanical robustness in case of collision [38]. In others the whole hand structure is exclusively constructed from soft material to exploit the compliance and adaptivity of the soft material to ease control [15].

In the following, we present an example, namely the design of the *KIT Sensorized Soft Hand* [54] and *KIT Finger-Vision Soft Hand* [28, 29], which illustrates the introduction of soft elements into the fingers of humanoid robotic hands. The hands are shown in Fig. 13.

Both hands share the same palm, including three motors. One motor drives the thumb, one the Index finger and the third motor drives the remaining fingers through an adaptive underactuation mechanism. The palm also houses an hybrid embedded system based on a combination of a microcontroller and a FPGA. Different fingers can be attached to the hand using a fixed interface, improving modularity of the design. The fingers for both hands share a similar mechanical design but differ in

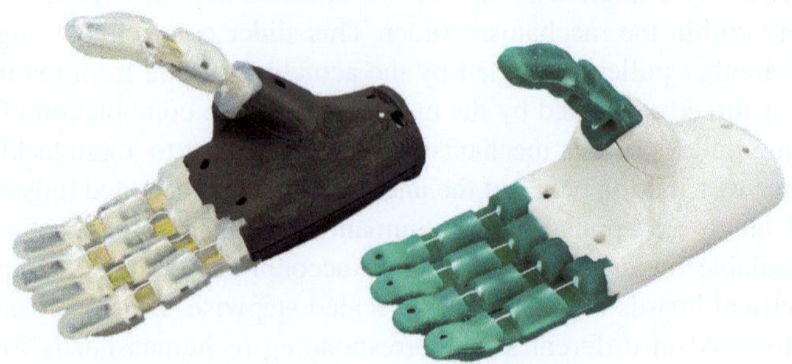

Fig. 13 The *KIT Sensorized Soft Hand* and *KIT Finger-Vision Soft Hand* [55]

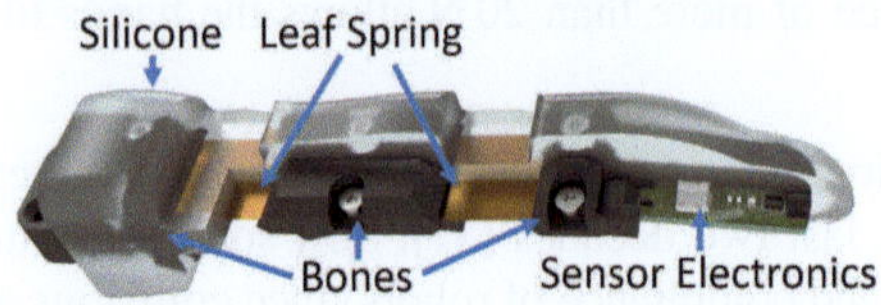

Fig. 14 The mechanical structure of the soft fingers. The silicone is cut after the proximal phalanx to expose the leaf spring and bones, the distal bone is also cut to expose electronics that can be embedded in the fingertip

the included sensors. While the *KIT Finger-Vision Soft Hand* features a camera at the tip of each finger, the *KIT Sensorized Soft Hand* includes a multimodal haptic sensor system.

The fingers are constructed from 3D-printed bones for the proximal, intermediate and distal phalanx connected by a metal leaf spring, as can be seen in Fig. 14. The finger is actuated by a tendon in flexion direction by a tendon and passively by the leaf spring in extension direction. While the bones completely prohibit bending of the finger at the phalanges, the finger can bend at the areas between the bones. Due to the leaf spring, the fingers bend preferably in the flexion/extension direction while being substantially more stiff in all other directions. The leaf spring has a thickness of 90 μm and forms the neutral phase during bending of the joints. electrical cables, connecting sensors in the finger tip to electronics in the palm can be directly taped onto the leaf spring. This way the cables will mainly be subjected to a bending motion with a defined bending radius and do not elongate during bending. *Flat flex cables (FFCs)* are especially well suited for this application. After the internal structure is assembled, the finger is inserted into a mold for casting. For casting, a two component silicone with a Shore A hardness of 13 is used.

3.2.2 Sensors and Embedded System

Very important for successful and reliable grasping and manipulation is the sensory information that can include tactile as well as visual information. The obtained information is required to provide feedback during the grasping process. Depending on the task, very high resolution tactile feedback in combination with visual feedback might be required, imagine the task of putting a thread through a sewing needle. In simpler tasks such as grasping of objects or manipulation, sensory information can be only necessary for detecting success and failure. While humans are not equipped with in-hand visual perception, in robotics visual feedback can be helpful for vision based control such as visual servoing, where in hand vision can provide an un-occluded and better perspective onto the target object.

In the human body, sensory information is transmitted via the spinal cord towards the brain where decision are made, however typically in robotics data is acquired locally and due to the principal of modularity can also be processed locally in hand, allowing a well defined and simple communication interface to the hand.

Tactile Perception Humans possess a sophisticated haptic perception system, enabling complex interaction with the environment. Likewise, haptic perception is a key enabling factor for dexterous grasping and manipulation in robotic hands. Yet, the robust integration of a rich sensor system into the confined space of robotic hands and fingers presents a unique mechatronic challenge [46]. The sensors need physical contact with the environment, making them especially prone to mechanical failure of the sensor structure or electrical connections. Additionally, the sensors need to be distributed throughout the mechanical structure of the hands since each part that is potentially in contact with the environment should ideally be innervated with sensors. This leads to a large number of required sensors and electrical signals that need to be routed through the mechanical structure of the hand.

Research on tactile sensors and their integration into robotic hands has gained a lot of traction in the last decade. Different sensor technologies have been explored to realize tactile sensors, including optical and visual, resistive, capacitive sensing, strain gauges, magnetic flux, vibrations, MEMS and the piezo-resistive/-electric effect [30, 39]. To illustrate the implementation of such a system into the fingers of a robotic hand, we will continue the example of the soft fingers of the *KIT Sensorized Soft Hand*.

The main idea behind the sensor system inside the soft fingers is to use readily available digital sensors and fabrication techniques to reduce fabrication effort and ensure reproducible results. Digital sensors offer the additional advantage that the analog signal of the sensing element is directly digitized inside the sensor and that the sensor usually offers a digital bus interface used for readout, minimizing wiring. This follows the paradigm of encapsulation and modularity, allows for scalability of the design and offers the high robustness inherent to industrial sensing devices. The sensor system is multi-modal, including both dedicated normal force and separate shear force sensors, a proximity sensor, an accelerometer and custom joint encoders. All sensors are mounted on two PCBs, both connected to a central processing system in the palm through FFCs. An overview of the system is depicted in Fig. 15.

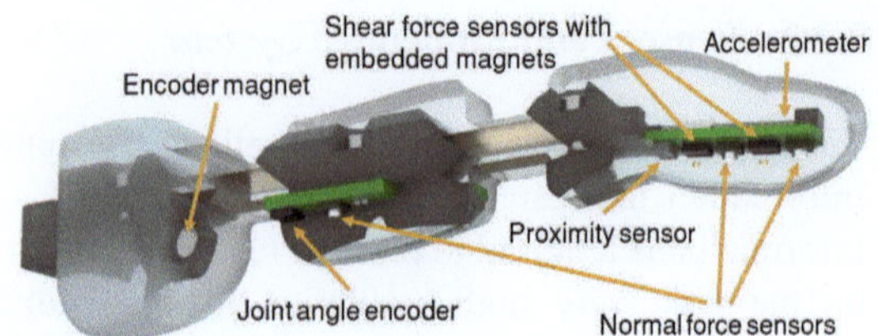

Fig. 15 Overview of the sensor system of the soft finger. The normal force sensor and joint angle encoder on the intermediate phalanx are duplicated the other side [54]

The shear force sensors are based on digital three-axis Hall effect sensors. To fabricate a shear force sensor, a magnet is placed in soft material above the sensor. When a force is exerted on the soft material, the magnet is displaced in relation to the sensor which causes a change in the measurement of magnetic flux measured by the sensor [33]. Co-planar movement of the magnet relative to the sensor represents a shear force while orthogonal movement represents the normal force component. While these sensors are able to measure shear forces well, they are not as sensitive to normal forces. Therefore, a second tactile sensing modality for normal forces is included in the sensing system.

The normal force sensors are based on digital miniature MEMS barometers. These sensors are cast into silicone which then acts as a transducer for the pressure applied to the surface of the finger onto the sensing element [50]. Depending on the hardness and thickness of the soft material, the sensitivity and sensing range can be adjusted. These sensors are able to detect weights as small as 0.5 g placed on the sensor and are hence well suited to detect initial contact between the fingers and the environment.

An infrared proximity sensor is located at the base of the fingertip to detect objects in close proximity. This allows sensing objects a few centimeters before they come in contact with the fingers and can for example be used in human-robot handover scenarios or to detect if the hand is close enough to an object to start grasping. Such proximity sensors can also be utilized as additional normal force sensors by measuring the deformation of the soft material above the sensor [40, 57]. Additionally, an accelerometer is placed on the back of the sensor PCB of the distal phalanx to detect sudden acceleration spikes. This can be utilized to detect collisions of the fingers with the environment or in an object placing task to detect contact between object and a supporting surface.

The soft joints of the fingers make joint angle measurement especially challenging as there is no fixed center of rotation and the joint is able to deflect in all three directions of space. While the leaf spring mostly eliminates bending of the joint to the sides, it allows for a considerable amount of twist in the joints. To measure movement in multiple axes, a strong magnet is placed at the joint on the proximal and distal phalanx. On the PCB of the intermediate phalanx, two 3D Hall effect sensors are placed opposite to these magnets. Each combination of joint flexion and twist causes a unique but nonlinear sensor reading that can then be either fitted to an analytical model or alternatively can be calibrated.

In-Hand Vision In addition to tactile perception, cameras offer an efficient and cheap way of obtaining scene information for grasping and manipulation tasks. Compared to tactile perception, cameras can provide pre-contact information. Cameras can be either installed in a static position in the working setup of the robot, mounted at the robot base or placed in head position. Integrating the camera in the end-effector or artificial hand of a robot comes with the advantage that the view of the contact point between end-effector and scene is visible well in the center of the camera frame. Further, if the hand is regarded as a tool, the tool-center-point is located at a fixed position in the image frame. This configuration is referred to as eye-in-hand visual servoing and is used for a variety of manipulation tasks where an exact placement of the end-effector is necessary. Depending on the purpose and manipulation strategy, the camera position inside the hand can either be chosen inside the hand palm or inside one of the finger phalanges, as the following example of the KIT Prosthetic Hand and the KIT Finger-Vision Soft Hand demonstrate.

When artificial hands are used as prosthetic hands, the controlling of the hand by the user can be quite challenging, since commonly used surface EMG electrodes are limited in resolution and can be unreliable in situations when electrodes are moved or skin properties change. To address this issue, quite a number of researchers have investigated the use of intelligent functions that are based on visual scene perception to implement semi autonomous control functions [14, 23, 24]. Visual perception can include shape estimation of objects that are grasped or recognition of specific objects to select from a predefined set of finger trajectories or forces.

The KIT Prosthetic Hand described in [53] is designed to comprehend the scene and thereby suggest a suitable grasp-type to the user. Therefor it has a 1.3 megapixel camera included in the palmar hand side that is used for object recognition. The camera is connected to the hand internal controller, the user can initiate object recognition via a command sent over the EMG interface. For the purpose of image processing and especially image classification tasks, convolutional neural networks are well suited, however they come with a quite high processing workload. Since the hand internal computation resources are very limited due to space and power constraints, special attention was given for a resource aware implementation of the image processing algorithms of the KIT Prosthetic Hand [27]. Since the user expects a grasp suggestion and thereby network result after a certain amount of time, real-time constraints can be derived: The maximum available time slot allows to estimate the highest possible processing workload per network inference. To obtain this value, the available processing time is multiplied with the throughput of the system, meaning the number of operations that the system is capable to perform of each second. To find a network architecture, that matches the available hardware resources and allows the highest accuracy, different optimization techniques can be used as for example genetic algorithms or Bayesian optimization. The processor (Arm Cortex M7) that is used in the first version of the KIT Prosthetic Hand is capable of approximately 2 million multiply-accumulate operations in the given timing budget that are mainly used for processing of convolutional networks. A best suited network architecture was obtained using a genetic algorithm, the network can recognize objects from 13 classed at an accuracy of 96.5%.

The very small size of the commercially available cameras allows to also integrate cameras inside the fingertips of a humanoid hand. This allows to obtain pre-touch information from the perspective of the expected contact point. This allows to detect objects, whether they are in reach of the hand, this information can be used in reactive grasping such as handover tasks, grasping of dynamic objects and also precision grasps. The KIT Finger-Vision Soft Hand implements the concept of in-finger vision. Here, each finger includes a miniature camera as shown in Fig. 13. To detect objects that are visible in the camera image, a binary segmentation algorithm is implemented. The segmentation algorithm is realize using a convolutional neural network in an encoder-decoder architecture. The output of the network is a pixel-wise semantic segmentation of all camera images. In a set of experiments the applicability of the concept to a set of typical grasping tasks is shown.

Embedded System The request for modular and encapsulated hand design comes with the need for a hand embedded controller. The controller is required to fulfill 4 different functions that include reading of sensor values, controlling actuators, processing data and providing an interface to other control units.

To allow processing of data as well as providing interfaces a data processing unit is an essential part of the controller, here the design choice that has to be made is the selection of suitable type of processing system. Commonly used processing architectures are microcontrollers, application processors, field programmable gate arrays (FPGAs) and System on a chip modules, that can combine logic circuitry and processors. While microcontrollers mostly are used without an operating system and execute relatively simple control algorithms in combination with specialized peripheral modules like times/counters and dedicated interfaces, application processors depend on external memory and an operating system. FPGAs come with the advantage of completely customizable configuration of logic circuits, that allow specialized and parallel interfaces and data processing algorithms. SoCs (system on a chip) can combine processors with reconfigurable logic or other types of hardware accelerators allowing to implement highly efficiency data processing algorithms in combination with complex control methods.

In Fig. 16 the right side image shows a SoC based hand controller for the KIT Prosthetic Hand (50th percentile female). The controller is based on a Xilinx Zynq SoC (XC7Z010) and additional external RAM (2Gb) and 32GB flash storage. The SoC has the advantage of containing a dual core Application processor as well an FPGA. This allows high flexibility for sensor interfaces, allows hardware-accelerated processing of sensor data as well as procedural programming on the processing system. However, the power consumption is comparably high compared to a microcontroller based system. The board includes interfaces for camera, display and relative encoders, as well as two brushed motor drivers. Especially in design of artificial hands, miniaturization of the controller is important, since the mechanisms and actuators already take up large parts of the- available space inside the hand.

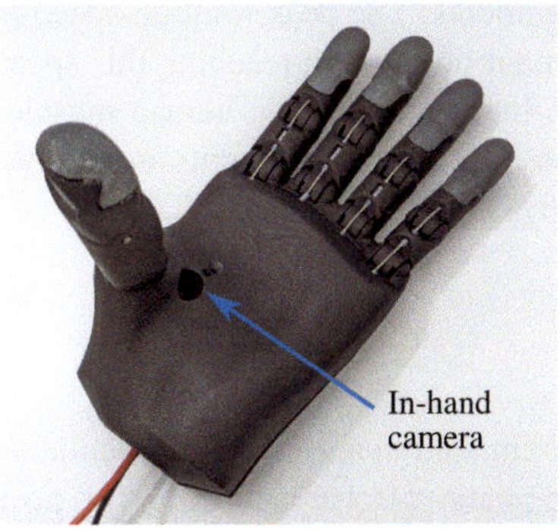

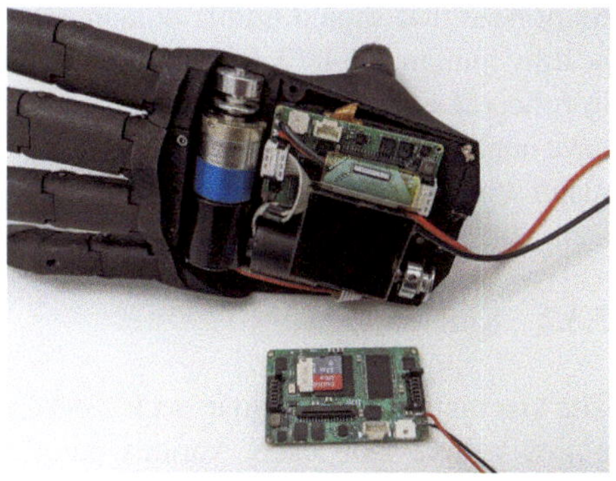

Fig. 16 KIT Prosthetic Hand (Version 2) with in-hand camera in the palm. On the right, the SoC-based embedded in-hand controller board is shown

Therefor, during the mechanical and electrical co-design placement of connectors, routing of cables and heat dissipation must be coordinated carefully. An other important aspect is a robust and easily to assemble and disassemble design, so that in case of any needed modification or mechanical/electrical failure, the system can be accessed and repaired at a reasonable effort.

3.3 Humanoid Head Design

In order for humanoid robots to perceive and interact with their environment, their perception capabilities are a crucial element. Thus, the perception system for such robots should provide sensory information necessary to execute various visuomotor tasks, such as detecting salient regions, and more complex sensorimotor tasks, such as hand-eye coordination, gestures, etc. as well as audio information. For the head, the appearance and the effect on the human observer is especially important, since the face is the center of attention during interactions.

The previously mentioned design principles are also followed in humanoid head design. This applies to modularity and encapsulation of the different components of subsystems of the head. High and robust integration with electrical and mechanical co-design are also very important.

3.3.1 Requirements

The human neck consists of 7 cervical vertebrae which allow an almost continuous bending of the neck in 3 directions The cervical spine can move in flexion (80–90°), extension 70° (pitch), lateral bending 20–45°(roll), and rotation 90° (yaw) in both directions [49]. This movement is complex since pure rotational movement cannot accurately reproduce it and thus it has to be to be approximated for a robot head that should mimic human neck movements. The peak rotation speed in healthy human is $348 \pm 92°/s$ [43] where the main problem in reaching this speed in robots is acceleration and thus the torque-to-inertia ratio while human muscles have negligible intern inertia. In addition, human eyes allow movements around two major axis in order to be able support gaze control and gaze control.

3.3.2 Kinematics

The kinematics of the human neck can be approximated in a robot by 2 articulation points with 2 DoF each. Various parallel kinematic mechanisms [36] for neck motion are possible, but rarely chosen for humanoid robots. The two 2 DoF of the eyes for direction and gaze stabilization can be realized directly or compensated by camera characteristics like image stabilization. The ability to mimic fast human head movements depends largely on the ratio of head inertia to available drive torque, since acceleration and deceleration phases account for most of the movement.

The kinematic comparison of the robot heads of ARMAR-III, ARMAR-4 and ARMAR-6 show that the first two have a focus on humanoid design and kinematics while the latter is focused on a simple kinematics and the use of certain camera systems. The kinematics of the ARMAR-III head with seven DoF arranged as lower pitch (1), roll (2), yaw (3), upper pitch (4) is shown in Fig. 17a. The complete head with eyes tilt, right eye pan and left eye pan can be seen in Fig. 17b. ARMAR-IV has a 9 DOF head with independent eye pan and tilt, a five DOF neck mechanism with lower and upper pitch, lower and upper roll and a yaw joint (Fig. 17c). In ARMAR-6 and ARMAR-DE a reduced kinematics setup with only pitch and yaw joints was chosen. The order of the joints is pitch-yaw in ARMAR-6 and yaw-pitch in ARMAR-DE (Fig. 17d).

3.3.3 Sensors and Actuation

Accurate position sensing is important for visual perception and hand-eye coordination. The obtained values from the absolute encoders are also used for joint angle control for angular velocity control relative encoders can provide additional feedback in the control loop. Additionally to the actuation of the neck, the cameras can be actuated similarly to the human movable eyes. This enlarges the field of view. ARMAR-III and ARMAR-IV has four digital color cameras each for wide and narrow angle. ARMAR-IV has four cameras and six microphones. The head is also equipped with microphones an IMU that is used for gaze stabilization and oculomotor control.

For actuation speed the possible acceleration is important. This is specified by actuation torques and inertia, where geared motors with typically used high transmission ratios have high internal inertia which is added to the heads inertia. Static forces depend on mass and center of mass. The motors for joint actuation used in humanoid robots are generally either brushed (DC) or brushless (BLDC)

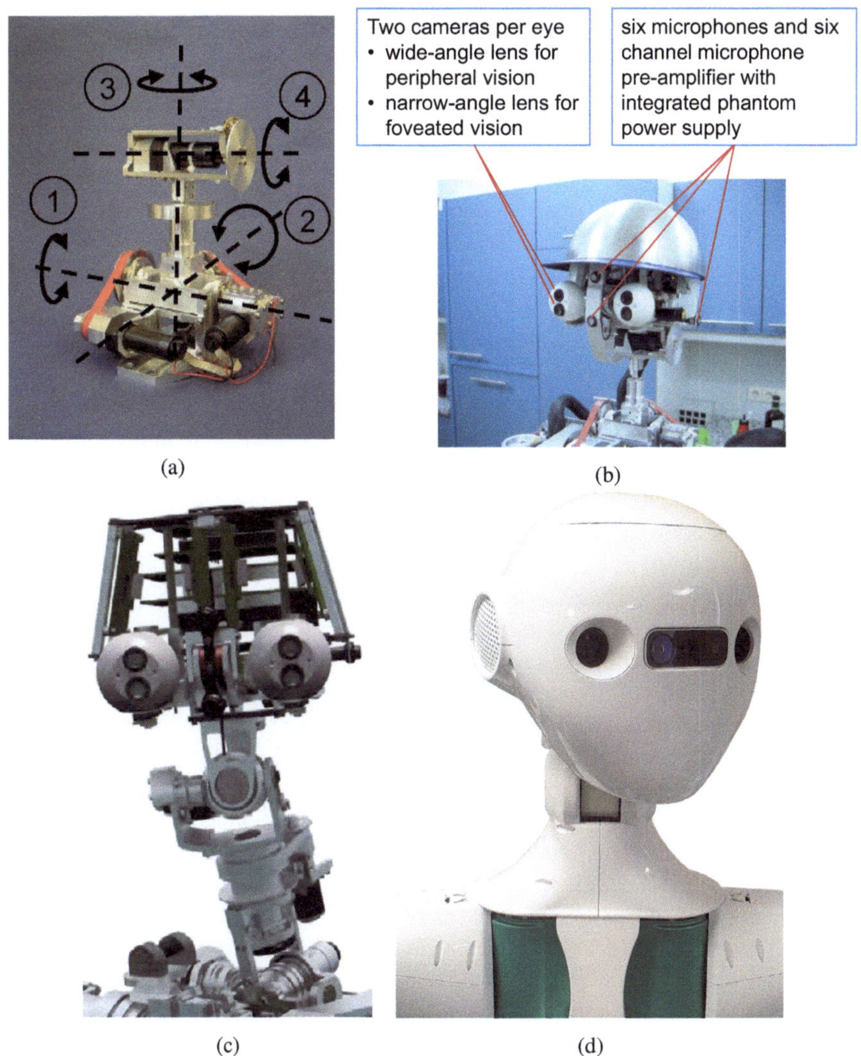

Fig. 17 The neck design of ARMAR-III [1] (**a**), the head of ARMAR-III [3] with seven DoF (**b**), the head design of ARMAR-4 [4] (**c**) and the head of ARMAR-DE (**d**)

direct current motors. Brushed motors are generally thinner but longer since space for the brushes is needed while BLDC motors are shorter with a larger diameter. The control electronics for BLDC motors is larger and has a minimum size and therefore small motors. Since the torque of small electric motors fitting into the neck and head volume is not sufficient, gearboxes with high transmission ratios are necessary. This can be multi-stage planetary gear boxes or Harmonic Drives which have the advantage of no backlash and where therefore chosen for all actuators in all robot heads. While the joints in ARMAR-6 and ARMAR-DE heads are directly driven, some of the higher number of joints in ARMAR-III and ARMAR-4 heads are driven by belts with remote actuator position.

4 Software Design

Developing software to control complex humanoid robots is a challenging task, that requires a sophisticated software design and architecture. Especially with multiple robots such as ARMAR-III, ARMAR-6, ARMAR-DE, ARMAR-7, re-using functionality is crucial. This can range from pre-processing sensor signals on embedded devices up to autonomous abilities such as grasping an object or navigating through a room. We developed ArmarX [52] as a unified software framework to fulfill these needs. ArmarX emerged over generations of ARMAR robots and is capable of seamlessly and concurrently integrating hardware and software. This allows to maintain and extend existing robots by both its software and hardware as well as its cognitive abilities allowing the integration of new robot components quickly. ArmarX is publicly available under an open-source license.[1]

To address the problems of high complexity and transferability, a layered architecture with clear interfaces was designed. The architecture is depicted in Fig. 18 and consists of three layers. 1. The *embedded devices* for highly specialized functionality, 2. the *hardware abstraction layer*, which provides integrating interfaces and abstracts from several bus systems such as USB, Ethernet or EtherCAT, and 3. the *high-level framework* for control and orchestration of multiple robot components which are transferable to other robots. Each of the layers has its individual properties and requirements and will be described in the following sections. One general paradigm that can be found across all layers is the realization of an event-driven closed control cycle as shown in Fig. 19.

For example, on the level of *embedded devices*, there are highly integrated devices with their own firmware-implemented controllers using available relative encoders. On the *hardware abstraction layer*, the SAC units are modeled using a virtual device with two bus participants, namely one motor and one sensing device. Virtual devices also support implementing controllers, using the sensor data of one bus participant and passing control commands to another bus participant. Finally on the *high-level layer*, previously independent components are combined such that multi-modal sensor data can be fused to enable e.g., visual servoing or navigation using the laser scanners.

---

[1] https://gitlab.com/ArmarX.

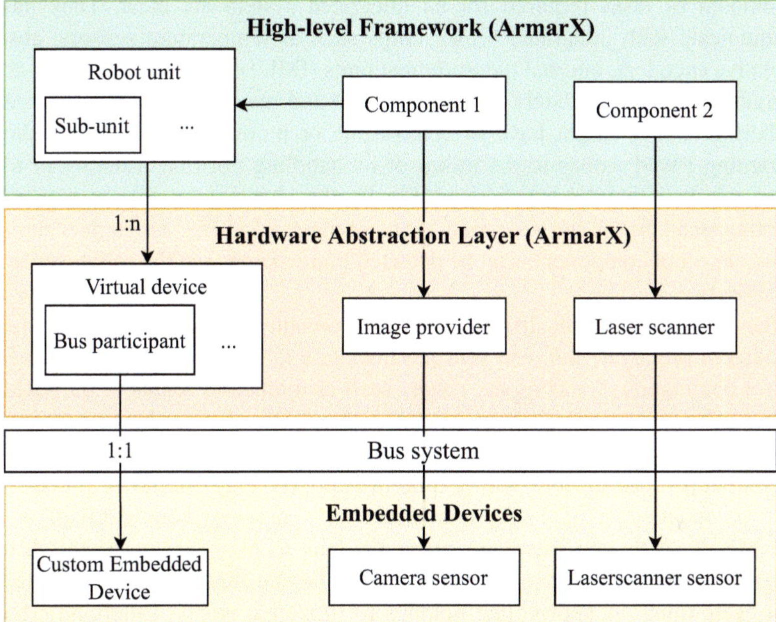

Fig. 18 The three layers of the ArmarX software architecture. **Bottom layer:** Embedded devices, including custom ones, but also commercial sensors such as cameras or laser scanners. **Middle:** Hardware abstraction layer, with unifying interfaces for bus participants (virtual devices, left), cameras (image provider, middle), laser scanners (right), etc. **Top:** High-level framework, with the unifying concept of a robot (robot unit, right), and several exemplary high level components accessing sensors and interacting with the robot unit. Note how several bus systems (e.g., Ethernet, USB, or EtherCAT) are abstracted from using the hardware abstraction layer

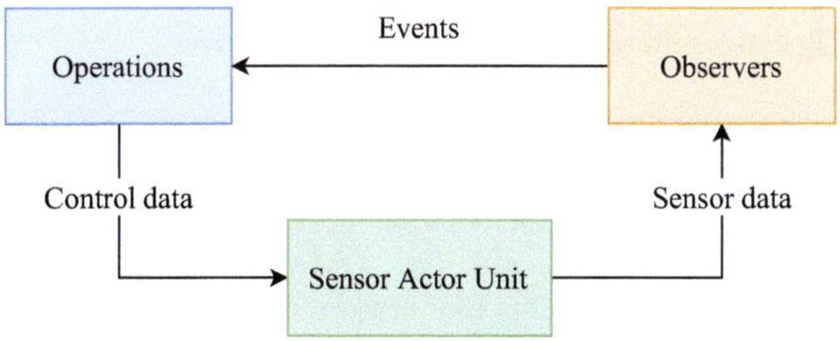

Fig. 19 Event-driven closed control cycle

4.1 Embedded Software Design

Embedded software is in most cases rather difficult to change or update once the electronic system is integrated into the robot. The most important design aspects for embedded software in a robot are therefore simplicity and reliability. Mostly the amount of tasks required for an integrated system are clear. They have to communicate with integrated sensor chips such as temperature sensors, absolute or relative encoders, internal measurement units (IMU) or torque sensors, be able to configure them appropriately and periodically and read the current sensor values. Furthermore, they might have to control one or more motors by either directly forwarding PWM values to the motor or by handling position, velocity or torque commands in internal control loops. Finally they have to be able to report if an error happened during the execution of any of their tasks. The last point ensures that failures in robot components can be detected immediately and the robot can be kept in a safe state.

The embedded systems do not only have to communicate with sensors or motors, but need to propagate the read sensor values to a main computer and receive new control commands. For complex robots such as humanoid robots (e.g., ARMAR-6) this communication needs to be fast and reliable due to the vast amount of data that needs to be transferred in every iteration: in ARMAR-6 there are 43 different embedded devices with a total of approximately 3400 B of process data. With an update rate of 1 kHz this results in a required bandwidth of at least 27.2 Mbit/s. Although there are many fieldbus technologies used for different industrial requirements, most of them can not handle this amount of data in such short cycle times. A appropriate fieldbus for this case is e.g., EtherCAT,[2] but for smaller robots with less DoF and therefore less periodically data slower fieldbus systems like PROFIBUS[3] or CAN[4] might be sufficient. The low-level interaction on the physical layer and data link layer of the OSI-Model[5] with these fieldbus systems should be handled by either dedicated integrated chips or FPGA systems since the required calculations might use a lot of processing power which is better used for performing the aforementioned tasks of the embedded system.

4.1.1 Best Practices for Reliable Embedded Software

In most cases the embedded systems are reasonable simple that there is no need for a real-time operating system (RTOS) but instead the software can be written as one single program which includes all necessary hardware abstractions for on-chip peripherals like UART, SPI, I2C, DMA, timers, GPIO and the drivers for the external peripheral systems like sensors or motors. This program will be executed once the embedded system is powered on and keep running until it gets shutdown by removing the power supply. So-called bare metal programs are today

---

[2] https://www.ethercat.org/en/technology.html.

[3] https://new.siemens.com/global/en/products/automation/industrial-communication/profibus.html.

[4] https://www.iso.org/standard/63648.html.

[5] https://www.itu.int/rec/T-REC-X.225-199511-I/en.

mostly written in C, C++ or Rust but for smaller projects do not require real-time capabilities (Micro)Python is getting more popular.[6] However for absolute control over aspects like memory layout or RAM usage non-interpreted languages are usually advantageous. In the following we provide a few practices for developing embedded software that is based around a single while-loop without a scheduler that handles different threads:

- *No dynamic memory allocations:* Memory should not be allocated during the run time of the program, but instead statically allocated. This has the benefit that the amount of required RAM is known at compile time and therefore prevents possible RAM overflow errors which are impossible to handle at run time.
- *Cooperative multitasking:* The simplest form of a scheduling system. Multiple tasks (e.g., reading values from different sensors) run one after another in a fixed order, but every task is divided into sub tasks, that take minimal execution time. The execution of these sub tasks is handled by a worker function for each task which selects the appropriate sub-task for the current situation. Since every worker always only executes one sub-task, all tasks can run almost in parallel and do not block each other. This principle can be refined by assigning less important tasks a frequency based on the system clock which controls how often the worker does nothing and returns immediately. The programmer has full control over the control flow of the system and does not rely on an underlying scheduler which might introduce non-determinant behavior.
- *Non blocking communication:* Communication over I2C, SPI, UART or other local bus systems should take up as little processing power as possible. It should especially not block other subroutines from being executed, due to for example waiting for a response from an external sensor. The most useful technique here is the usage of DMA (Direct Memory Access) which handles together with dedicated chip hardware the communication completely parallel to the main CPU and only triggers a hardware interrupt once the requested communication step has finished. The programmer can then for example set a certain flag that the data has been successfully received which itself will be handled in the next iteration of the cooperative multitasking loop. Alternatively one could design the PCB itself in a way that more complex communication (e.g., EtherCAT or CAN) gets handled by a specialized integrated chip and the micro controller only triggers this communication by sending the payload to the chip and gets notified once the communication has finished. This practice goes hand in hand wit cooperative multitasking.
- *Error reporting:* This point is valid for all kind of embedded software. Since the program is running on a system without easy physical access (e.g., because the micro-controller is embedded deep into a mechanical device), it is very difficult to handle malfunctions if there is no way of reporting errors from the embedded system to the main control system. Unlike programs running on a

---

[6] https://www.qt.io/embedded-development-talk/embedded-software-programming-languages-pros-cons-and-comparisons-of-popular-languages.

personal computer there is no easy way to debug the system once it is assembled. Important type of errors which are useful to be reported via status words or bit-wise error fields include communication errors with integrated devices like sensors, forwarded error bits from those devices, external signals (like STO) and whether limits of for example temperature sensors are reached.

The design of embedded software depends heavily on the complexity of the underlying embedded system. Smaller robots can be controlled from one large embedded microprocessor, which has the capabilities of running an operating system and can be used and programmed like a small computer. But every robotic system with a high number of DoF needs modular, specialized micro-controller-based embedded systems with programs that need to be as simple and reliable as possible. Some strategies to reach that goal were presented here.

4.2 ArmarX—Hardware Abstraction Layer

The hardware abstraction layer (HAL) is an essential part of the core functionalities of ArmarX and is the foundation all higher components of ArmarX build upon. As shown in Fig. 18, the HAL provides bus-agnostic interfaces for sensors and actors available on the robot to the whole ArmarX ecosystem (potentially spanning across multiple computers inside the robot).

We distinguish two cases for tethering new embedded devices. The simplest case is connecting commercial embedded devices such as cameras or laser scanners via general purpose interfaces (e.g., Ethernet, USB, ...). For these devices, generic software interfaces are already available in ArmarX, e.g., for image providers (cameras), or point cloud providers (laser scanners). The standard procedure, however, is to daisy chain several (custom) embedded devices into a dedicated bus system, in order to save space and to meet hard real-time constraints such as new control targets for actuators. When constructing a robot, we also replicate the physical bus in software by writing device specific software libraries, which integrate embedded devices such as actors or sensors. When starting operation, the physical bus is queried for the bus participants, to find and instantiate the corresponding class describing the state and functionality of each embedded device. These small software libraries for each embedded device allow for a very modular design, as the functionality can be shared between multiple generations of robots, as it is the case with ARMAR-6, ARMAR-DE, and ARMAR-7. New embedded devices can easily be integrated. To build a new robot, we thus first start by bundling the libraries representing the used embedded devices for a concrete robot (Fig. 20).

On a higher level of abstraction, we interpret groups of embedded devices on the bus as virtual devices, which are represented in software by objects being assigned to one or more embedded device with the ability to fully control the whole group and thus provide high-level interfaces or high-level control for the remaining framework. An example for this are the wheels of a mobile robotic

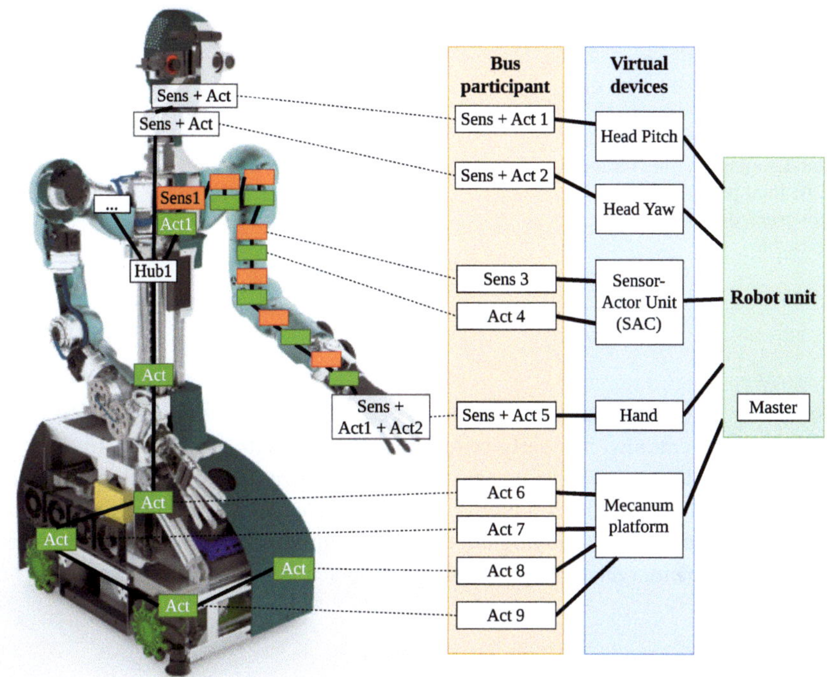

Fig. 20 Schematic of the bus layout in ARMAR-6. Shown is how physical components are mapped to their software counterpart. Virtual devices group one or more bus participants to implement higher-level functionalities. These are then instantiated by the robot unit

platform, which are participants of the bus system each on their own. For each wheel (a motor connected to an embedded device on the bus), a software equivalent would be instantiated to provide the low-level functionality in software, such as motor velocity control. On a higher level of abstraction, the mobile platform as a whole is represented through a virtual device in order to provide high-level control by coordinating the set of embedded devices the virtual device was assigned to, e.g., to implement Cartesian platform velocity control by controlling individual wheel velocities.

To ensure *safe operation* of the robot, a continuous monitoring of the system state stops the robot's movement in case of failure. This can be due to software issues e.g., if a crucial component enters a failure state (or crashes) or hardware issues such as the robot being close to self-collision. The implemented state machine is shown in Fig. 21. At startup, the robot enters *safe stop* mode in which the joint velocities of each joint are controlled to be zero. Entering *operational* mode requires manual intervention of the operator in order to enable the robot to actuate its motors. In case the robot itself detects a software or hardware failure, it returns to safe stop mode. As it might only be a temporary failure, e.g., due to a connection loss to a camera which can be re-established automatically, the robot might recover and

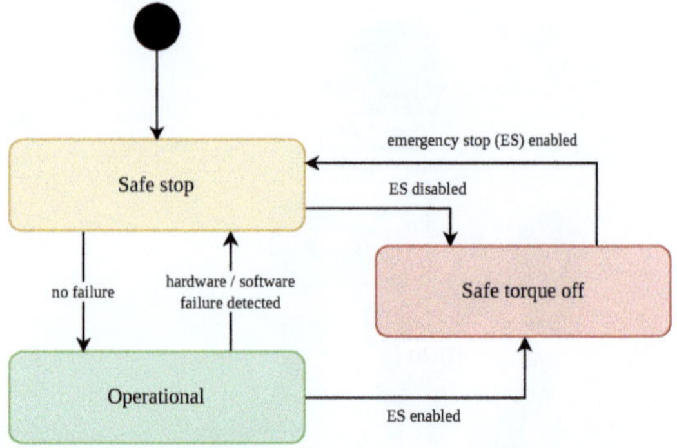

Fig. 21 State machine of the different operational modes. Operational: All control modes are allowed. Safe stop: Joints are zero-velocity controlled, no other control modes allowed. Safe torque off: Joint power supply interrupted, no control possible

transition to operational mode again. In case of unforeseen or unmodeled failures, the operator can also manually bring the robot into safe stop mode. In emergency cases, it is also possible to use any of the hardware emergency stop buttons that are either attached to the robot or connected wirelessly. These directly interrupt the motor's power supply, bringing the robot into *safe torque off* mode. After releasing the hardware emergency stop buttons, the robot will transition into safe stop mode again.

4.3 ArmarX—High-Level Framework

On the high-level layer, robot-specific modules need to be combined and made available in a robot-agnostic way. To do so, all virtual devices that are necessary for the operation of the robot are instantiated by the robot unit. The robot unit also manages the concrete bus as master. As virtual devices might differ between robots, some functionality is exposed via hardware-abstracting sub-unit interfaces that are common to all robots, e.g., a kinematic unit to control individual joints or a platform unit to move the robot within its local coordinate frame. In addition, the robot unit also manages the real-time controllers for complex task and joint-space control of the whole kinematic chain. These multi-joint controllers are linked to a subset of the kinematic chain. The robot unit ensures that only one controller is active per joint. This allows realizing unimanual, bimanual or even whole-body control. High-level components allow to integrate multi-modal sensor data to build a coherent environment model e.g., by performing vision-based object localization and 6D pose estimation. This enables the realization of complex skills such as vision-based grasping, human-aware navigation and active perception, and many other complex tasks.

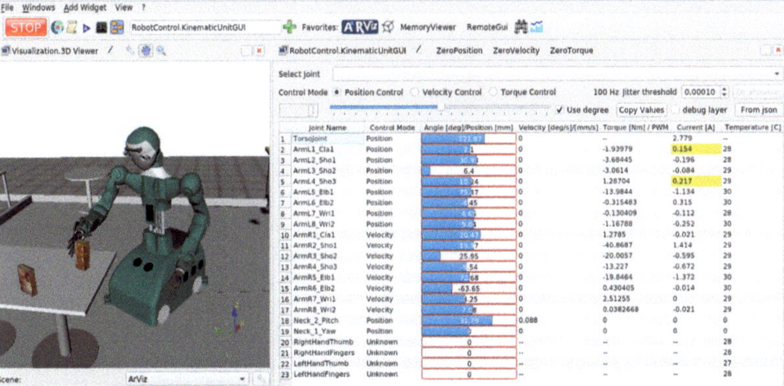

Fig. 22 The ArmarX graphical user interface (GUI). On the left, the virtual scene with ARMAR-6 is shown. The kinematic unit GUI on the right can be used to control all joints of the robot in either position, velocity or torque mode

4.3.1 Visualization and Introspection

For concurrent hardware and software integration, visualization and introspection tools are essential. For this, the ArmarX GUI can be used to disclose the robot's state. As shown in Fig. 22, the GUI shows information about the robot such as the current joint positions, velocities and torques. In addition, the GUI can be used to control each joint individually. The robots internal state, its location in the environment, as well as the robots model of the environment are visualized on the left side.

5 Conclusions

The development of the mechatronics system for robots is a very complex and time-consuming task. Expert knowledge from the various disciplines of mechatronics is needed to design a robust, highly integrated, high-performance system. Thus, it is important to find solutions to preserve such expert knowledge and made it available for future generation of robot designers. In addition to general guidelines, it would be helpful to provide concrete solutions for the realization of mechatronics components based on previous experience. This requires the formalization of the design process on all levels. To support this process, we developed an ontology-based expert system [32]. Based on user requirements such as spatial dimensions, performance, cost, weight and sensor types, the system proposes concept solutions for the design of mechatronic components. These concept solutions include all required purchased parts, their arrangement to each other and the resulting properties of the overall system. The expert system draws on an ontological knowledge base that preserves the design knowledge from previous developments.

Cross-References

This chapter contains cross-references to content published in other chapters belonging to the four "Robotics Goes MOOC" books: KNO2, KNO3, DES5, DES8, INT1, INT2. The full list of chapters abbreviations is available in the Preface.

Acknowledgments The research leading to the development of the ARMAR humanoid robots has received funding from the German Research Foundation (DFG: Deutsche Forschungsgemeinschaft) within the German Humanoid Research project SFB 588 and the European Union within the EU projects PACO-PLUS, Xperience, GRASP, WALK-MAN and SecondHands as well as from the German Federal Ministry of Education and Research (BMBF) in the Competence Center ROBDEKON and Innovation Cluster INOPRO. We would like to thank all members of the H^2T lab and all students who contributed to this research in its different phases.

References

1. A. Albers, S. Brudniok, J. Ottnad, C. Sauter, K. Sedchaicharn, Upper body of a new humanoid robot-the design of Armar iii, in *2006 6th IEEE-RAS International Conference on Humanoid Robots* (IEEE, 2006), pp. 308–313
2. M.A. Arbib, Coordinated control programs for movements of the hand. Exp. Brain Res. **10**, 111–129 (1985)
3. T. Asfour, K. Welke, P. Azad, A. Ude, R. Dillmann, The Karlsruhe humanoid head, in *Humanoids 2008 - 8th IEEE-RAS International Conference on Humanoid Robots* (2008), pp. 447–453
4. T. Asfour, J. Schill, H. Peters, C. Klas, J. Bücker, C. Sander, S. Schulz, A. Kargov, T. Werner, V. Bartenbach, Armar-4: a 63 DOF torque controlled humanoid robot, in *2013 13th IEEE-RAS International Conference on Humanoid Robots (Humanoids)* (IEEE, 2013), pp. 390–396
5. T. Asfour, L. Kaul, M. Wächter, S. Ottenhaus, P. Weiner, S. Rader, R. Grimm, Y. Zhou, M. Grotz, F. Paus et al., Armar-6: a collaborative humanoid robot for industrial environments, in *2018 IEEE-RAS 18th International Conference on Humanoid Robots (Humanoids)* (IEEE, 2018), pp. 447–454
6. T. Asfour, M. Waechter, L. Kaul, S. Rader, P. Weiner, S. Ottenhaus, R. Grimm, Y. Zhou, M. Grotz, F. Paus, Armar-6: a high-performance humanoid for human-robot collaboration in real-world scenarios. IEEE Robot. Autom. Mag. **26**(4), 108–121 (2019). https://doi.org/10.1109/MRA.2019.2941246
7. B. Barry et al., *Software Engineering Economics*, vol. 197 (Prentice-Hall, Englewood Cliffs, 1981)
8. J.T. Belter, A.M. Dollar, Novel differential mechanism enabling two DOF from a single actuator: application to a prosthetic hand, in *IEEE International Conference on Rehabilitation Robotics (ICORR)*, Seattle (2013), pp. 1–6. https://doi.org/10.1109/ICORR.2013.6650441
9. B.W. Boehm, A spiral model of software development and enhancement. Computer **21**(5), 61–72 (1988)
10. C. Brown, H. Asada, Inter-finger coordination and postural synergies in robot hands via mechanical implementation of principal components analysis, in *IEEE/RSJ International Conference on Intelligent Robots and Systems (IROS)* (2007), pp. 2877–2882
11. J. Carryer, R. Ohline, T. Kenny, *Introduction to Mechatronic Design* (Prentice Hall, Upper Saddle River, 2011)

12. M. Catalano, G. Grioli, E. Farnioli, A. Serio, C. Piazza, A. Bicchi, Adaptive synergies for the design and control of the Pisa/IIT SoftHand. Int. J. Robot. Res. **33**(5), 768–782 (2014). https://doi.org/10.1177/0278364913518998

13. X. Chen, X. Zhang, Y. Huang, L. Cao, J. Liu, A review of soft manipulator research, applications, and opportunities. J. Field Robot. (2021). https://doi.org/10.1002/rob.22051

14. J. DeGol, A. Akhtar, B. Manja, T. Bretl, Automatic grasp selection using a camera in a hand prosthesis, in *2016 38th Annual International Conference of the IEEE Engineering in Medicine and Biology Society (EMBC)* (IEEE, 2016), pp. 431–434

15. R. Deimel, O. Brock, A novel type of compliant and underactuated robotic hand for dexterous grasping. Int. J. Robot. Res. **35**(1–3), 161–185 (2016)

16. A.M. Dollar, R.D. Howe, Joint coupling design of underactuated hands for unstructured environments. Int. J. Robot. Res. **30**(9), 1157–1169 (2011). https://doi.org/10.1177/0278364911401441

17. K. Forsberg, H.M. Co-Principals, 4 system engineering for faster, cheaper, better, in *INCOSE International Symposium*, vol. 9 (1999), pp. 924–932. Wiley Online Library

18. N. Fukaya, S. Toyama, T. Asfour, R. Dillmann, Design of the TUAT/Karlsruhe Humanoid Hand, in *IEEE/RSJ International Conference on Intelligent Robots and Systems (IROS)*, Takamatsu (2000)

19. N. Fukaya, T. Asfour, R. Dillmann, S. Toyama, Development of a five-finger dexterous hand without feedback control: the TUAT/Karlsruhe humanoid hand, in *IEEE/RSJ International Conference on Intelligent Robots and Systems (IROS)*, Tokyo (2013)

20. I. Gaiser, S. Schulz, A. Kargov, H. Klosek, A. Bierbaum, C. Pylatiuk, R. Oberle, T. Werner, T. Asfour, G. Bretthauer, R. Dillmann, A new anthropomorphic robotic hand, in *Humanoids 2008 - 8th IEEE-RAS International Conference on Humanoid Robots* (2008), pp. 418–422. https://doi.org/10.1109/ICHR.2008.4755987

21. I. Gaiser, R. Wiegand, O. Ivlev, A. Andres, H. Breitwieser, S. Schulz, G. Bretthauer, Compliant robotics and automation with flexible fluidic actuators and inflatable structures. IntechOpen (2012)

22. J. Gausemeier, S. Moehringer, New guideline vdi 2206-a flexible procedure model for the design of mechatronic systems, in *DS 31: Proceedings of ICED 03, the 14th International Conference on Engineering Design*, Stockholm (2003)

23. G. Ghazaei, A. Alameer, P. Degenaar, G. Morgan, K. Nazarpour, Deep learning-based artificial vision for grasp classification in myoelectric hands. J. Neural Eng. **14**(3), 036025 (2017)

24. F. Giordaniello, M. Cognolato, M. Graziani, A. Gijsberts, V. Gregori, G. Saetta, A.G.M. Hager, C. Tiengo, F. Bassetto, P. Brugger et al., Megane pro: myo-electricity, visual and gaze tracking data acquisitions to improve hand prosthetics, in *2017 International Conference on Rehabilitation Robotics (ICORR)* (IEEE, 2017), pp. 1148–1153

25. M. Grebenstein, A. Albu-Schäffer, T. Bahls, M. Chalon, O. Eiberger, W. Friedl, R. Gruber, S. Haddadin, U. Hagn, R. Haslinger, H. Höppner, S. Jörg, M. Nickl, A. Nothhelfer, F. Petit, J. Reill, N. Seitz, T. Wimböck, S. Wolf, T. Wüsthoff, G. Hirzinger, The DLR hand arm system, in *2011 IEEE International Conference on Robotics and Automation* (2011), pp. 3175–3182

26. M. Grebenstein, M. Chalon, W. Friedl, S. Haddadin, T. Wimböck, G. Hirzinger, R. Siegwart, The hand of the DLR hand arm system: designed for interaction. Int. J. Robot. Res. **31**(13), 1531–1555 (2012). https://doi.org/10.1177/0278364912459209

27. F. Hundhausen, D. Megerle, T. Asfour, Resource-aware object classification and segmentation for semi-autonomous grasping with prosthetic hands, in *2019 IEEE-RAS 19th International Conference on Humanoid Robots (Humanoids)* (IEEE, 2019), pp. 215–221

28. F. Hundhausen, J. Starke, T. Asfour, A soft humanoid hand with in-finger visual perception, in *IEEE/RSJ International Conference on Intelligent Robots and Systems (IROS)* (2020), pp. 8722–8728

29. F. Hundhausen, R. Grimm, L. Stieber, T. Asfour, Fast reactive grasping with in-finger vision and in-hand FPGA-accelerated cnns, in *IEEE/RSJ International Conference on Intelligent Robots and Systems (IROS)* (2021)

30. Z. Kappassov, J.A. Corrales, V. Perdereau, Tactile sensing in dexterous robot hands – review. Robot. Auton. Syst. **74**(1) (2015). https://doi.org/10.1016/j.robot.2015.07.015
31. A. Kargov, T. Werner, C. Pylatiuk, S. Schulz, Development of a miniaturised hydraulic actuation system for artificial hands. Sens. Actuators A: Phys. **141**(2), 548–557 (2008). https://doi.org/10.1016/j.sna.2007.10.025
32. O. Karrenbauer, S. Rader, T. Asfour, An ontology-based expert system to support the design of humanoid robot components, in *2018 IEEE-RAS 18th International Conference on Humanoid Robots (Humanoids)* (IEEE, 2018), pp. 1–8
33. P.J. Kyberd, P.H. Chappell, A force sensor for automatic manipulation based on the hall effect. Meas. Sci. Technol. **4**(3), 281–287 (1993). https://doi.org/10.1088/0957-0233/4/3/005
34. T. Laliberte, L. Birglen, C. Gosselin, Underactuation in robotic grasping hands. Mach. Intell. Robot. Control **4**(3), 1–11 (2002)
35. J.M.F. Landsmeer, The coordination of finger-joint motions. J. Bone Joint Surg. **45**(8), 1654–1662 (1963)
36. J.P. Merlet, C. Gosselin, T. Huang, Parallel mechanisms, in *Springer Handbook of Robotics* (Springer, Berlin, 2016), pp. 443–462
37. A.A. Mohd Faudzi, J. Ooga, T. Goto, M. Takeichi, K. Suzumori, Index finger of a human-like robotic hand using thin soft muscles. IEEE Robot. Autom. Lett. **3**(1), 92–99 (2018)
38. L.U. Odhner, L.P. Jentoft, M.R. Claffee, N. Corson, Y. Tenzer, R.R. Ma, M. Buehler, R. Kohout, R.D. Howe, A.M. Dollar, A compliant, underactuated hand for robust manipulation. Int. J. Robot. Res. **33**(5) (2014). https://doi.org/10.1177/0278364913514466
39. M. Park, B.G. Bok, J.H. Ahn, M.S. Kim, Recent advances in tactile sensing technology. Micromachines **9**(7) (2018). https://doi.org/10.3390/mi9070321
40. R. Patel, R. Cox, N. Correll, Integrated proximity, contact and force sensing using elastomer-embedded commodity proximity sensors. Auton. Robots **42**(7), 1443–1458 (2018). https://doi.org/10.1007/s10514-018-9751-4
41. C. Piazza, G. Grioli, M. Catalano, A. Bicchi, A century of robotic hands. Annu. Rev. Control Robot. Auton. Syst. **2**(1), 1–32 (2019). https://doi.org/10.1146/annurev-control-060117-105003
42. S. Rader, L. Kaul, P. Weiner, T. Asfour, Highly integrated sensor-actuator-controller units for modular robot design (2017), pp. 1160–1166. https://doi.org/10.1109/AIM.2017.8014175
43. U. Röijezon, M. Djupsjöbacka, M. Björklund, C. Häger-Ross, H. Grip, D.G. Liebermann, Kinematics of fast cervical rotations in persons with chronic neck pain: a cross-sectional and reliability study. BMC Musculoskelet. Disord. **11**(1), 1–10 (2010)
44. F.A. Salem, A.A. Mahfouz, Mechatronics design and implementation education-oriented methodology; a proposed approach. J. Multidiscip. Sci. Technol. **1**(3), 34–45 (2014)
45. M. Santello, M. Flanders, J.F. Soechting, Postural hand synergies for tool use. J. Neurosci. **18**(23), 10105–10115 (1998). https://doi.org/citeulike-article-id:423192
46. A. Saudabayev, H.A. Varol, Sensors for robotic hands: a survey of state of the art. IEEE Access **3**(0) (2015). https://doi.org/10.1109/ACCESS.2015.2482543
47. H.F. Schulte, The characteristics of the McKibben artificial muscle. National Academy of Sciences - National Research Council (1961)
48. G. Smit, D.H. Plettenburg, F.C.T. van der Helm, The lightweight delft cylinder hand: first multi-articulating hand that meets the basic user requirements. IEEE Trans. Neural Syst. Rehabil. Eng. **23**(3), 431–440 (2015)
49. E.E. Swartz, R. Floyd, M. Cendoma, Cervical spine functional anatomy and the biomechanics of injury due to compressive loading. J. Athl. Train. **40**(3), 155 (2005)
50. Y. Tenzer, L.P. Jentoft, R.D. Howe, The feel of mems barometers: inexpensive and easily customized tactile array sensors. IEEE Robot. Autom. Mag. **21**(3) (2014). https://doi.org/10.1109/MRA.2014.2310152
51. N. Vahrenkamp, T. Asfour, R. Dillmann, Efficient inverse kinematics computation based on reachability analysis. Int. J. Humanoid Robot. **9**(4), 1250035 (2012)
52. N. Vahrenkamp, M. Wächter, M. Kröhnert, K. Welke, T. Asfour, The robot software framework ArmarX. Inf. Technol. **57**(2), 99–111 (2015). https://doi.org/10.1515/itit-2014-1066

53. P. Weiner, J. Starke, F. Hundhausen, J. Beil, T. Asfour, The KIT prosthetic hand: design and control, in *IEEE/RSJ International Conference on Intelligent Robots and Systems (IROS)*, Madrid (2018), pp. 3328–3334
54. P. Weiner, F. Hundhausen, R. Grimm, T. Asfour, Detecting grasp phases and adaption of object-hand interaction forces of a soft humanoid hand based on tactile feedback, in *IEEE/RSJ International Conference on Intelligent Robots and Systems (IROS)* (2021), pp. 3956–3963
55. P. Weiner, J. Starke, S. Rader, F. Hundhausen, T. Asfour, Designing prosthetic hands with embodied intelligence: the KIT prosthetic hands. Front. Neurorobot. **16**, 815716 (2022)
56. T. Werner, A. Kargov, I. Gaiser, A. Bierbaum, J. Schill, S. Schulz, G. Bretthauer, Eine fluidisch angetriebene anthropomorphe roboterhand. at - Automatisierungstechnik **58**(12), 681–687 (2010). https://doi.org/10.1524/auto.2010.0877
57. N. Yamaguchi, S. Hasegawa, K. Okada, M. Inaba, A gripper for object search and grasp through proximity sensing, in *2018 IEEE/RSJ International Conference on Intelligent Robots and Systems (IROS)* (2018), pp. 1–9. https://doi.org/10.1109/IROS.2018.8593572
58. C. Zheng, M. Bricogne, J. Le Duigou, B. Eynard, Survey on mechatronic engineering: a focus on design methods and product models. Adv. Eng. Inf. **28**(3), 241–257 (2014). https://doi.org/10.1016/j.aei.2014.05.003. Multiview Modeling for Mechatronic Design

Redundant Robots

Anthony A. Maciejewski and Biyun Xie

Abstract The vast majority of robots in use today operate in very structured environments, e.g., in factory assembly lines, and possess only those limited motion capabilities required to perform specific tasks. While these robots can outperform humans in terms of speed, strength, and accuracy for these tasks, they are no match for the dexterity of human motion. Part of a human's inherent advantage over industrial robots is due to the large number of degrees of freedom in the human body. Articulated, i.e., jointed, motion systems that possess more degrees of freedom than the minimum required to perform a specified task are referred to as kinematically redundant. In an effort to mimic the dexterity of biological systems, researchers have built a number of kinematically redundant robotic systems, e.g., anthropomorphic arms, multi-fingered hands, dual-arm manipulators, and walking machines. While these systems vary in their appearance and intended applications, they all require motion control strategies that coordinate large numbers of joints to achieve the high degree of dexterity possible with redundant systems. This chapter will discuss the issues that arise when designing such strategies, frequently drawing on the use of the singular value decomposition, including the characterization of redundancy, the quantification of dexterity, and the development of efficient and numerically stable motion control algorithms that simultaneously optimize multiple criteria.

A. A. Maciejewski (✉)
Department of Electrical and Computer Engineering, Colorado State University, Fort Collins, CO, USA
e-mail: aam@colostate.edu

B. Xie
Department of Electrical and Computer Engineering, University of Kentucky, Lexington, KY, USA
e-mail: Biyun.Xie@uky.edu

B. Siciliano (ed.), *Robotics Goes MOOC*,
https://doi.org/10.1007/978-3-319-75823-7_2

1   Introduction

By definition, kinematically redundant robots are robots that possess more joints (or degrees of freedom) than the minimum required to complete a specified task [1]. Thus robots may not be intrinsically redundant, but redundant with respect to a particular task. Because six degrees of freedom are required to arbitrarily position and orient an object in space, robots with seven or more joints are commonly referred to as being intrinsically redundant. For example, the robots in Fig. 1 are all referred to as redundant robots. However, it is important to note that there are many tasks that do not require a full six-dimensional task specification, e.g., for arc welding only two-dimensional pointing of the end effector is required, so a traditional six degree-of-freedom robot will be redundant with respect to that task. The advantage of redundancy is that it implies choices, i.e., there are many different ways that the robot can complete the assigned task, and so additional desirable criteria can be optimized. These criteria are frequently associated with dexterity, e.g., avoiding singularities, joint limits, and obstacles in the workspace, or efficiency, e.g., minimizing velocity, torque, or energy. In addition, redundancy by definition allows one to consider fault tolerance. This chapter presents motion planning techniques for effectively using redundancy to achieve all these desirable criteria, under the constraint of completing the desired task.

2   Task-Oriented Kinematics

2.1   Position Level Forward Kinematics

In order to control the motion of a robot, the kinematic structure of the particular robot to be controlled must first be defined. This structure is usually described by a notation like that developed by Denavit and Hartenberg. Using this notation, each degree of freedom within an articulated robot is assigned a unique coordinate system with the relationship between adjacent coordinate systems defined by four parameters. The degrees of freedom of the robot, either rotary or prismatic, are referred to as joints while the interconnecting portions are called links.

The four parameters used to specify the relationships between coordinate systems are the length of the link a, the twist of the link α, the distance between links d, and the angle between links θ. The definition of these parameters for both rotary and prismatic degrees of freedom is illustrated in Fig. 2. More complicated joints, such as ball and socket joints, can be modelled mathematically as combinations of simple joints. It should be pointed out that these parameters only specify the kinematic structure of a robot and do not describe the physical appearance of its links.

Given the above specification of coordinate frames, the relationship between adjacent coordinate frames is given by a rotation of θ, followed by translations of d and a, and a final rotation of α. By combining these transformations, it can be easily

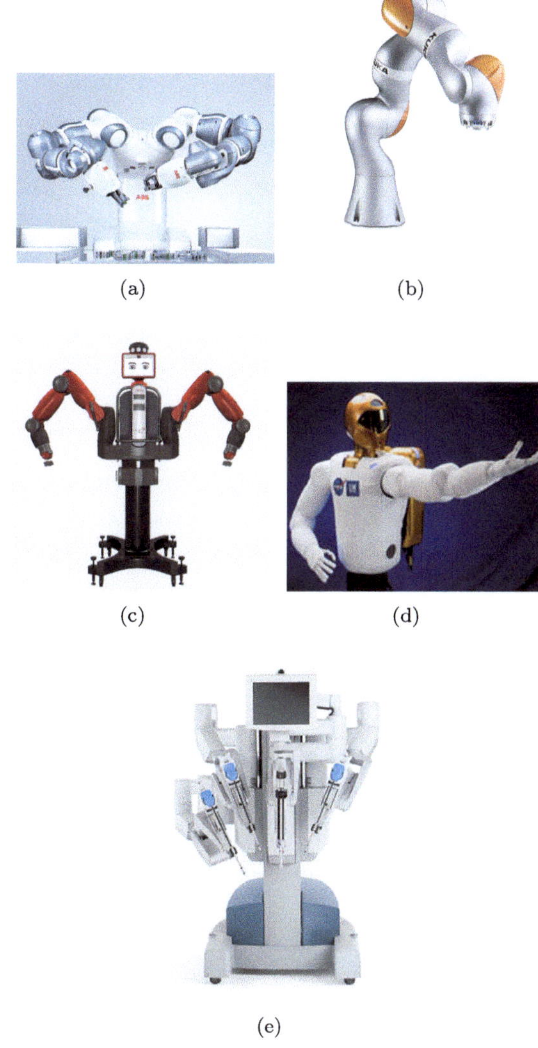

Fig. 1 Several examples of redundant robots include: (**a**) ABB's YuMi, (**b**) KUKA's IIWA, (**c**) Rethink Robotics's Baxter, (**d**) NASA's Robonaut2, and (**e**) Intuitive Surgical's Da Vinci

shown that the relationship between adjacent coordinate frames i and $i - 1$ denoted by $^{i-1}\mathbf{A}_i$ is given by the homogeneous transformation matrix

$$
^{i-1}\mathbf{A}_i =
\left[
\begin{array}{ccc|c}
\multicolumn{3}{c|}{\mathbf{R}_{3\times3}} & \boldsymbol{p}_{3\times1} \\
\hline
0 & 0 & 0 & 1
\end{array}
\right]
\tag{1}
$$

                                                      A. A. Maciejewski and B. Xie

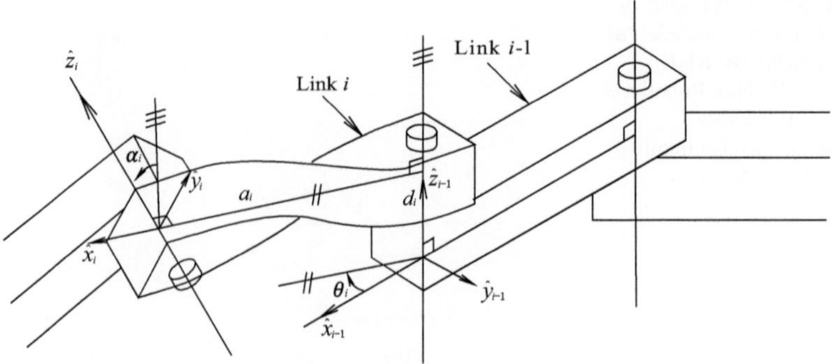

Fig. 2 The Denavit-Hartenberg parameters of a robot are shown

where

$$\mathbf{R} = \begin{bmatrix} \cos\theta_i & -\cos\alpha_i \sin\theta_i & \sin\alpha_i \sin\theta_i \\ \sin\theta_i & -\cos\alpha_i \cos\theta_i & -\sin\alpha_i \cos\theta_i \\ 0 & -\sin\alpha_i & \cos\alpha_i \end{bmatrix} \tag{2}$$

and

$$\boldsymbol{p} = \begin{bmatrix} a_i \cos\theta_i \\ a_i \sin\theta_i \\ d \end{bmatrix}. \tag{3}$$

The above homogeneous transformation matrix has an easily visualized physical interpretation. The submatrix $\mathbf{R}$, which is sometimes called the rotation matrix, is composed of the direction cosines between the two related coordinate frames; thus it also represents the three-dimensional rotation that is required to align the two coordinate frames, so that it is frequently written as

$$\mathbf{R} = \begin{bmatrix} \hat{n} | \hat{o} | \hat{a} \end{bmatrix}. \tag{4}$$

where $\hat{n}$, $\hat{o}$, and $\hat{a}$, are the axes of the ith coordinate frame described in the $(i-1)$th coordinate frame. The vector $\boldsymbol{p}$ is a position vector that specifies the difference in position between the origins of the two coordinate systems.

By multiplying adjacent link transformations, the homogeneous transformation between any two coordinate systems i and j may be computed as follows:

$$^{i}\mathbf{A}_j = {}^{i}\mathbf{A}_{i+1} \, {}^{i+1}\mathbf{A}_{i+2} \cdots {}^{j-1}\mathbf{A}_j. \tag{5}$$

Let the position and orientation of a robot's end effector be represented by a six-dimensional vector x given by

$$x = \begin{bmatrix} p_x \\ p_y \\ p_z \\ \alpha \\ \beta \\ \gamma \end{bmatrix},$$
(6)

where α, β, and γ are a minimal description of the end-effector orientation. The value of x can be computed using (5) given the values of the joint variables, represented by the vector q, whose elements are either θ for rotary joints or d for prismatic joints. This computation is typically referred to as forward kinematics, i.e.,

$$x = f(q).$$
(7)

Note, however, that when applying robots for the completion of some useful task, the reverse problem needs to be solved. That is, given the position and orientation of the robot's end effector, one must find the required joint values

$$q = f^{-1}(x).$$
(8)

This problem, referred to as inverse kinematics, is not as easily solved. This is especially true for kinematically redundant robots because, in general, there are an infinite number of solutions, which are frequently referred to as self-motion manifolds. One approach to characterizing all these solutions is to cast the problem in a differential form using Jacobian-based kinematics.

2.2 Jacobian-Based Kinematics

The relationship between the end-effector velocity, denoted $\dot{x}$, and the joint velocity, denoted $\dot{q}$, is given by

$$\dot{x} = J\dot{q}$$
(9)

where J is the Jacobian matrix [2]. While the end-effector velocity can be of arbitrary dimension m, based on the requirements of the task that the robot is to perform, for general tasks it is frequently represented as

$$\dot{x} = \begin{bmatrix} v \\ \omega \end{bmatrix},$$
(10)

where v is a three-dimensional vector of the desired translational velocity and ω is a three-dimensional vector of the desired orientational velocity. The joint velocity vector $\dot{q}$ is an n-dimensional vector, n being the number of degrees of freedom, where for kinematically redundant robots $n > m$.

Mathematically, the elements of $\mathbf{J}$ are the partial derivatives of the elements of x with respect to the elements of q, however, computing these derivatives is not the most efficient way to compute $\mathbf{J}$. An elegant and efficient method is to geometrically describe the ith column of $\mathbf{J}$ as the velocity created at the end effector due to the velocity of joint i, that is

$$j_i = \left[\frac{p_i \times \hat{a}_i}{\hat{a}_i} \right] \tag{11}$$

for rotational joints and

$$j_i = \left[\frac{\hat{a}_i}{\mathbf{0}} \right] \tag{12}$$

for translational joints, where $\hat{a}_i$ and p_i are in the third and fourth columns, respectively, of the homogeneous transformation matrix $^n\mathbf{A}_{i-1}$. This formulation of the Jacobian results in a minimum amount of computation because the majority of the work has already been done in generating the homogeneous transformations required to compute the forward kinematics.

The objective of defining a smooth linear transformation between the two sets of variables, i.e., the joint variables describing the degrees of freedom of the robot and the convenient task-oriented set of variables, is thus achieved by the Jacobian. In this manner the cumbersome nonlinear relationship between position specification of the two sets of variables is avoided. It is clear from (9) that the desired motion specified by $\dot{x}$ can be achieved by applying the joint velocities given by

$$\dot{\theta} = \mathbf{J}^{-1}\dot{x} \tag{13}$$

if $\mathbf{J}$ is square and nonsingular. One can also describe the relationship between the end effector and the joint variables at the acceleration level by differentiating (9) to obtain

$$\ddot{x} = \mathbf{J}\ddot{\theta} + \dot{\mathbf{J}}\dot{\theta} \tag{14}$$

where once again a solution can be expressed in the form

$$\ddot{\theta} = \mathbf{J}^{-1}(\ddot{x} - \dot{\mathbf{J}}\dot{\theta}) \tag{15}$$

if $\mathbf{J}$ is square and nonsingular.

In both cases, the solution for either the joint angle velocities or accelerations is highly dependent on the properties of $\mathbf{J}$. However, for redundant robots the number of degrees of freedom is by definition greater than the dimension of the specified end-effector velocity. Thus the Jacobian is rectangular and the inverse is not defined, however, there do exist generalized inverses that provide useful solutions to (9). These generalized inverses are best described using the singular value decomposition (SVD), which is the topic of the next section.

3 Singular Value Decomposition

3.1 *Definition in Terms of the Jacobian*

Considering the Jacobian to be composed of columns, as in (11) and (12), provides an intuitive interpretation for how the end effector moves when individual joints are moved. The joints of robots, however, are very seldom moved one at a time. Most frequently some combination of all of the joints must be moved to achieve a desired trajectory. To characterize the Jacobian transformation under these conditions consider the set of all possible different combinations of joint velocities of unit magnitude for the simple planar three degree-of-freedom robot shown in Fig. 3. This can be represented as a sphere in joint velocity space as illustrated in Fig. 4. Due to the directionally dependent scaling of the Jacobian transformation, the velocity at the end effector resulting from all of these possible inputs will in general be described by an ellipse. This ellipse graphically depicts the directions in which it is

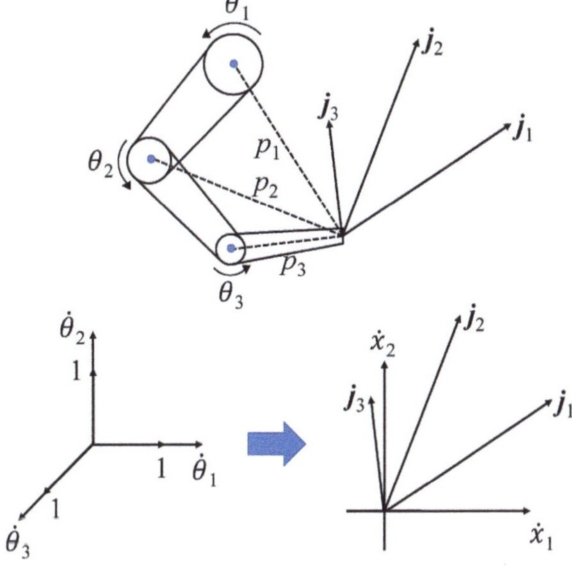

Fig. 3 A physical interpretation of the columns of the Jacobian matrix that describes the kinematic transformation between the joint velocities and the velocity of the end effector

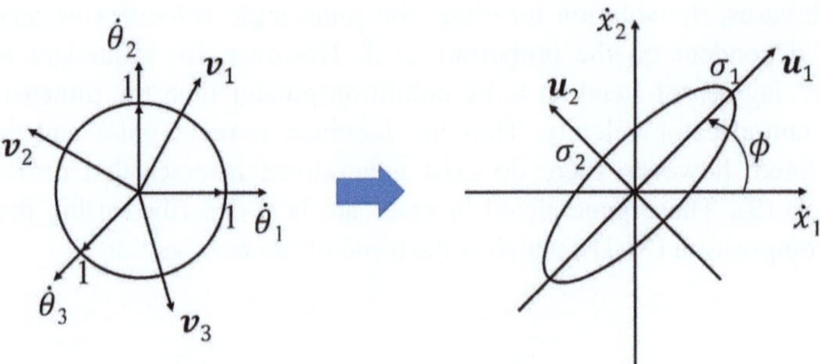

Fig. 4 A geometric interpretation of the singular value decomposition as applied to the Jacobian transformation illustrated in Fig. 3

easier for the end effector to move, i.e., the largest scaling between the input joint velocity input and the end-effector velocity output.

It should be noted that the choice of a coordinate frame for the end effector is rather arbitrary and it can be changed to any convenient one. Because the axes of the ellipse describing the end-effector velocities for all possible joint inputs have a physical significance they represent an ideal choice. Unit vectors along these new axes will be denoted $\hat{u}_1$ and $\hat{u}_2$ for the major axis and minor axis, respectively. This new coordinate frame can be viewed as a simple rotation of the old coordinate frame by an angle ϕ so that vectors defined in one frame can be transformed to the other using the rotation matrix $\mathbf{U}$ given by

$$\mathbf{U} = \begin{bmatrix} \hat{u}_1 & \hat{u}_2 \end{bmatrix} = \begin{bmatrix} \cos\phi & \sin\phi \\ -\sin\phi & \cos\phi \end{bmatrix}. \tag{16}$$

Having thus described a new coordinate frame for the end effector, one might be interested in which particular combination of joint inputs results in motion of the end effector along the vectors $\hat{u}_1$ and $\hat{u}_2$. By rotating the coordinate frame for the joint space one can define a new coordinate frame given by the unit vectors $\hat{v}_1$, $\hat{v}_2$, and $\hat{v}_3$ such that an input along $\hat{v}_1$ results in motion of the end effector along $\hat{u}_1$ and any input along $\hat{v}_2$ results in motion along $\hat{u}_2$. This rotation can be mathematically represented in matrix form as

$$\mathbf{V} = \begin{bmatrix} \hat{v}_1 & \hat{v}_2 & \hat{v}_3 \end{bmatrix}. \tag{17}$$

A joint velocity along $\hat{v}_3$ results in a change in the configuration of the robot without producing any motion at the end effector. This is the null vector of the Jacobian and is a characteristic of redundant robots because they possess more joints than are required to satisfy the desired end-effector velocity.

If the original problem described by (9) is formulated with respect to these new coordinate frames the following equation results

$$\mathbf{U}^\mathrm{T}\dot{\mathbf{x}} = \mathbf{D}\mathbf{V}^\mathrm{T}\dot{\boldsymbol{\theta}} \tag{18}$$

where $\mathbf{U}^\mathrm{T}\dot{\mathbf{x}}$ represents the desired end-effector velocity in the $\hat{\boldsymbol{u}}_1, \hat{\boldsymbol{u}}_2$ coordinate frame, $\mathbf{V}^\mathrm{T}\dot{\boldsymbol{\theta}}$ represents the joint velocity in the $\hat{\boldsymbol{v}}_1, \hat{\boldsymbol{v}}_2, \hat{\boldsymbol{v}}_3$ coordinate frame and $\mathbf{D}$ is a diagonal matrix of the form

$$\mathbf{D} = \begin{bmatrix} \sigma_1 & 0 & 0 \\ 0 & \sigma_2 & 0 \end{bmatrix}. \tag{19}$$

The values σ_1 and σ_2 are known as the singular values and are identically equal to the length of the semi-major and semi-minor axes, respectively, of the ellipse in Fig. 4. These three matrices $\mathbf{U}$, $\mathbf{V}$, and $\mathbf{D}$ are the singular value decomposition of the Jacobian transformation, i.e.,

$$\mathbf{J} = \mathbf{U}\mathbf{D}\mathbf{V}^\mathrm{T}. \tag{20}$$

It is also common to write the singular value decomposition as the summation of vector outer products, which for an arbitrary Jacobian would result in

$$\mathbf{J} = \sum_{i=1}^{\min(m,n)} \sigma_i \hat{\boldsymbol{u}}_i \hat{\boldsymbol{v}}_i^\mathrm{T} \tag{21}$$

where m and n are the number of rows and columns of $\mathbf{J}$ and the singular values are typically ordered from largest to smallest.

The singular values, by specifying how a transformation scales different vectors between the input space and the output space, play a crucial role in determining how well-posed the inverse kinematic problem is. In particular, the condition number of a transformation [3], denoted by κ, is defined as

$$\kappa = \frac{\sigma_{max}}{\sigma_{min}}. \tag{22}$$

The condition number is important because it provides a bound on the worst-case magnification of relative errors. To see why this is true consider the case where the solution of (9) is desired for a given $\dot{\mathbf{x}}$. If the uncertainty in $\dot{\mathbf{x}}$ is denoted by $\delta\dot{\mathbf{x}}$ then the range of possible values of $\dot{\mathbf{x}} + \delta\dot{\mathbf{x}}$ defines a circle in end-effector space as illustrated in Fig. 5. The transformation of this circle into joint space results in an ellipse with major and minor axes aligned with $\hat{\boldsymbol{v}}_1$ and $\hat{\boldsymbol{v}}_2$ and of magnitude equal

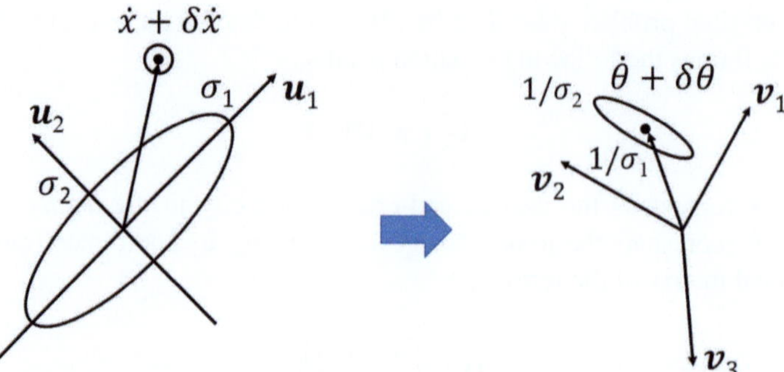

Fig. 5 The transmission of relative uncertainty in the velocity of the end effector to uncertainty in the calculated velocity of the joints. The range of this uncertainty if given by the ratio of the largest to the smallest singular values and is the definition of the condition number

to the reciprocals of their respective singular values. The relative uncertainty in the solution is therefore bounded by

$$\frac{\|\delta\dot{\boldsymbol{\theta}}\|}{\|\dot{\boldsymbol{\theta}}\|} \leq \kappa \frac{\|\delta\dot{\boldsymbol{x}}\|}{\|\dot{\boldsymbol{x}}\|}. \tag{23}$$

It is easy to verify that the worst case denoted by the equality will occur if $\dot{x}$ is entirely along $\hat{u}_1$ and $\delta\dot{x}$ is entirely along $\hat{u}_2$. Thus by monitoring the maximum and minimum singular values one can assess the amount of confidence that should be placed in a computed result.

3.2 Singularities and Measures of Dexterity

From the above discussion on singular values it should be clear that the minimum singular value plays an important role in determining the conditioning of the Jacobian and therefore, in obtaining usable solutions to (9). Because the conditioning of (9) becomes worse as σ_{min} decreases one might want to consider the worst case, for example what happens when σ_{min} is equal to zero? In this case $\mathbf{J}$ is referred to as singular and an exact solution to (9) will not exist for an arbitrary desired end-effector velocity [4]. To examine the physical significance of this situation, consider the configuration of the robot shown in Fig. 6. The robot is approaching the singular configuration that occurs when the robot is completely stretched out. One can see from the figure that the effect of joint rotations about each of the three joint axes result in end-effector velocities that are in approximately the same direction, i.e., in mathematical terms they are becoming linearly dependent. It should also be physically intuitive that it becomes increasingly easier to move in the direction of $\hat{u}_1$ while velocities along $\hat{u}_2$ become increasingly more difficult. At the actual

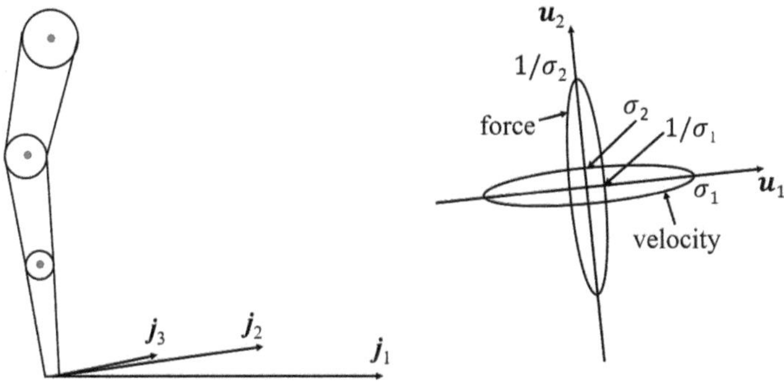

Fig. 6 The physical effects of approaching a singularity. The achievable velocity of the end effector becomes restricted to the nonsingular direction. However, because of the duality of the static force and velocity relationships defined by the Jacobian, the singularity also results in the ability to balance large forces in the singular direction

singularity there is no combination of joint velocities that can result in an end-effector velocity along $\hat{u}_2$. However, it should be pointed out that the singular configuration may be desirable due to its static force properties. In particular, the static force at the end effector, denoted f, is related to the torque at the joints, denoted τ, by the transpose of the Jacobian, i.e.,

$$\tau = \mathbf{J}^\mathrm{T} f \tag{24}$$

so that at the singularity the robot can balance forces along $\hat{u}_2$ without any torque at the joints.

Because the singular values quantify how easy it is for a robot to move its end effector in the various directions of the task space, most measures of a robot's dexterity are some function of the singular values of the Jacobian matrix. It is important to point out that when both the end-effector linear velocity and orientational velocity are considered, a scaling is needed for the rows of the Jacobian to eliminate the difference in units. Analogously, when the robot includes both prismatic joints and rotational joints, a scaling is needed for the columns of the Jacobian. The most common of these is perhaps the manipulability measure proposed by Yoshikawa [5, 6] which is defined as the square root of the determinant of the matrix $\mathbf{J}\mathbf{J}^\mathrm{T}$ which is simply the product of the singular values of $\mathbf{J}$. Other proposed measures include the trace of the Jacobian, the minimum singular value of the Jacobian [3], the compatibility index [7], and isotropy (all equal singular values) [8].

3.3 Computing the Singular Value Decomposition

All of the above dexterity measures have a physical significance and justification for their use, but the key point here is that they are all closely linked to the SVD. Calculation of the SVD has a reputation for being numerically expensive, however, with the right formulation it is computationally feasible for use in real-time control. Traditionally, the computation of the SVD of an arbitrary matrix is an iterative procedure so that the exact number of computations cannot be known a priori [9]. However, the control of robotic systems is not based on the solution of arbitrary matrix equations but involves the solution of the inverse kinematics based on the Jacobian matrix. The current Jacobian for a robot is a perturbation of a previously known matrix for which perturbation bounds on the singular values and singular vectors can be established. This knowledge of the previous state can be exploited during the current computation of the SVD in order to reduce the overall computational burden. The following algorithm, based exclusively on Givens rotations, illustrates how to take advantage of these incremental perturbations [10].

Givens rotations are orthogonal transformations of the form

$$
Q = \begin{bmatrix}
1 & & & & \cdot & & & & & \cdot & & & \\
& \ddots & & & \cdot & & & & & \cdot & & & \\
& & 1 & & \cdot & & & & & \cdot & & & \\
\cdot & \cdot & \cdot & \cos(\theta) & \cdot & \cdot & \cdot & -\sin(\theta) & \cdot & \cdot & \cdot & & i \\
& & & \cdot & 1 & & & & \cdot & & & & \\
& & & \cdot & & \ddots & & & \cdot & & & & \\
& & & \cdot & & & 1 & & \cdot & & & & \\
\cdot & \cdot & \cdot & \sin(\theta) & \cdot & \cdot & \cdot & \cos(\theta) & \cdot & \cdot & \cdot & & k \\
& & & \cdot & & & & & 1 & & & & \\
& & & \cdot & & & & & \cdot & \ddots & & & \\
& & & \cdot & & & & & \cdot & & 1 &
\end{bmatrix} \quad\quad i \quad\quad\quad\quad\quad k
$$

$$(25)$$

where all other elements not shown are zero. This transformation can be geometrically interpreted as a plane rotation of θ in the $i - k$ plane. In contrast to Householder reflections, Givens rotations only affect two rows or columns of the matrix with which they are multiplied, a property that is useful when designing parallel computing structures based on this transformation.

The most important property of Givens rotations is their ability to orthogonalize the two rows or columns on which they operate, which can be used to compute the SVD of the Jacobian. In particular consider an orthogonal matrix $\mathbf{V}$, composed of successive Givens rotations, such that

$$\mathbf{JV} = \mathbf{B} \tag{26}$$

where the columns of $\mathbf{B}$ are orthogonal. If the columns of $\mathbf{B}$ are orthogonal then it can be written as the product of an orthogonal matrix $\mathbf{U}$ and a diagonal matrix $\mathbf{D}$

$$\mathbf{B} = \mathbf{UD} \tag{27}$$

by letting the columns of $\mathbf{U}$ be equal to normalized versions of the columns of $\mathbf{B}$,

$$\mathbf{u}_i = \frac{\mathbf{b}_i}{\|\mathbf{b}_i\|} \tag{28}$$

and defining the diagonal elements of $\mathbf{D}$ to be equal to the norm of the columns of $\mathbf{B}$

$$d_{ii} = \|\mathbf{b}_i\|. \tag{29}$$

By substituting (27) into (26) and solving for $\mathbf{J}$ one obtains

$$\mathbf{J} = \mathbf{UDV}^{\mathrm{T}} \tag{30}$$

which is the SVD of $\mathbf{J}$.

The critical step in the above procedure for calculating the SVD is determining the orthogonal matrix $\mathbf{V}$ that will orthogonalize the columns of $\mathbf{J}$. This matrix is usually formed as a product of Givens rotations, each of which is designed to orthogonalize two columns. Considering the current ith and kth columns of $\mathbf{J}$, multiplication by a Givens rotation results in the new columns, $\mathbf{j}'_i$ and $\mathbf{j}'_k$ given by

$$\mathbf{j}'_i = \mathbf{j}_i \cos(\theta) + \mathbf{j}_k \sin(\theta) \tag{31}$$

$$\mathbf{j}'_k = \mathbf{j}_k \cos(\theta) - \mathbf{j}_i \sin(\theta). \tag{32}$$

The constraint that these columns be orthogonal results in

$$\mathbf{j}'^{\mathrm{T}}_i \mathbf{j}'_k = 0 = \mathbf{j}^{\mathrm{T}}_i \mathbf{j}_k (\cos^2(\theta) - \sin^2(\theta)) + (\mathbf{j}^{\mathrm{T}}_k \mathbf{j}_k - \mathbf{j}^{\mathrm{T}}_i \mathbf{j}_i) \sin(\theta) \cos(\theta). \tag{33}$$

The terms in the Givens rotation matrix to achieve orthogonality can be computed by using

$$p = \boldsymbol{j}_i^{\mathrm{T}} \boldsymbol{j}_k \tag{34}$$

$$q = \boldsymbol{j}_i^{\mathrm{T}} \boldsymbol{j}_i - \boldsymbol{j}_k^{\mathrm{T}} \boldsymbol{j}_k \tag{35}$$

$$v = \sqrt{4p^2 + q^2} \tag{36}$$

so that for $q \geq 0$

$$\cos(\theta) = \sqrt{\frac{v+q}{2v}} \quad \text{and} \quad \sin(\theta) = \frac{p}{v\cos(\theta)} \tag{37}$$

and for $q < 0$

$$\sin(\theta) = \mathrm{sgn}(p)\sqrt{\frac{v-q}{2v}} \quad \text{and} \quad \cos(\theta) = \frac{p}{v\sin(\theta)} \tag{38}$$

where

$$\mathrm{sgn}(p) = \begin{cases} 1 & \text{if } p \geq 0 \\ -1 & \text{if } p < 0 \end{cases}. \tag{39}$$

The two sets of formulas are given so that ill-conditioned equations resulting from the subtraction of nearly equal numbers can always be avoided.

The above discussion shows how to determine a single Givens rotation that will orthogonalize two columns of a given matrix. It still needs to be shown how the matrix $\mathbf{V}$ can be computed from these elementary rotations. If the Givens rotation to orthogonalize columns i and k is denoted by $\mathbf{V}_{ik}$ then the product of a set of $n(n-1)/2$ rotations

$$\prod_{i=1}^{n-1} \left(\prod_{k=i+1}^{n} \mathbf{V}_{ik} \right) \tag{40}$$

is referred to as a sweep. Unfortunately, a single sweep will not, in general, orthogonalize all of the columns of a matrix because subsequent rotations can destroy the orthogonality produced by previous ones. However, the procedure can be shown to converge so that $\mathbf{V}$ can be obtained as the product of l sweeps, where l is not known a priori. Convergence of the algorithm is based on completing an entire sweep with all of the columns being orthogonal. Orthogonality is measured by the parameter α defined as

$$\alpha = \frac{(j_i^T j_k)^2}{(j_i^T j_i)(j_k^T j_k)} \tag{41}$$

dropping below a preset threshold. If for two columns α is below the threshold then the rotation is not performed. The above algorithm, by virtue of being composed exclusively of Givens rotations, can be highly parallelized.

The algorithm for computing the SVD using Givens rotations outlined above uses successive sweeps in order to make the columns of a matrix more orthogonal. The more orthogonal the columns are to begin with, the fewer the number of sweeps required for convergence. If one considers the current Jacobian to be a perturbation of the previous Jacobian

$$\mathbf{J}(t + \Delta t) = \mathbf{J}(t) + \Delta\mathbf{J}(t) \tag{19}$$

the SVD of which is known and given by

$$\mathbf{J}(t) = \mathbf{U}(t)\mathbf{D}(t)\mathbf{V}^T(t) \tag{20}$$

then the matrix $\mathbf{J}(t + \Delta t)\mathbf{V}(t)$ will have nearly orthogonal columns provided the perturbation $\Delta\mathbf{J}(t)$ is small relative to $\mathbf{J}(t)$. The foundation of the above lies in the fundamentally well-behaved nature of the SVD of a matrix. The perturbation bounds on singular values are very well-known and it is easy to show that

$$|\sigma_i(\mathbf{J}(t + \Delta t)) - \sigma_i(\mathbf{J}(t))| \leq \|\Delta\mathbf{J}(t)\| \tag{21}$$

The perturbation bounds on the rotation of subspaces defined by singular vectors are not as widely known but are also well-behaved. Motivated by these perturbation bounds, after computing $\mathbf{J}(t + \Delta t)\mathbf{V}(t)$ one only needs to perform a single sweep to completely orthogonalize the columns of $\mathbf{J}(t + \Delta t)$, thereby significantly reducing the computational expense and removing the need for convergence tests.

4 Differential Inverse Kinematics

4.1 Pseudoinverse Solution

The inverse kinematics problem given by (9) can be solved using (13) only if $\mathbf{J}$ is square and nonsingular. However, for redundant robots $\mathbf{J}$ is always rectangular because there are more joints than required to satisfy the desired end-effector velocity. In these cases there may exist no exact solution, one solution, or an infinite number of solutions depending on $\mathbf{J}$ at the current configuration and the desired $\dot{x}$. To obtain meaningful solutions to non-square (and possibly singular) systems of equations that occur with redundant robots, pseudoinverses are commonly suggested. The pseudoinverse solution can be considered as the best possible approximate solution because it results in the least squares solution of minimum norm. Mathematically, it is the value of $\dot{\theta}$ that minimizes the residual, i.e.,

$$\min_{\dot{\theta}} \| \dot{x} - \mathbf{J}\dot{\theta} \| \tag{42}$$

and if this $\dot{\theta}$ is not unique then the $\dot{\theta}$ of smallest Euclidean norm, i.e., minimum $\|\dot{\theta}\|$, is selected. This physically means that the end-effector velocity will be as close to the desired velocity as possible. However, for redundant robots there will, in general, be an infinite number of joint velocities that minimize the residual. The pseudoinverse solution is unique in that it also minimizes the norm of the joint velocity under the constraint of minimizing the residual. This in effect means that there is no unnecessary joint motion in achieving the desired end-effector velocity. The pseudoinverse solution is very easily obtained from the singular value decomposition by simply taking the reciprocal of all nonzero singular values. In particular, the pseudoinverse of $\mathbf{J}$, denoted by $\mathbf{J}^+$ is given by

$$\mathbf{J}^+ = \sum_{i=1}^{r} \frac{1}{\sigma_i} \hat{v}_i \hat{u}_i^T \tag{43}$$

where r is the rank of $\mathbf{J}$, which is by definition the number of nonzero singular values. For $\mathbf{J}$ that are of full rank, the pseudoinverse is frequently written as

$$\mathbf{J}^+ = (\mathbf{J}^T\mathbf{J})^{-1}\mathbf{J}^T \quad \text{if } m > n = r \tag{44}$$

or

$$\mathbf{J}^+ = \mathbf{J}^T(\mathbf{J}\mathbf{J}^T)^{-1} \quad \text{if } r = m > n \tag{45}$$

although the pseudoinverse should not be computed in this manner if $\mathbf{J}$ is ill conditioned.

The ability of the pseudoinverse to provide meaningful solutions to any linear system of equations regardless of whether it is underdetermined, overdetermined, or even singular accounts for its popularity. However, there are still fundamental drawbacks with using true pseudoinverse solutions for any equations describing the motion of robots. This is due to the fact that such equations are not isolated mathematical formulas but they represent how the robot's configuration evolves over time. As such, one is repeatedly solving the inverse kinematic equations that are slightly perturbed from the previous set of equations. From a physical point of view there must be continuity in the solutions of these equations and it is on this point that pseudoinverse solutions fail. As an example, consider the SVD of the Jacobian for the robot in Fig. 6, where the diagonal matrix of singular values is given by (19). While σ_2 remains nonzero the pseudoinverse of **D** will be given by

$$\mathbf{D}^+ = \begin{bmatrix} \frac{1}{\sigma_1} & 0 \\ 0 & \frac{1}{\sigma_2} \\ 0 & 0 \end{bmatrix}. \tag{46}$$

however, when the robot moves into the singular configuration and σ_2 becomes zero then the pseudoinverse becomes

$$\mathbf{D}^+ = \begin{bmatrix} \frac{1}{\sigma_1} & 0 \\ 0 & 0 \\ 0 & 0 \end{bmatrix}. \tag{47}$$

The difficulty therefore is not *at* the singularity, where the pseudoinverse provides a perfectly reasonable solution, it is the *discontinuous transition* between singular and non-singular configurations. Unfortunately it is also at this transition that the equations are most ill conditioned. Furthermore, even if the equations could be solved exactly, note that the use of (46) for $\sigma_2 \to 0$ results in solution norms that approach infinity, i.e., joint velocities that are not physically possible.

The large norm of pseudoinverse solutions that is characteristic of proximity to singularities may at first seem paradoxical for a technique that is guaranteed to yield the minimum norm solution. This apparent paradox, however, can be resolved by considering the least-squares criterion of achieving the end-effector velocity to take priority over minimizing the joint velocity. Thus the solution of minimum norm is chosen from among only those solutions that have already satisfied the least-squares criterion. This explains the discontinuity encountered in pseudoinverse solutions at singularities because it represents a switch in the optimization criterion. The decrease in rank encountered at a singularity means that a component of joint angle velocity can no longer be used to improve the least-squares criterion and is therefore set to zero in order to satisfy the minimum norm criterion. Thus one method of removing this discontinuity and also limiting the maximum solution norm is to consider both criteria simultaneously.

4.2 *Damped Least Squares*

Properties of Damped Least-Squares Solutions The concept of considering a solution's norm along with the accuracy to which it solves a set of linear equations was first proposed by Levenberg where he coined the phrase "damped least squares". Since that time the use of damped least squares, also referred to as singularity robustness, has become one of a number of common techniques for obtaining solutions to ill-conditioned equations. In terms of the inverse kinematics problem specified by (9) the damped least-squares criterion requires a solution that minimizes the sum

$$\|\dot{x} - \mathbf{J}\dot{\theta}\|^2 + \lambda^2\|\dot{\theta}\|^2 \tag{48}$$

where λ is a weighting factor, sometimes referred to as the damping factor, which can be used to set the relative importance of the minimum residual criterion, i.e., end-effector tracking error, versus the norm of the joint velocity [11, 12]. The solution that satisfies this criterion for a particular value of λ will be denoted as $\dot{\theta}^{(\lambda)}$. The damped least-squares criterion results in the augmented system of equations

$$\begin{bmatrix} \mathbf{J} \\ \lambda\mathbf{I} \end{bmatrix}\dot{\theta} = \begin{bmatrix} \dot{x} \\ \mathbf{0} \end{bmatrix} \tag{49}$$

where the solution can be obtained by solving the consistent set of equations

$$(\mathbf{J}^{\mathrm{T}}\mathbf{J} + \lambda^2\mathbf{I})\dot{\theta} = \mathbf{J}^{\mathrm{T}}\dot{x} \tag{50}$$

which results in $\dot{\theta}^{(\lambda)}$. It is easily shown that this $\dot{\theta}^{(\lambda)}$ represents the solution with the minimal residual over all $\dot{\theta}$ whose norm does not exceed that of $\dot{\theta}^{(\lambda)}$.

The properties of the damped least-squares solution are perhaps best revealed by using the SVD. By writing $\dot{x}$ in terms of the basis specified by the output singular vectors

$$\dot{x}_i = \hat{u}_i^{\mathrm{T}}\dot{x} \tag{51}$$

the components of the damped least-squares solution can be written as

$$\dot{\theta}_i^{(\lambda)} = \frac{\sigma_i}{\sigma_i^2 + \lambda^2}\hat{v}_i\dot{x}_i. \tag{52}$$

It is important to note that inclusion of the vector norm criterion into the least-squares process only affects the magnitude of the singular values; the singular vectors remain unchanged. From (52) one can see that for σ_i much larger than λ the damped least-squares solution has little effect because

$$\frac{\sigma_i}{\sigma_i^2 + \lambda^2} \approx \frac{1}{\sigma_i} \tag{53}$$

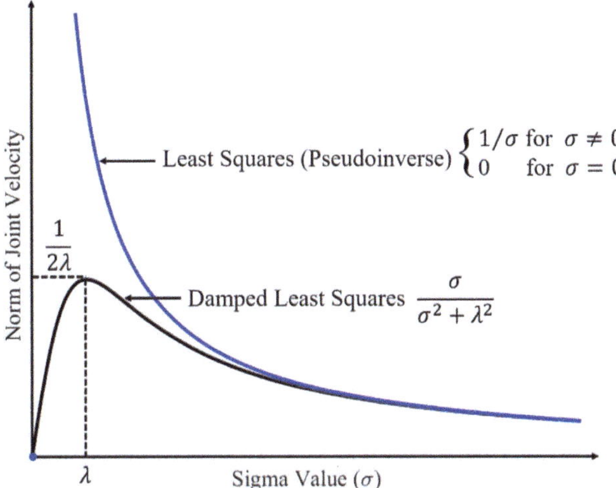

Fig. 7 A comparison of the damped least-squares solution to that of the pseudoinverse as a function of the singular value. The maximum possible value of the joint velocity is determined by the damping factor λ. The damped least-square solution also resolves the discontinuity of the pseudoinverse solution at a singularity ($\sigma = 0$)

which results in approximately the same value as the standard least squares. However for σ_i that are on the order of λ, the λ term in the denominator "damps" the potentially high norm of that component of the solution so that

$$\max(\|\dot{\boldsymbol{\theta}}_i^{(\lambda)}\|) = \frac{\dot{x}_i}{2\lambda} \tag{54}$$

which occurs when $\sigma_i = \lambda$. For σ_i smaller than λ, $\dot{\boldsymbol{\theta}}_i^{(\lambda)}$ approaches zero as σ_i approaches zero which demonstrates the desired continuity in the solution in spite of the change in the rank of $\mathbf{J}$. A plot of the norm of a component of the damped least-squares solution as compared to the standard least squares as a function of σ_i is presented in Fig. 7.

The norm of the complete solution is obtained by combining each of the components from (52) which results in

$$\|\dot{\boldsymbol{\theta}}^{(\lambda)}\|^2 = \sum_{i=1}^{r} \dot{x}_i^2 \left[\frac{\sigma_i}{\sigma_i^2 + \lambda^2} \right]^2 \tag{55}$$

where r is the rank of $\mathbf{J}$. The above equation clearly demonstrates that for positive λ the norm is monotonically decreasing and approaches zero as λ approaches infinity. This decrease in norm, however, is unavoidably accompanied by an increase in the resultant residual. The residual, as a function of λ, is given by

$$\|\dot{x} - \mathbf{J}\dot{\theta}^{(\lambda)}\|^2 = \sum_{i=1}^{r} \dot{x}_i^2 \left[\frac{\lambda^2}{\sigma_i^2 + \lambda^2} \right]^2 + \sum_{i=r+1}^{m} \dot{x}_i^2. \tag{56}$$

The second term of the residual represents that portion of $\dot{x}$ that is outside of the range space of $\mathbf{J}$ and is therefore not a function of λ. The first term illustrates how the damping factor λ affects the resultant residual, the magnitude of which is a monotonically increasing function with a minimum at $\lambda = 0$.

The characteristics of the damped least-squares solution are similar to those obtained using the truncated SVD. The truncated SVD solution of a system of linear equations described by (9), denoted here by $\dot{\theta}^{(k)}$, is defined as

$$\dot{\theta}^{(k)} = \sum_{i=1}^{k} \frac{1}{\sigma_i} v_i u_i^{\mathrm{T}} \dot{x} \tag{57}$$

where k is an integer less than or equal to the rank r. The truncated SVD reduces the solution norm by removing all components of the solution that correspond to small singular values while retaining all of those associated with larger singular values. The parameter k is used to define small and large such that σ_i for $i \leq k$ are large and σ_i for $i > k$ are considered small. It can be shown that $\dot{\theta}^{(k)}$ is the minimum residual solution for all $\dot{\theta}$ in the k-dimensional subspace spanned by v_i for $i \leq k$. For cases where $\sigma_k \gg \sigma_{k+1}$ such that there is a large gap between the large and small singular values and λ is chosen to be midway between σ_k and σ_{k+1}, the results for the two types of solutions will be approximately the same.

Determining an Appropriate Damping Factor One of the difficulties in successfully applying the damped least-squares formulation is determining the optimum value for the damping factor λ. If there exist hard constraints on the maximum allowable joint velocity then an iterative technique based on the monotonic behavior of (55) can be applied. While this requires knowledge of the singular value decomposition, it is possible to approximate this solution with just an estimate of the minimum singular value. The justification for this is due to the fact that bounds on the norm of the joint angle velocity can be determined solely on the basis of the minimum singular value of $\mathbf{J}$.

Note that one can satisfy a constraint on the maximum joint angle velocity over all unit norm commanded end-effector velocities, denoted by $\dot{\theta}_{max}$, by specifying a damping factor based on (54), without even considering the minimum singular value. However, this value of λ will be much too conservative for those cases where the minimum singular value is greater than the damping factor, which can be seen from Fig. 7. Intuitively, if the value of $\|\dot{\theta}\|$ for which the residual is equal to zero is less than $\dot{\theta}_{max}$ then $\lambda = 0$, otherwise λ would take on the value that results in $\|\dot{\theta}\| = \dot{\theta}_{max}$. In physical terms, if the joint velocity that exactly tracks the desired end-effector trajectory is physically achievable then it should be used, otherwise the optimal solution requires that the joint velocity norm be at its limit. The value of λ that achieves this can be approximated based on the minimum singular value, obtained from (55) which results in

$$\lambda(\dot{\theta}_{max}) = \sqrt{\frac{\sigma_{min}}{\dot{\theta}_{max}} - \sigma_{min}^2} \tag{58}$$

for those cases where $\sigma_{min} > \lambda$.

Implementation of Damped Least-Squares Solutions with Minimum Singular Values Estimation The damped least-squares solution of (9) is given by (52), however, one may not want to explicitly compute the SVD. In this case, the solution can be obtained by solving (50), but it is preferable to compute it by using the mathematically equivalent form

$$\dot{\boldsymbol{\theta}}^{(\lambda)} = \mathbf{J}^{\mathrm{T}}(\mathbf{J}\mathbf{J}^{\mathrm{T}} + \lambda^2\mathbf{I})^{-1}\dot{\boldsymbol{x}} \tag{59}$$

especially when dealing with redundant manipulators. For redundant manipulators $m < n$ so that solutions based on the m by m matrix $\mathbf{J}\mathbf{J}^{\mathrm{T}} + \lambda^2\mathbf{I}$ require fewer calculations than the n by n matrix $\mathbf{J}^{\mathrm{T}}\mathbf{J} + \lambda^2\mathbf{I}$. The solution of (59) is best performed in two parts by first solving

$$(\mathbf{J}\mathbf{J}^{\mathrm{T}} + \lambda^2\mathbf{I})z = \dot{\boldsymbol{x}} \tag{60}$$

for z and then substituting it into (59) to obtain

$$\dot{\boldsymbol{\theta}}^{(\lambda)} = \mathbf{J}^{\mathrm{T}}z \tag{61}$$

to avoid computing an explicit inverse.

The solution of (60) presents an ideal place to obtain an estimate of the minimum singular value of $\mathbf{J}$ [13, 14]. The solution for an additional right-hand side of (60), chosen for minimum singular value estimation, represents a minimal amount of additional computation. For this reason (60) is modified to the partitioned matrix equation

$$(\mathbf{J}\mathbf{J}^{\mathrm{T}} + \lambda^2\mathbf{I})\left[z \ \vdots \ \tilde{\boldsymbol{u}}'_m \right] = \left[\dot{\boldsymbol{x}} \ \vdots \ \tilde{\boldsymbol{u}}_m \right] \tag{62}$$

where $\tilde{\boldsymbol{u}}_m$ is a unit vector designed to optimize the estimate of the minimum singular value. The vector $\tilde{\boldsymbol{u}}_m$ can be written in terms of the basis specified by the output singular vectors of $\mathbf{J}$ as

$$\tilde{\boldsymbol{u}}_m = \sum_{i=1}^{m} a_i \hat{\boldsymbol{u}}_i \tag{63}$$

were the a_i denote the component along the respective singular vector. The solution for $\tilde{u}'_m$ can then be obtained from (62) as

$$\tilde{u}'_m = (\mathbf{J}\mathbf{J}^\mathrm{T} + \lambda^2 \mathbf{I})^{-1}\tilde{u}_m = \sum_{i=1}^{m} \frac{a_i}{\sigma_i^2 + \lambda^2}\hat{u}_i. \tag{64}$$

If $\tilde{u}_m$ has a strong component in the direction of $\hat{u}_m$ so that $a_m \approx 1$, then

$$\|\tilde{u}'_m\| \approx \frac{1}{\sigma_m^2 + \lambda^2} \tag{65}$$

and for a known value of λ an estimate of the minimum singular value can be obtained. This technique is an application of the singular value estimation procedure used in numerical analysis packages involving matrix computations and is based on the inverse iteration method for computing eigenvalues.

In order to obtain a good estimate of the minimum singular value, the vector $\tilde{u}_m$ must contain a significant component within the subspace spanned by the singular vectors associated with small singular values. In particular, for those cases where there is only one minimum singular value, $\tilde{u}_m$ must be nearly identical (within a sign) to $\hat{u}_m$. This vector is maintained by first setting $\tilde{u}_m$ exactly to $\hat{u}_m$ before starting the trajectory. Then at every computation cycle time when (62) is solved the vector $\tilde{u}_m$ is replaced by a normalized version of $\tilde{u}'_m$. In this manner, as the singular vector associated with the minimum singular value rotates, $\tilde{u}_m$ is able to rotate along with it, always maintaining a strong component along that direction.

The estimate for the minimum singular value of $\mathbf{J}$ can now be used to set an appropriate damping factor in (58). There is an apparent circular dilemma in that λ must be known in order to estimate σ_m from (65) yet the reason for calculating σ_m is to be able to set λ to an appropriate value. This dilemma is resolved by setting the damping factor using the previous estimate of the minimum singular value. Because singular values are well conditioned, the bounds on their rates of change guarantee that the change in the minimum singular value will be small during Δt so that the above value of the damping factor can be used.

Numerical Filtering of Singular Components One of the limitations of the damped least-squares approach is a result of the uniform damping factor applied to all singular values. From a physical point of view it would be desirable if only those components of the end effector velocity that are difficult to achieve would be damped when a robot is near a singular configuration. The effect of such a formulation would be similar to a continuous version of the truncated SVD solution where error introduced by the damping of acceptable singular values is eliminated. The advantages of such a formulation are the removal of unnecessary end effector tracking errors due to the presence of small singular values even though the desired end effector velocity has no component in the singular directions. Such a solution requires replacing the identity matrix in (62) with a more general matrix designed to selectively filter different components to varying degrees.

The estimate $\tilde{u}_m$ can be used to achieve a solution where the component associated with the smallest singular value is damped more than the others. By replacing (62) with

$$(\mathbf{J}\mathbf{J}^{\mathrm{T}} + \alpha^2 \tilde{u}_m \tilde{u}_m^{\mathrm{T}} + \lambda^2 \mathbf{I}) \left[z \,\vdots\, \tilde{u}'_m \right] = \left[\dot{x} \,\vdots\, \tilde{u}_m \right] \tag{66}$$

the filter gain α can be used to provide further damping of the component with the smallest singular value in addition to the overall damping factor λ. By using (66) in place of (62) the solution norm is now given by

$$\|\dot{\boldsymbol{\theta}}^{(\alpha,\lambda)}\|^2 \approx \sum_{i=1}^{m-1} \dot{x}_i^2 \left[\frac{\sigma_i}{\sigma_i^2 + \lambda^2} \right]^2 + \dot{x}_m^2 \left[\frac{\sigma_m}{\sigma_m^2 + \alpha^2 + \lambda^2} \right]^2 \tag{67}$$

where the accuracy of the approximation depends on the accuracy to which $\tilde{u}_m$ matches $\hat{u}_m$. As a robot approaches a singularity, it is the last term in (67) that induces the high joint velocities. Without the filter gain α, the overall damping factor λ would be forced to increase, unnecessarily damping the well-behaved components in the summation term and resulting in an unnecessary end effector tracking error. By introducing an additional mechanism for selectively applying damping, the filter gain α can be set based on the estimate of the minimum singular value obtained from (65) which then allows a reduction in the overall damping factor. Note that these advantages are present regardless of the value of $\dot{x}_m$.

The computation of an appropriate overall damping factor λ is based on establishing an effective singular value for the specific commanded end effector velocity. This effective singular value is based on modeling the commanded end effector velocity $\dot{x}$ as being composed of two components, one within the possibly singular subspace defined by $\tilde{u}_m$ and denoted by $\dot{x}_s$, with the other component being in the remaining orthogonal subspace denoted by $\dot{x}_o$. These two components are easily calculated using

$$\dot{x}_s = \tilde{u}_m \tilde{u}_m^{\mathrm{T}} \dot{x} \tag{68}$$

and

$$\dot{x}_o = \dot{x} - \dot{x}_s. \tag{69}$$

The additional information required for establishing an effective singular value is obtained from the vector z calculated from the previous computation interval. From (66) one can write z as

$$z \approx \left[\sum_{i=1}^{m-1} \frac{\dot{x}_i \hat{u}_i}{\sigma_i^2 + \lambda^2} \right] + \frac{\dot{x}_s}{\sigma_m^2 + \alpha^2 + \lambda^2} \tag{70}$$

where once again the accuracy of the approximation depends on the accuracy to which $\tilde{u}_m$ matches $\hat{u}_m$. All of the values in the last term of (70) are known so that the portion of z that results from $\dot{x}_o$, denoted here by z_o, can be calculated from

$$z_o = z - \frac{\dot{x}_s}{\sigma_m^2 + \alpha^2 + \lambda^2} \tag{71}$$

A measure of the effective singular value outside of the singular subspace, denoted by σ_o, can then be obtained by modeling the summation term in (70) by a one-dimensional space resulting in

$$\|z_o\| = \frac{\|\dot{x}_o\|}{\sigma_o^2 + \lambda^2} \tag{72}$$

or

$$\sigma_o^2 = \frac{\|\dot{x}_o\|}{\|z_o\|} - \lambda^2. \tag{73}$$

A measure of the overall effective singular value for the current commanded end effector velocity can then be obtained by comparing the components $\dot{x}_s$ and $\dot{x}_o$. In particular, the overall effective singular value, denoted by σ_e is given by

$$\frac{\|\dot{x}\|}{\sigma_e} = \frac{\|\dot{x}_s\|}{\sigma_m} + \frac{\|\dot{x}_o\|}{\sigma_o}. \tag{74}$$

The entire motivation for developing (74) is to provide a means of incorporating information about how the current commanded end effector velocity will affect the resulting solution. Equation (74) provides an estimate of what the resulting joint angle velocity norm would be if no damping were applied. This effective singular value is used in place of the minimum singular value (58) for calculating an appropriate overall damping factor.

4.3   *Truncated SVD*

The use of damped least-squares provides the optimal solution for tracking a given end-effector trajectory under the constraint of minimizing the joint norm. While this solution is optimal with respect to the joint norm, it may be undesirable to implement due to the iterative nature of calculating the appropriate damping factor λ. The characteristics of the damped least-squares solution are very similar to those obtained using the truncated SVD solution, which does not require any iterations to satisfy a joint norm constraint of $\dot{\theta}_{max}$. By modifying the truncated SVD to be continuous instead of a stepwise function of k, a solution that is virtually identical to the damped least-squares solution can be obtained. This type of solution will be denoted $\dot{\boldsymbol{\theta}}^{(c)}$ and is defined by

$$\dot{\boldsymbol{\theta}}^{(c)} = \sum_{i=1}^{k} \frac{\dot{x}_i}{\sigma_i} \boldsymbol{v}_i + \frac{(c-k)\dot{x}_{k+1}}{\sigma_{k+1}} \boldsymbol{v}_{k+1} \tag{75}$$

where c is a real number less than or equal to the rank and k is the greatest integer less than or equal to c. The norm of this type of solution is given by

$$\|\dot{\boldsymbol{\theta}}^{(c)}\|^2 = \sum_{i=1}^{k} \left(\frac{\dot{x}_i}{\sigma_i}\right)^2 + \left(\frac{(c-k)\dot{x}_{k+1}}{\sigma_{k+1}}\right)^2 \tag{76}$$

because singular vectors are mutually orthogonal unit vectors.

The advantages of using this form of a solution are that, when the SVD is available, it is extremely easy to compute. The norms of the truncated singular value solutions $\|\dot{\boldsymbol{\theta}}^{(i)}\|$, obtained from (57), are computed as i is incremented until either i is equal to the rank or the norm is greater than $\dot{\theta}_{max}$. The latter case is the one of interest because it represents reaching the physical constraint on the joint velocity. In this case k is now known, namely, $i-1$, and the desired quantity $(c-k)$ for which $\|\dot{\boldsymbol{\theta}}^{(c)}\|$ is equal to $\dot{\theta}_{max}$ can be easily obtained from (76). This value is then used to obtain the continuous truncated singular value solution defined by (75).

4.4 A Simple Illustrative Example

It is useful to compare the different types of damped least squares solutions and the continuous truncated SVD on a simple example. Consider the two-link planar robot depicted in Fig. 8. The robot is shown in its initial configuration along with the desired square trajectory for the end effector. As illustrated, the link lengths are 110 and 100 units for the first and second links, respectively. The desired trajectory is 200 units square, being offset by 10 units from the base of the robot. By commanding the robot to travel along the square ABCD, the reach singularity is encountered along a significant part of the trajectory, including the entire segment from B to C for which the desired end effector position is kinematically unachievable except for the midpoint of segment BC. In addition, the trajectory is designed to force the robot to go through the internal boundary singularity at the midpoint between points D and A.

The simulation results for the three cases of a constant damping factor, a variable damping factor, and the numerical filtering formulation are presented in Fig. 9. The results for all three cases are superior to any formulation relying on traditional pseudoinverse solutions because the transition from a full rank to a singular configuration for pseudoinverse solutions results in unacceptably high joint

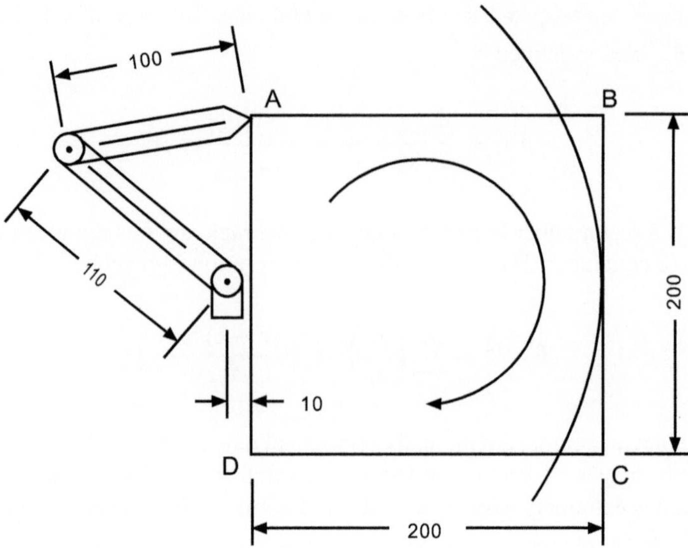

Fig. 8 Two-link planar robot and desired trajectory used for the two-dimensional example. A portion of the maximum reach of the robot is shown with an arc

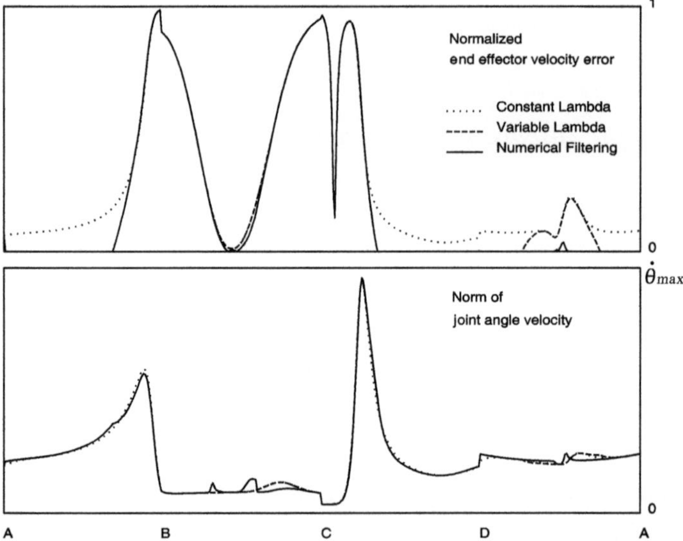

Fig. 9 Results of the two-dimensional example showing the end effector velocity error and joint velocity for the three cases of constant damping factor, variable damping factor, and numerical filtering

velocities, oscillations, and poor end effector tracking. The damped least-squares solution, however, results in smooth transitions and velocity bounds that can be set a priori. The normalized end effector tracking error is much as would be expected. The largest errors occur at points B and C along the trajectory because they are physically unachievable. The dip in the error at the midpoint between B and C is due to the fact that the commanded end effector velocity is directed increasingly along the singular vector associated with the non-zero singular value. The notch in the end effector velocity error occurring just after point C marks the position at which the commanded end effector position matches the reach of the robot and it can now begin to come out of the singularity. The overshoot in the end effector tracking as the robot comes out of its reach singularity is due to the fact that the commanded end effector velocity has a strong component in the direction of the now small singular value. It is at this point that the peak joint velocity is reached in getting the robot out of the singular configuration.

The constant damping factor case has the undesirable characteristic of end effector tracking errors along the entire trajectory, even when the robot is far from a singular configuration. In addition, a very large error occurs near the internal boundary singularity midway between points D and A even though the robot is only singular at a single point and at that point there is no component of desired end effector velocity in the direction of the singular vector associated with the zero singular value. The use of a variable damping factor based on an estimate of the minimum singular value is designed to alleviate some of the difficulties associated with the constant damping factor solution, namely, that of end effector errors in well-conditioned configurations. The variable damping factor simulation results illustrate that, as expected, the end effector velocity error can be reduced to near zero outside of the proximity of singular configurations. These results are particularly evident in the neighborhood of points A and D where the error is reduced to zero without an appreciable change in the joint velocity. By using this formulation the desirable properties of predictable performance at singularities can be coupled with improved performance away from singularities. Note, however, that there is still a significant amount of end effector tracking error in the neighborhood of the internal boundary singularity. This error is due to the fact that all components of the desired end effector velocity are equally damped. Thus the non-singular component is damped by the maximum damping factor because the minimum singular value is zero. This degree of damping is clearly unnecessary as can be seen from the still relatively low joint velocity norm.

By applying the numerical filtering technique the robot can treat the various components of the commanded end effector velocity with different damping factors, thus further improving the performance at singular configurations. The most dramatic difference in performance with numerical filtering as opposed to treating all components equally occurs in the neighborhood of singularities. In particular, at the internal boundary singularity midway between points D and A, the end effector error is significantly reduced. Note that this improvement occurs without violating any of the constraints on the joint angle velocities. The reason for the improvement lies in the fact that the numerically filtered solution more accurately reflects the physical situation of the robot. Proximity to a singularity only reflects loss of motion in those directions associated with the corresponding singular vectors. Treating all directions equally only induces unnecessary errors in the nonsingular directions.[1] The portion of this trajectory through the internal boundary singularity is meant to emphasize the importance of this fact, namely that manipulators can physically pass through singular configurations without inducing high joint velocities or end effector tracking errors.

5   Redundancy Resolution via Optimization

5.1   Gradient Projection

It has been shown in the previous section that damped least squares is instrumental in best achieving the primary task described by $\dot{x}$ even if the robot is at or near singular configurations. Yet in the case of redundant degrees of freedom, the damped least squares solution is typically only one of an infinite number of solutions that can satisfy both $\dot{x}$ and the $\dot{\theta}_{max}$ constraint on the joint velocity. These extra degrees of freedom that are not required to achieve the primary goal $\dot{x}$ are available for achieving secondary goals specified by the user. The secondary goals are typically used to control the configuration of the robot while it completes the desired task specified by the primary goal, and may include criteria such as avoiding joint limits and maximizing dexterity. This section presents formulations that generalize the damped least squares solution, retaining all of the characteristics described above, as well as providing the added versatility of permitting the user to specify additional constraints on how a motion is to occur for situations where redundant degrees of freedom are present.

The general solution to a system of linear equations described by (9) is given by

$$\dot{\theta} = \mathbf{J}^+ \dot{x} + (\mathbf{I} - \mathbf{J}^+ \mathbf{J})z \tag{77}$$

where $(\mathbf{I} - \mathbf{J}^+ \mathbf{J})$ is an $n \times n$ projection onto the null space of $\mathbf{J}$ given by

$$(\mathbf{I} - \mathbf{J}^+ \mathbf{J}) = \sum_{i=r+1}^{n} \hat{v}_i \hat{v}_i^{\mathrm{T}} \tag{78}$$

and z is an arbitrary vector in $\dot{\theta}$ space. That is, the ith element in the vector z can be described as the desired joint velocity for the ith joint. The homogeneous portion of this solution is described by a projection operator $(\mathbf{I} - \mathbf{J}^+ \mathbf{J})$ that maps

---

[1] Using the continuous version of the truncated SVD is able to remove even the small amount of end effector tracking error because it considers all of the singular values and vectors.

the arbitrary vector z into the null space of the Jacobian. That is, the projection operator allows the user to choose secondary goals, described by the vector z, and blend them into the desired manner of motion without affecting the primary goal of task completion. In effect what occurs is that the projection operator removes all of the degrees of freedom that would disturb achievement of the primary goal. Any remaining degrees of freedom are used to achieve the goal specified by the vector z. In cases where there are no redundant degrees of freedom, the projection operator is the null matrix and the general formulation reduces to the simple pseudoinverse solution.[2]

We will illustrate the use of a homogenous solution for the simple case of joint limit avoidance. The goal of staying in a configuration where all the joints are as close to their center position as possible can be described by minimizing the function [15]

$$H(\theta) = \sum_{i=1}^{n} \left[(\theta_i - \theta_{i_c})/(\theta_{i_{max}} - \theta_{i_{min}}) \right]^2 \tag{79}$$

where θ_i is the ith joint angle, θ_{i_c} is its center position, $\theta_{i_{min}}$ is its lower limit, and $\theta_{i_{max}}$ is its upper limit. If the vector z in (77) is selected as

$$z = -\alpha \nabla H(\boldsymbol{\theta}) \tag{80}$$

where α is a real positive scalar gain and ∇H is the gradient of H, the homogeneous solution can be used to minimize H, i.e., the robot will achieve the given primary goal in a manner that keeps the joints as close as possible to their center position.

5.2 Task Priority

The above section shows how one can achieve secondary goals that can be described in the joint space using a function of the form $H(\theta)$ without affecting the primary goal. There are, however, some secondary goals that can be more intuitively described in Cartesian worldspace coordinates. A typical example of such a goal is that of obstacle avoidance. The distance between the robot and other objects in the world is much more easily defined in worldspace coordinates rather than joint space ones. Clearly, obstacle avoidance is a useful secondary goal that again should be satisfied under the constraints imposed by the primary goal [16, 17]. The equation for achieving this type of solution is given by

$$\dot{\boldsymbol{\theta}} = \mathbf{J}^+ \dot{x} + (\mathbf{J}_2(\mathbf{I} - \mathbf{J}^+\mathbf{J}))^+ (\dot{x}_2 - \mathbf{J}_2\mathbf{J}^+\dot{x}) \tag{81}$$

---

[2] This all assumes that the damping factor is zero, i.e., the norm of the pseudoinverse solution is less than $\dot{\theta}_{max}$.

where

$\dot{x}$ = primary goal velocity,
$\mathbf{J}$ = Jacobian for primary reference frame,
$\dot{x}_2$ = secondary goal velocity, and
$\mathbf{J}_2$ = Jacobian for secondary reference frame.

For obstacle avoidance while performing a specified task, $\dot{x}$ represents the desired end effector velocity required to perform the task and $\mathbf{J}$ is the corresponding Jacobian. The secondary goal is dynamically determined by first identifying the point on the robot that is currently closest to an obstacle, referred to as the *obstacle avoidance point*. Then, the Jacobian for this point, denoted $\mathbf{J}_2$, is computed and it is assigned a desired velocity, denoted $\dot{x}_2$, in a direction that is directly away from the obstacle surface.

Each of the terms in (81) has an easily visualized physical interpretation. As stated previously, the pseudoinverse solution $(\mathbf{J}^+\dot{x})$ of the desired end-effector trajectory composes the particular portion of the solution that in the redundant case guarantees the exact desired end-effector velocity with the minimum joint velocity norm. The added homogeneous solution sacrifices the minimum norm solution to satisfy a different goal, that of obstacle avoidance. The matrix composed of the obstacle Jacobian times the projection operator, $\mathbf{J}_2(\mathbf{I} - \mathbf{J}^+\mathbf{J})$, represents the degrees of freedom available to move the obstacle point while creating no motion at the end-effector. This matrix is used to transform the desired obstacle point motion from Cartesian obstacle velocity space into the best available solution in joint velocity space again through the use of the pseudoinverse. The vector describing the desired obstacle point motion is composed of the commanded motion, $\dot{x}_2$, obtained from environmental information, modified by subtracting the motion caused at the obstacle point due to satisfaction of the end-effector velocity constraint, $\mathbf{J}_2\mathbf{J}^+\dot{x}$.

Numerical considerations regarding rank are especially important because the robot tends to move such that either the obstacle point cannot move any farther without changing the end-effector position or the robot configuration results in a new point that becomes the obstacle point. Some of these issues can be mitigated by only considering obstacle points that are sufficiently close to an obstacle, which would be compatible with limited range proximity sensors. In order to not introduce a discontinuity, the obstacle avoidance term should be tapered as a function of distance.

To effectively control the degree of influence that an obstacle has on the resultant robot motion, the solution for the joint angle rates can be modified to

$$\dot{\boldsymbol{\theta}} = \mathbf{J}^+\dot{x} + \alpha(d)\left[\mathbf{J}_2(\mathbf{I} - \mathbf{J}^+\mathbf{J})\right]^+ (\beta(d)\hat{\dot{x}}_2 - \mathbf{J}_2\mathbf{J}^+\dot{x}) \tag{82}$$

where d is the distance between the obstacle surface and the obstacle avoidance point, $\hat{\dot{x}}_2$ is now a unit vector indicating direction, $\beta(d)$ is the speed of the obstacle avoidance velocity, and $\alpha(d)$ is a gain term for the amount of the homogeneous

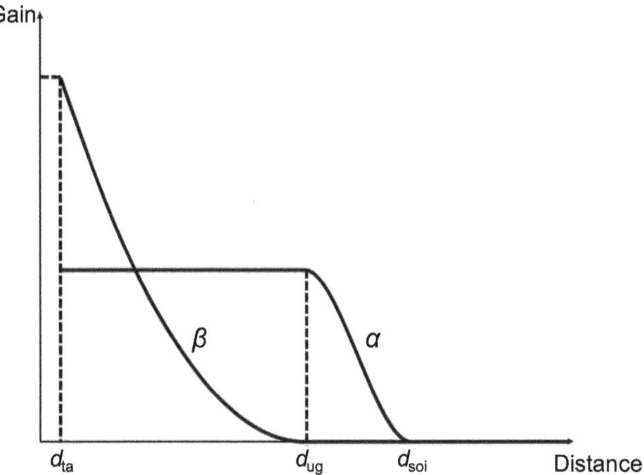

Fig. 10 The form of the homogeneous term gain, α, and the obstacle avoidance speed, β, as a function of obstacle distance

solution to be included in the total solution. The functions α and β are described by polynomials of the form shown in Fig. 10.

From Fig. 10 it can be seen that there are three distances that characterize changes in the value of the functions α and β. These distances are defined as the task abort distance, d_{ta}, the unity gain distance, d_{ug}, and the sphere of influence distance, d_{soi}. These distances define three zones that encompass each obstacle. If the robot is further from the obstacle than the d_{soi}, then the obstacle has no influence on the robot and the homogeneous solution can be used to satisfy some other desirable criterion. Between the d_{soi} and the d_{ug}, there is a smooth transition where avoidance of the obstacle is included in the motion specification of the robot. Inside of the d_{ug}, the influence of the obstacle in repelling the robot is inversely related to the distance. If the robot reaches the d_{ta}, then a collision is considered imminent and the specified task is suspended. This distance is determined by such factors as the maximum velocity and the physical dimension of the robot, thus allowing the robot to slow down and stop before physical contact occurs.

Because the use of a single worst-case obstacle point for motion planning may result in oscillations for some configurations or environments, one should use multiple secondary goals to minimize switching between homogeneous solutions. This is done by modifying (82) to

$$\dot{\theta} = \mathbf{J}^+ \dot{x} + \sum_{i=1}^{l} \alpha_i(d_i) \mathbf{h}_i \tag{83}$$

where $\mathbf{h}_i$ is the ith homogeneous solution for the ith obstacle point, α_i is its corresponding gain, d_i is the distance to that obstacle, and l is the number of obstacle points that are within d_{soi}. When one blends the homogeneous solutions in this manner, the robot has the ability to make smooth transitions between the varying priorities of the secondary goals due to the robot's motion or moving obstacles.

5.3 *Augmented Jacobian*

In the previous sections it was shown how secondary goals could be achieved under the constraint of satisfying the primary goal of the end-effector velocity, $\dot{x}$. These goals were either of the form

$$\dot{\theta} = z \tag{84}$$

or

$$\mathbf{J}_2\dot{\theta} = \dot{x}_2. \tag{85}$$

One additional way to find a common solution to both the primary goal of (9) and secondary goals of the form (84) or (85) is to combine them into a single matrix equation:

$$\begin{bmatrix} \mathbf{J} \\ \mathbf{J}_2 \\ \mathbf{I} \end{bmatrix} \dot{\theta} = \begin{bmatrix} \dot{x} \\ \dot{x}_2 \\ z \end{bmatrix}. \tag{86}$$

The resultant matrix is frequently referred to as an Augmented Jacobian [18, 19]. Because this matrix is not square or of full rank, there is not an exact solution to (86). However, one can use least-squares to determine a best approximate solution. Because following the desired trajectory $\dot{x}$ is the primary goal, one would like that error to be minimal. This can be done by suitably weighting the augmented rows of (86) so it becomes

$$\begin{bmatrix} \mathbf{W}_1\mathbf{J} \\ \mathbf{W}_2\mathbf{J}_2 \\ \mathbf{W}_3\mathbf{I} \end{bmatrix} \dot{\theta} = \begin{bmatrix} \mathbf{W}_1\dot{x} \\ \mathbf{W}_2\dot{x}_2 \\ \mathbf{W}_3 z \end{bmatrix} \tag{87}$$

where $\mathbf{W}_1$, $\mathbf{W}_2$, and $\mathbf{W}_3$ are weighting matrices so that the least-squares solution allocates the error appropriately to the desired secondary goals. The elements of the matrix $\mathbf{W}_1$ can also be used to set a priority between the dimensions of the task space. For example, the position and orientation components of $\dot{x}$ may have different accuracy requirements. If the norm of $\mathbf{W}_1$ is much greater than the norms of $\mathbf{W}_2$ and $\mathbf{W}_3$, then the solution of the weighted augmented Jacobian equation will result in solutions that are virtually the same as using the task priority formulation.

As an example, consider the case above where $z = \nabla H(\boldsymbol{\theta})$ with H specified by (79) in order to keep all joints near their center position. This will not necessarily keep a joint from violating a strict joint limit constraint. However, one can modify the performance criterion to be

$$H(\theta) = \sum_{i=1}^{n} \frac{(\theta_{i_{max}} - \theta_{i_{min}})^2}{(\theta_{i_{max}} - \theta_i)(\theta_i - \theta_{i_{min}})}. \tag{88}$$

and the weighting matrix $\mathbf{W}_3$ be a diagonal matrix with diagonal elements w_i defined as

$$w_i = \begin{cases} 1 + |\partial H(q)/\partial q_i|, & \text{if } \Delta|\partial H(q)/\partial q_i| \geq 0 \\ 1, & \text{if } \Delta|\partial H(q)/\partial q_i| \leq 0 \end{cases}. \tag{89}$$

where $\Delta|\partial H(q)/\partial q_i|$ is the rate of change of $|\partial H(q)/\partial q_i|$. This means that a joint will not violate its constraint because the weighting on a joint that approaches its limit will approach infinity, so that it takes priority over all other constraints.

Note that, as discussed above, there will typically be many $\mathbf{J}_2$'s in the augmented Jacobian, i.e., one for every obstacle point. Likewise, there will typically be many identity matrices, one for each desirable property that can be described by the optimization of a function of the robot configuration. Finally, note that the properties of the damped least squares solution can be obtained by selecting $z = \mathbf{0}$ for the right-hand side of one of the secondary tasks.

5.4 Extended Jacobian and Algorithmic Singularities

The gradient projection approach presented in Sect. 5.1 changes the configuration of the robot to improve the desired optimization criterion $H(\theta)$ by moving in the appropriate direction of the gradient. However, it does not guarantee that the robot will be at the optimal value of $H(\theta)$ under the constraint of the current desired end effector location. The configuration with the optimal value of $H(\theta)$ is one where there is no component of $\nabla H(\theta)$ in the null space of $\mathbf{J}$, i.e.,

$$\hat{\boldsymbol{v}}_i^{\mathrm{T}} \nabla H(\theta) = \mathbf{0} \quad \text{for } i = n - m + 1 \text{ to } n. \tag{90}$$

In order to track this extremal value one must make sure that its gradient is equal to zero, i.e.,

$$\nabla(\hat{\boldsymbol{v}}_i^{\mathrm{T}} \nabla H(\theta)) = 0. \tag{91}$$

One can now augment the differential forward kinematics with $n - m$ constraints to obtain

$$\begin{bmatrix} \mathbf{J} \\ \nabla^{\mathrm{T}}(\hat{\boldsymbol{v}}_i^{\mathrm{T}} \nabla H(\theta)) \end{bmatrix} \dot{\boldsymbol{\theta}} = \begin{bmatrix} \dot{\boldsymbol{x}} \\ \mathbf{0} \end{bmatrix}. \tag{92}$$

This $n \times n$ matrix is referred to as the extended Jacobian [20]. If it is not singular, then its inverse exists and it maintains the robot in a configuration that corresponds to the extremal value of $H(\theta)$ under the constraint of the desired end effector location $\boldsymbol{x}$.

One of the advantages of the extended Jacobian method is that it guarantees cyclic behavior. In other words, for every closed trajectory $\boldsymbol{x}(t)$ the corresponding $\boldsymbol{\theta}(t)$ is also closed. However, one serious disadvantage is that the extended Jacobian method introduces algorithmic singularities [21]. Algorithmic singularities are configurations where the robot is not kinematically singular, i.e., $\mathbf{J}$ is not singular, however, the extended Jacobian is singular. In other words, the additional constraints of optimizing $H(\theta)$ are not compatible with the desired end effector trajectory. The characteristics of algorithmic singularities are perhaps best described in terms of the self-motion manifold, i.e., the set of all configurations associated with a given end effector location, which is described in the next subsection.

5.5 Self-Motion Manifolds

Recall that the forward kinematics of a robot is given by (7) and for redundant robots, $n > m$, where $n - m$ is the degree of redundancy. The self-motion manifold(s) is (are) the set of all solutions that result from solving the inverse-kinematic problem given by (8) [22]. The upper limit on the number of self-motion manifolds for redundant spherical, positional, and spatial manipulators is 2, 4, and 16, respectively. At the velocity level, self motion corresponds to:

$$\mathbf{J}\dot{\boldsymbol{\theta}} = \mathbf{0}. \tag{93}$$

For the case where $n - m = 1$ and the robot is in a non-singular configuration, the null space is one dimensional and given by the nth input singular vector, $\hat{\boldsymbol{v}}_n$, which is tangent to a self-motion manifold associated with this location. One can use $\hat{\boldsymbol{v}}_n$ to step along the self-motion manifold(s) by maintaining a fixed desired end-effector location, $\boldsymbol{x}_d$. Numerically, this can be done by identifying an initial configuration $\boldsymbol{\theta}_0$ where $\boldsymbol{x}_d = f(\boldsymbol{\theta}_0)$, and repeatedly solving

$$\Delta\boldsymbol{\theta} = \gamma\hat{\boldsymbol{v}}_n + \mathbf{J}^+ \Delta\boldsymbol{x}_e \tag{94}$$

where $\Delta\boldsymbol{\theta}$ is the change in the joint angles, γ is a real positive scalar that represents the step size along the manifold, $\mathbf{J}^+ \Delta\boldsymbol{x}_e$ is an error correction term and $\Delta\boldsymbol{x}_e$ is the end-effector error, i.e., the difference between $f(\theta + \Delta\theta)$ and x_d [23]. If there are multiple self-motion manifolds, this procedure must be performed on each of them with an appropriate initial θ_0. The characteristics of the individual manifolds can be significantly different in terms of their shape and size. One can use the above procedure to identify the globally optimal configuration for satisfying a desirable secondary criterion function $H(\theta)$ for a given x_d. However, it should be noted that if one is tracking a desired trajectory $\dot{x}$, for example by using the extended Jacobian approach described above, this globally optimal configuration may instantaneously jump from one location on a self-motion manifold to another. This behavior is associated with algorithmic singularities.

6 Fault Tolerance

6.1 Local Fault Tolerance

This section will consider the use of kinematic redundancy to implement fault-tolerant operation of robots [24–26]. For this application, the differential formulation of the kinematics introduced in Sect. 2.2 is most useful. Recall, that the columns of the Jacobian matrix can be represented as a collection of columns

$$\mathbf{J}_{m \times n} = \begin{bmatrix} j_1 & j_2 & \cdots & j_n \end{bmatrix} \tag{95}$$

where j_i, given by either (11) or (12), represents the end-effector velocity due to the velocity of joint i. For an arbitrary single joint failure at joint f, assuming that the failed joint can be locked, the resulting m by $n-1$ Jacobian will be missing the fth column, where f can range from 1 to n. This Jacobian will be denoted by a preceding superscript so that in general

$$^{f}\mathbf{J}_{m \times n} = \begin{bmatrix} j_1 & j_2 & \cdots & j_{f-1} & 0 & j_{f+1} & \cdots & j_n \end{bmatrix}. \tag{96}$$

One definition of local fault tolerance is based on the worst-case dexterity following an arbitrary locked joint failure [27]. Because $^{f}\sigma_m$ denotes the minimum singular value of $^{f}\mathbf{J}$ then $^{f}\sigma_m$ is a measure of the worst-case dexterity if joint f fails [28]. If all joints are equally likely to fail, then a measure of the worst-case failure tolerance, denoted $\mathcal{K}$, is given by

$$\mathcal{K} = \min_{f=1}^{n}(^{f}\sigma_m). \tag{97}$$

To insure that robot performance is optimal prior to a failure, an optimally failure tolerant Jacobian is further defined as having all equal singular values due to the desirable properties of isotropic robot configurations. Under these conditions, to guarantee that the minimum $^f\sigma_m$ is as large as possible they should all be equal. Physically, this can be interpreted as attempting to balance the use of all joints so that they contribute equally to the velocity of the end-effector. Thus, the worst-case dexterity of an isotropic robot that experiences a single joint failure is governed by the inequality

$$\mathcal{K} = \min_{f=1}^{n}(^f\sigma_m) \leq \sigma\sqrt{\frac{n-m}{n}} \tag{98}$$

where σ denotes the norm of the original Jacobian. The best case of equality occurs if the robot is in an optimally failure tolerant configuration. The above inequality makes sense from a physical point of view because it represents the ratio of the degree of redundancy to the original number of degrees of freedom. This inequality can be used to determine the degree of redundancy required to maintain a minimum amount of dexterity in the event of any single joint failure.

Using the above definition of an optimally failure tolerant configuration one can identify the structure of the Jacobian required to obtain this property [29]. In terms of the SVD of $\mathbf{J}$, the matrix of output singular vectors, $\mathbf{U}$, simply represents a rotation of the end-effector coordinate frame so that it does not affect the configuration of the robot and can be arbitrarily set to identity. In addition, because the robot is initially in an isotropic configuration only the matrix $\mathbf{V}^\mathrm{T}$ needs to be considered. If $\mathbf{V}^\mathrm{T}$ is partitioned between the m and $m+1$ rows in the following manner

$$\mathbf{V}^\mathrm{T} = \left[\begin{array}{cccc}
v_{1,1} & v_{2,1} & \cdots & v_{n,1} \\
v_{1,2} & v_{2,2} & \cdots & v_{n,2} \\
\vdots & \vdots & \ddots & \vdots \\
v_{1,m} & v_{2,m} & \cdots & v_{n,m} \\
\hline
v_{1,m+1} & v_{2,m+1} & \cdots & v_{n,m+1} \\
\vdots & \vdots & \ddots & \vdots \\
v_{1,n} & v_{2,n} & \cdots & v_{n,n}
\end{array}\right]. \tag{99}$$

then the $n-m$ rows v_{m+1}^T to v_n^T will span the null space and the upper m rows v_1^T to v_m^T will be equivalent to $\mathbf{J}$ so that the ith column of $\mathbf{J}$ can be taken to be

$$\boldsymbol{j}_i = \begin{bmatrix} v_{i,1} \\ v_{i,2} \\ \vdots \\ v_{i,m} \end{bmatrix}. \tag{100}$$

The magnitude of the contribution of an individual joint i to the motion at the end-effector is given by the norm of $\boldsymbol{j}_i$. Because the definition of an optimally failure tolerant configuration requires that each joint contribute equally to the motion at the end-effector, this translates into a constraint that all of the norms of the columns of the Jacobian be equal. In the following discussion, the scalar R_i will be used to denote the portion of joint i's motion that is transformed into the range space of $\mathbf{J}$, so that

$$R_i = \|\boldsymbol{j}_i\|^2 = \sum_{k=1}^{m} v_{i,k}^2. \tag{101}$$

The optimally failure tolerant criteria can be alternatively described as requiring each joint to contribute equally to the null space of the Jacobian transformation. Physically, this means that the redundancy of the robot is uniformly distributed among all the joints so that a failure at any joint can be compensated for by the remaining joints. An individual joint's contribution to the null space, denoted by N_i, is given by

$$N_i = \sum_{k=m+1}^{n} v_{i,k}^2. \tag{102}$$

It is easy to see that the condition that all of the R_i be equal is equivalent to all of the N_i being equal because

$$R_i + N_i - 1 \tag{103}$$

due to the fact that $\mathbf{V}$ is an orthogonal matrix. While these conditions are mathematically equivalent, one or the other may be computationally preferable depending on the degree of redundancy relative to the dimension of the end-effector space.

To summarize, an optimally failure tolerant Jacobian is defined as being isotropic, i.e. $\sigma_i = \sigma$ for all i, and having a maximum worst-case dexterity following a failure, i.e., one for which ${}^f\sigma_m = \sigma\sqrt{\frac{n-m}{n}}$ for all f. The second condition is equivalent to having the columns of the Jacobian have equal norms.

The simplest example of an optimally failure tolerant configuration is given by the following Jacobian

$$\mathbf{J} = \begin{bmatrix} -\sqrt{\frac{2}{3}} & \sqrt{\frac{1}{6}} & \sqrt{\frac{1}{6}} \\ 0 & -\sqrt{\frac{1}{2}} & \sqrt{\frac{1}{2}} \end{bmatrix}. \tag{104}$$

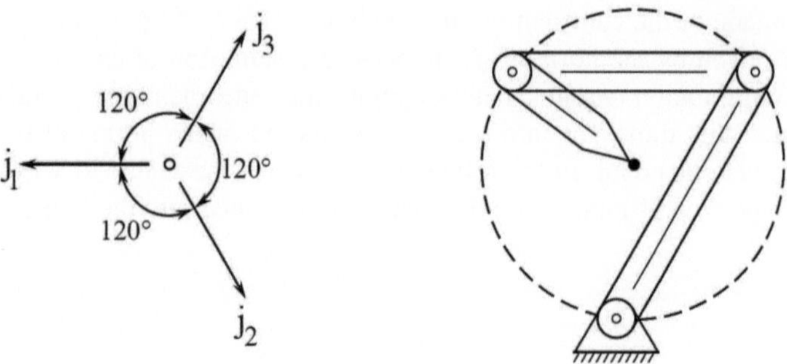

Fig. 11 Optimally failure tolerant configuration for a planar 3R robot generated from an optimally failure tolerant Jacobian

One planar 3R robot configuration that possesses this Jacobian is illustrated in Fig. 11. The null space at this configuration is given by

$$
v_3 = \begin{bmatrix} \sqrt{\frac{1}{3}} \\ \sqrt{\frac{1}{3}} \\ \sqrt{\frac{1}{3}} \end{bmatrix}
\tag{105}
$$

which illustrates that each joint contributes equally to the null space motion thus distributing the redundancy proportionally to all degrees of freedom. Geometrically, it is easy to see that the three vectors j_1, j_2, and j_3 are all 120° apart which results in a balanced coverage of the planar task space. If the three possible joint failures are considered, one can show that

$$
{}^f\sigma_2 = \sqrt{\frac{1}{3}}
\tag{106}
$$

for $f = 1$ to 3, which satisfies the optimally failure tolerant criterion. The end-effector motion that suffers most from a failure at one of the joints is, intuitively, in the direction of the column of the Jacobian that is associated with that joint, so that

$$
{}^f\boldsymbol{u}_2 = \frac{\boldsymbol{j}_f}{\|\boldsymbol{j}_f\|}.
\tag{107}
$$

Configurations of a robot that result in the worst-case fault tolerance, i.e., $\mathcal{K} = 0$, are referred to as fault intolerant configurations. For the planar 3R case being considered, this means that after a joint failure, the remaining joints are collinear with the end effector. For example, from Fig. 12a it is clear that a failure in joint one will result in a singular configuration whenever $\theta_3 = k\pi$ because this failure is physically equivalent to having a two-link robot that is at a reach singularity. A

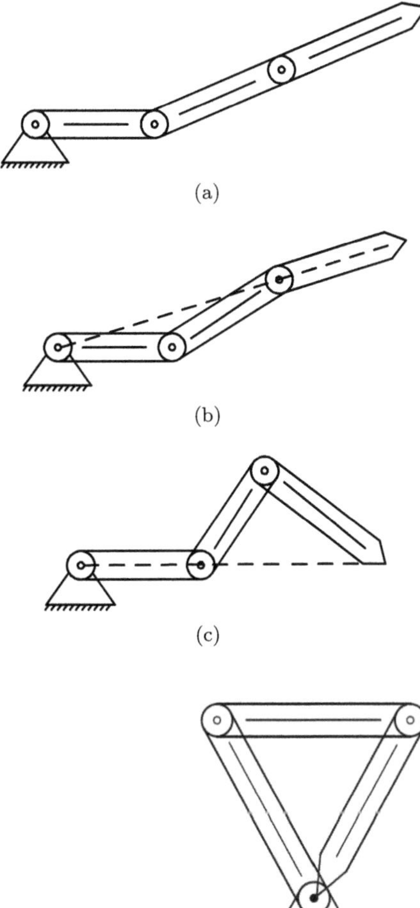

Fig. 12 The planar 3R robot configurations shown represent examples of the three types of fault intolerant configurations for this robot. These configurations correspond to situations in which a single locked joint failure will effectively result in a 2R robot in a singular configuration. This occurs for a failure in joint one in (**a**), joint two in (**b**), and joint three in (**c**)

Fig. 13 A configuration where a joint failure in joint two or three will result in a singular configuration. This is because all of the redundancy is located in the first joint

failure in joint two will result in a singular configuration whenever the origins for link one, link three, and the end effector are collinear, which is illustrated in Fig. 12b. A similar geometric argument identifies the family of configurations represented in Fig. 12c as being intolerant to failures in joint three. It is important to note that every location in the task space can be reached in a configuration like case Fig. 12c. This means that, unlike kinematic singularities, one cannot guarantee fault tolerance by simply restricting the task space of the robot. Figure 13 illustrates a case where the configuration is fault intolerant to failures in either joint two or joint three. This is because the null vector is $[1 \ 0 \ 0]^{T}$, which means that all of the redundancy is located in the first joint.

                                              A. A. Maciejewski and B. Xie

It is instructive to consider the effects of adding another degree of freedom to this planar Jacobian and applying the conditions for optimal failure tolerance. For this case one can identify an entire family of optimally failure tolerant Jacobians that can be represented by the matrix

$$
\mathbf{V}^{\mathrm{T}} = \left[\begin{array}{cccc}
\sqrt{\tfrac{1}{2}} & 0 & \sqrt{\tfrac{1}{2}}\cos\alpha & -\sqrt{\tfrac{1}{2}}\sin\alpha \\[4pt]
0 & \sqrt{\tfrac{1}{2}} & \sqrt{\tfrac{1}{2}}\sin\alpha & \sqrt{\tfrac{1}{2}}\cos\alpha \\[4pt]
\hline
\sqrt{\tfrac{1}{2}} & 0 & -\sqrt{\tfrac{1}{2}}\cos\alpha & \sqrt{\tfrac{1}{2}}\sin\alpha \\[4pt]
0 & \sqrt{\tfrac{1}{2}} & -\sqrt{\tfrac{1}{2}}\sin\alpha & -\sqrt{\tfrac{1}{2}}\cos\alpha
\end{array} \right]. \tag{108}
$$

where $\sin\alpha$ and $\cos\alpha$ are the sine and cosine of an arbitrary parameter α describing this family. This family of robot configurations is illustrated in Fig. 14. The worst-case dexterity for an arbitrary joint failure in this case is given by

$$
{}^{f}\sigma_2 = \sqrt{\frac{1}{2}} \tag{109}
$$

for $f = 1$ to 4. If one considers optimal failure tolerance to both single and double joint failures then the optimal Jacobian is given by $\alpha = k\pi/4$ where k is any odd integer.

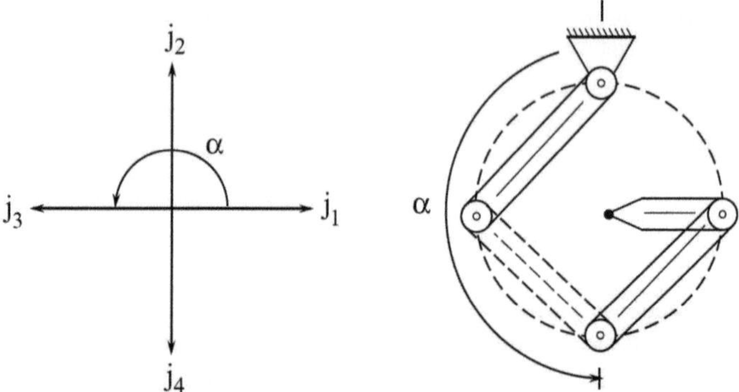

Fig. 14 A family of optimally failure tolerant configurations for a planar 4R robot corresponding to a family of optimally failure tolerant Jacobians

In order to specify an arbitrary end-effector velocity in three-dimensional space, the simplest redundant robot must possess four joints. An optimally failure tolerant Jacobian for such a robot is given by

$$\mathbf{V}^{\mathrm{T}} = \left[\begin{array}{cccc} -\sqrt{\frac{3}{4}} & \sqrt{\frac{1}{12}} & \sqrt{\frac{1}{12}} & \sqrt{\frac{1}{12}} \\ 0 & -\sqrt{\frac{2}{3}} & \sqrt{\frac{1}{6}} & \sqrt{\frac{1}{6}} \\ 0 & 0 & -\sqrt{\frac{1}{2}} & \sqrt{\frac{1}{2}} \\ \hline \frac{1}{2} & \frac{1}{2} & \frac{1}{2} & \frac{1}{2} \end{array}\right]. \tag{110}$$

It is easy to see that the effects of the four joints of this robot symmetrically span the three-dimensional end-effector space. The worst-case dexterity for an arbitrary joint failure is given by

$$^{f}\sigma_3 = \frac{1}{2} \tag{111}$$

for $f = 1$ to 4. A robot in a configuration with this optimal Jacobian is shown in Fig. 15 [30].

In addition to using $\mathcal{K}$ for designing robots that possess an optimally fault tolerant configuration, one can use $\mathcal{K}$ as the desired secondary goal and employ the gradient projection technique to always maintain maximum fault tolerant performance.

6.2 Global Fault Tolerance

The subsection above focused on a local optimization of fault tolerance that allowed one to maintain a desired end effector velocity even after a joint fails and is locked.

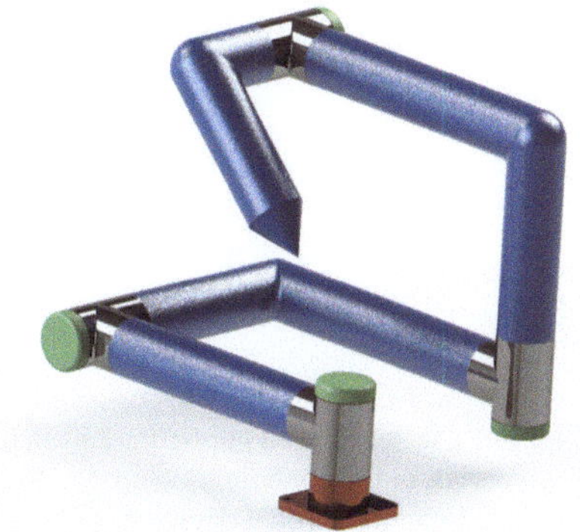

Fig. 15 Optimally failure tolerant configuration for a spatial 4R robot generated from an optimally failure tolerant Jacobian

In this subsection, we consider a different type of fault tolerance, i.e., guaranteeing that an entire trajectory is achievable after a joint fails and is locked [31–33]. Approaches to guaranteeing this behavior are frequently based on the self-motion manifolds of the robot. As a simple example, consider the 3 DOF planar robot shown in Fig. 16 for which the self-motion manifolds are one-dimensional curves. For this robot a projection of the self-motion curves onto the (θ_2, θ_3) plane is shown in Fig. 17. Each curve represents the family of joint variable combinations that place the end-effector at a constant radius from the base.

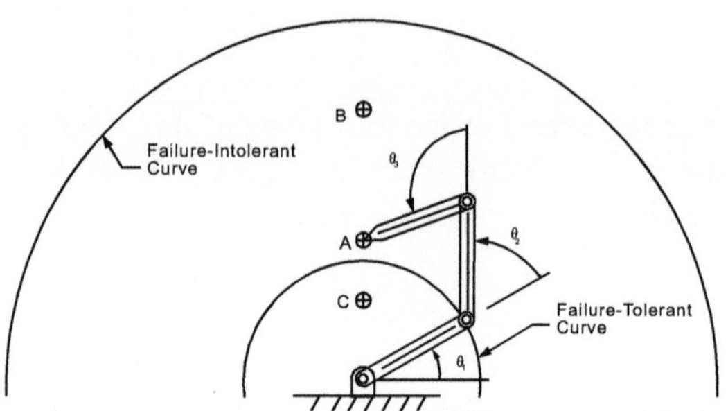

Fig. 16 A three degree-of-freedom planar robot with equal link lengths is shown with the curves in the task space having maximum and minimum failure tolerance capabilities. The points A, B, and C are representative task space points that are analyzed for their global failure tolerance properties

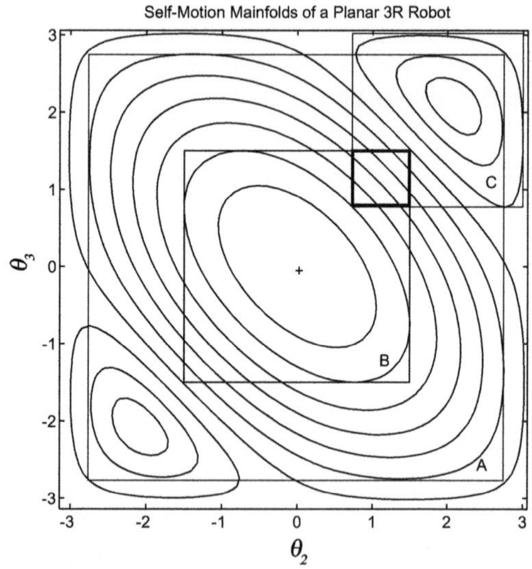

Fig. 17 The set of joint configurations that keep the robot's end-effector at a single 2D position form curves in the configuration space of the robot. The curves shown are the self-motion curves for the 3 link planar robot depicted in Fig. 16

From the figure, one can see that some regions of the task space have larger self-motion manifolds than others. For instance, consider the points corresponding to the reach singularity that occurs when the arm is fully extended and the end-effector is at the boundary of its task space. Each of these points is reachable in only a single joint configuration that corresponds to the self-motion manifold also being a point. Obviously, a task space boundary point will not be reachable after any joint failure unless the failure occurs with the joint at exactly the value required to reach that point. In contrast, the task space points exactly one link length from the base have self-motion manifolds that span the entire range of joint values for all three joints. In this unique case the failed robot will always be able to reach the entire set of points one link length from the base regardless of which joint fails, or the configuration in which it fails. This family of points is in fact the single joint failure tolerant task space for this robot. A critical task consisting of points that lie solely in this family may be completed regardless of any single failure.

To guarantee that a robot is capable of returning to a critical task space position and/or orientation $x(t_i)$, the motion range for each of the n joints must be constrained to lie within the range spanned by the self-motion manifold associated with that position and/or orientation. The minimum and maximum joint values of the ith joint, denoted $\theta_{i_{min}}$ and $\theta_{i_{max}}$, respectively, can be determined from the minimum and maximum values of θ_i over the entire self-motion manifold. This effectively superscribes an n-dimensional bounding box aligned with the joint axes around the self-motion manifold. It is important to note that these joint restrictions are implemented as simple software joint limits, referred to as artificial joint limits, and that after a joint failure they can be removed. The size of the self-motion manifold bounding box is a measure of the inherent failure tolerance of the task space position and/or orientation for which it was computed. If the robot fails while operating within the bounding box of a given desired end-effector position and/or orientation x^*, then it will always be able to position and/or orient its end-effector at x^* regardless of where the end-effector is when the failure occurs.

As a specific example, consider again the 3 DOF robot for which the bounding boxes associated with the self-motion surfaces for the three task space points labeled A, B, and C in Fig. 16 have been drawn in Fig. 17. Note that although θ_1 and its associated boundaries are not shown, they also need to be considered. If we want to guarantee that task point A is reachable after any single joint failure, the joint values must be restricted to the range of the bounding box for A's associated self-motion manifold. Now, if in addition we want to guarantee that task point B is reachable after any single joint failure, the joint values must be further restricted to B's bounding box (which is the intersection of boxes A and B). Note that by doing so we have restricted allowable configurations for being at task point A to only those inside of B's bounding box. This family of configurations is represented by the two small curves at the upper right and lower left corners of B's bounding box in Fig. 17. Notice that if the robot is in a configuration near the center of bounding box B when a joint failure occurs, then the artificial joint restrictions must be released for the robot to reach task point A. Finally, consider trying to add task point C to this scenario. The intersection of the three bounding boxes is indicated with a bold line in

Fig. 17. By design if a joint fails while operating in this region then by relaxing these artificial joint limits, the robot can reach all three task points. However, with these artificial joint restrictions in place, the family of configurations for point A has been reduced to only the curve in the upper right corner of the bold box. The family of configurations for point B has been reduced to a single point at the lower left corner of the bold box. Unfortunately, none of task point C's self-motion manifold lies within the bold box and therefore it is unreachable with these artificial constraints in place.

For one degree of redundancy, one can always map out a self motion manifold by stepping along the null vector of the Jacobian, which is tangent to the manifold, using (94). For systems with more than two degrees of redundancy, an estimate of the size of the self-motion manifold may be obtained by modifying (94) to

$$\dot{\boldsymbol{\theta}} = \pm(\mathbf{I} - \mathbf{J}^{+}\mathbf{J})\hat{\boldsymbol{e}}_i + \mathbf{J}^{+}\Delta\boldsymbol{x}_e \qquad (112)$$

where $\hat{\boldsymbol{e}}_i$ is a unit vector along the ith joint axis. In practice, the first term provides motion along the self-motion manifold until the tangent hyperplane of the self-motion manifold becomes orthogonal to the joint axis direction $\hat{\boldsymbol{e}}_i$, while the second term eliminates errors that could have accumulated during the iterative procedure. Because this technique is effectively a local optimization, it is subject to being trapped by local extrema and so only provides a guaranteed lower bound on the bounding box.

Thus the problem of maintaining the fault tolerance of a given critical position and/or orientation reduces to that of maintaining joint limits specified by the bounding box of the self-motion manifold for that position and/or orientation using the technique given in Sect. 5.1 above. Constraining the motion of a robot's joints prior to a failure will, in general, render a significant portion of the original task space unreachable. However, these joint restrictions are crucial for eliminating the possibility of a joint failing in a catastrophic configuration, i.e., one that precludes reaching a critical task space position and/or orientation. Thus, while it may appear counterintuitive, imposing appropriate software joint limits prior to a failure can actually increase the size of the task space that can be guaranteed reachable after an arbitrary failure.

Figure 18 illustrates this point using the simple 3 DOF robot from Fig. 16. If one denotes $\mathcal{W}_0$ as the pre-failure task space imposed by artificial joint limits, and $\mathcal{W}_i$ as the guaranteed task space resulting from a failure in joint i, then evaluating these workspaces after all possible joint failures and intersecting these regions with the original constrained task space results in the guaranteed failure tolerant task space, $\mathcal{W}_F$. This represents the portion of the task space in which any path can be tracked both before and after any joint failure. Note that the original unconstrained failure tolerant task space in Fig. 16 consisted of only a circle having a one link length radius.

Fig. 18 Restricting the range of motion of each joint yields a reduced pre-failure task space, $\mathcal{W}_0$. However, restrictions can eliminate the possibility that a joint will fail in a configuration where the partially failed system is unable to reach critical regions of the task space. The failure tolerant task space, $\mathcal{W}_F$, of the 3 DOF planar robot described in Fig. 16 is shown (This figure is courtesy of Ashraf Bader)

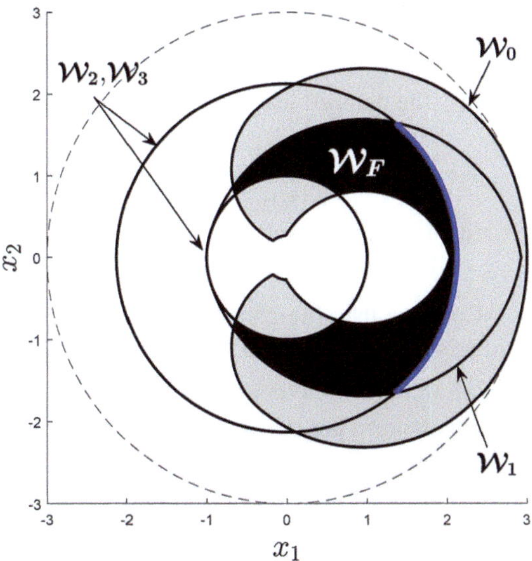

7 Summary and Conclusions

Kinematically redundant robots are by definition capable of more dexterous motion than non-redundant robots used for the same task. However, to take advantage of this inherent potential one must apply appropriate motion planning algorithms. One quantification of dexterity is associated the properties of the Jacobian, which can be elegantly described by using the singular value decomposition. The SVD not only provides a quantification of dexterity, which can be optimized with redundant robots, but also a foundation for the computation of numerically stable inverse kinematic algorithms. This chapter has provided an introduction to these algorithms, which provide a unified framework for incorporating a range of desirable criteria that can be accomplished in addition the requirements of the robot's task. Finally, this framework also incorporates fault tolerance, i.e., the ability of the robot to continue completing its assigned task while experiencing failures in some of its joints.

8 Exercises

1. Consider the equal link length 3 DOF planar robot in the configuration shown in Fig. 13. Assume that the physical joint limits on all joints are $-\pi$ and π and that the desired end effector velocity is $\dot{\boldsymbol{x}} = [1\ 0]^{\mathrm{T}}$. Compute the differential

inverse kinematics, i.e., the value of $\dot{\boldsymbol{\theta}}$, using the following techniques:

(a) The pseudoinverse of the Jacobian, $\mathbf{J}^{+}$.
(b) The damped least-squares inverse of the Jacobian, $\mathbf{J}^{(\lambda)}$ for $\lambda = 1$ and $\lambda = 10^{-6}$.
(c) The gradient projection optimization with simple joint limit avoidance, i.e., Eqs. (77) to (80).
(d) The augmented Jacobian approach to simple joint limit avoidance, i.e., Eqs. (86) and (87) with $\mathbf{W}_1 = \mathbf{I}$, for $\mathbf{W}_3 = \mathbf{I}$ and $\mathbf{W}_3 = 10^{-6}\mathbf{I}$.

Compare the difference between all of these solutions in terms of end effector tracking error $\|\dot{\boldsymbol{x}} - \mathbf{J}\dot{\boldsymbol{\theta}}\|$ and joint norm $\|\dot{\boldsymbol{\theta}}\|$.

2. Show that $\dot{\boldsymbol{\theta}}^{(\lambda)}$ represents the solution with the minimal residual over all $\dot{\boldsymbol{\theta}}$ whose norm does not exceed that of $\dot{\boldsymbol{\theta}}^{(\lambda)}$.

3. Show that $\dot{\boldsymbol{\theta}}^{(k)}$ is the minimum residual solution for all $\dot{\boldsymbol{\theta}}$ in the k-dimensional subspace spanned by $\boldsymbol{v}_i$ for $i \leq k$.

4. By comparing (63) with (64), show that $\hat{\boldsymbol{u}}'_m$ will always have an even larger relative component along the minimum singular vector than the original estimate $\hat{\boldsymbol{u}}_m$.

5. Consider a three degree-of-freedom manipulator operating in two-dimensional Cartesian space. The Jacobian for this manipulator is given by

$$\mathbf{J} = \begin{bmatrix} 1 & 0 & 0 \\ 0 & \cos\phi & \sin\phi \end{bmatrix}$$

where $0 \leq \phi \leq \pi/2$.

(a) What is the null vector of $\mathbf{J}$?
(b) Write expressions for $\dot{\theta}_1$, $\dot{\theta}_2$, and $\dot{\theta}_3$ in terms of $\dot{x}_1$, $\dot{x}_2$, and ϕ if $\dot{\boldsymbol{\theta}}$ is computed using the pseudoinverse of $\mathbf{J}$.
(c) Write expressions for $\dot{\theta}_1$, $\dot{\theta}_2$, and $\dot{\theta}_3$ in terms of $\dot{x}_1$, $\dot{x}_2$, and ϕ if $\dot{\boldsymbol{\theta}}$ is computed using the damped least squares inverse of $\mathbf{J}$.

6. Assume a Jacobian whose singular values are 10, 5, and 1 and that $\|\dot{\boldsymbol{x}}\| = 1$.

(a) What is the maximum value of $\dot{\boldsymbol{\theta}}$ if $\lambda = 2$?
(b) What value of λ is required to guarantee that $\|\dot{\boldsymbol{\theta}}\| \leq 1/2$?
(c) Assume that you solve $\mathbf{J}\dot{\boldsymbol{\theta}} = \dot{\boldsymbol{x}}$ using damped least squares with $\lambda = 0$ and your answer has a norm of one, i.e., $\|\dot{\boldsymbol{\theta}}\| = 1$. If you rotate $\dot{\boldsymbol{x}}$ by ninety degrees and re-solve $\mathbf{J}\dot{\boldsymbol{\theta}} = \dot{\boldsymbol{x}}$ using damped least squares with $\lambda = 0$, what are the range of possible values for the norm of your solution $\|\dot{\boldsymbol{\theta}}\|$?

7. Consider the following Jacobian matrix:

$$\mathbf{J} = \begin{bmatrix} 1 & -1 \\ 1 & -1 \end{bmatrix}$$

(a) Which of the following vectors are possible pseudoinverse solutions?

$$\begin{bmatrix} 1 & 1 \end{bmatrix} \qquad \begin{bmatrix} 0 & 1 \end{bmatrix} \qquad \begin{bmatrix} 1 & 0 \end{bmatrix} \qquad \begin{bmatrix} 1 & -1 \end{bmatrix}$$

For those that are not, find the closest pseudoinverse solution

(b) Assume that you do not want to change the current maximum singular value of $\mathbf{J}$ but that you want to add a column that will make the minimum singular value change from zero to one. What should that column be?

8. Assume that you would like to minimize $\|\mathbf{J}\dot{\theta} - \dot{x}\|$ subject to $\|\dot{\theta}\| \leq 1$ for an arbitrary $\mathbf{J}$ and $\dot{x}$.

   (a) What is the best value of λ^2 that will guarantee that the constraint is not violated?
   (b) Repeat part (a) if you know that $\|\dot{x}\| = 1$ and

$$\mathbf{J} = \begin{bmatrix} 1 & 0 \\ 0 & 0.1 \end{bmatrix}.$$

   (c) Repeat part (a) if you know that $\dot{x} = [1 \ 0]^{\mathrm{T}}$.

9. Consider a Jacobian given by

$$\mathbf{J} = \begin{bmatrix} 1 & 0 \\ 0 & \epsilon \end{bmatrix}$$

where $0 < \epsilon < 1$, a desired end-effector velocity given by

$$\dot{x} = \sqrt{2}\begin{bmatrix} \cos\phi & \sin\phi \end{bmatrix}$$

where $0 \leq \phi \leq \pi/2$, and the constraint that $\|\dot{\theta}\| \leq 1$.

   (a) Write an expression for $\dot{\theta}$ as a function of $\dot{x}$ and ϵ for the truncated SVD (TSVD) and the continuous version of the TSVD (CTSVD).
   (b) Write an expression for the maximum difference in end-effector tracking error for the following pairs of solutions. Make sure to give ALL values of ϕ and ϵ where the maxima occur.
   (c) Truncated SVD vs. Continuous Truncated SVD
   (d) Damped Least Squares vs. Truncated SVD
   (e) Damped Least Squares vs. Continuous Truncated SVD

10. What is the self-motion manifold for the end-effector location of the manipulator shown in Fig. 13?

Cross-References

This chapter contains cross-references to content published in other chapters belonging to the four "Robotic Goes MOOC" books: KNO3, DES3, INT1. The full list of chapters abbreviations is available in the Preface.

References

1. J. Léger, J. Angeles, Off-line programming of six-axis robots for optimum five-dimensional tasks. Mech. Mach. Theory **100**, 155–169 (2016)
2. D.E. Whitney, Resolved motion rate control of manipulators and human prostheses. IEEE Trans. Man-Mach. Syst. **10**(2), 47–53 (1969)
3. C.A. Klein, B.E. Blaho, Dexterity measures for the design and control of kinematically redundant manipulators. Int. J. Robot. Res. **6**(2), 72–83 (1987)
4. D.R. Baker, C.W. Wampler, On the inverse kinematics of redundant manipulators. Int. J. Robot. Res. **7**(2), 3–21 (1988)
5. T. Yoshikawa, Manipulability of robotic mechanisms. Int. J. Robot. Res. **4**(2), 3–9 (1985)
6. T. Yoshikawa, Dynamic manipulability of robot manipulators. Trans. Soc. Instrum. Control Eng. **21**(9), 970–975 (1985)
7. S. Chiu, Control of redundant manipulators for task compatibility, in *Proceedings of the 1987 IEEE International Conference on Robotics and Automation* (1987), pp. 1718–1724
8. J. Angeles, The design of isotropic manipulator architectures in the presence of redundancies. Int. J. Robot. Res. **11**(3), 196–201 (1992)
9. G.H. Golub, C. Reinsch, Singular value decomposition and least squares solutions, in *Linear Algebra* (Springer, Berlin, 1971), pp. 134–151
10. A.A. Maciejewski, C.A. Klein, The singular value decomposition: computation and applications to robotics. Int. J. Robot. Res. **8**(6), 63–79 (1989)
11. C.W. Wampler, Manipulator inverse kinematic solutions based on vector formulations and damped least-squares methods. IEEE Trans. Syst. Man Cybern. **16**(1), 93–101 (1986)
12. S. Chiaverini, O. Egeland, R.K. Kanestrom, Weighted damped least-squares in kinematic control of robotic manipulators. Adv. Robot. **7**(3), 201–218 (1992)
13. A.A. Maciejewski, C.A. Klein, Numerical filtering for the operation of robotic manipulators through kinematically singular configurations. J. Robot. Syst. **5**(6), 527–552 (1988)
14. S. Chiaverini, Estimate of the two smallest singular values of the jacobian matrix: application to damped least-squares inverse kinematics. J. Robot. Syst. **10**(8), 991–1008 (1993)
15. A. Liegeois, Automatic supervisory control of the configuration and behavior of multibody mechanisms. IEEE Trans. Syst. Man Cybern. **7**(12), 868–871 (1977)
16. A.A. Maciejewski, C.A. Klein, Obstacle avoidance for kinematically redundant manipulators in dynamically varying environments. Int. J. Robot. Res. **4**(3), 109–117 (1985)
17. Y. Nakamura, H. Hanafusa, T. Yoshikawa, Task-priority based redundancy control of robot manipulators. Int. J. Robot. Res. **6**(2), 3–15 (1987)
18. O. Egeland, Task-space tracking with redundant manipulators. IEEE J. Robot. Autom. **3**(5), 471–475 (1987)
19. H. Seraji, Configuration control of redundant manipulators: theory and implementation. IEEE Trans. Robot. Autom. **5**(4), 472–490 (1989)
20. J. Baillieul, Kinematic programming alternatives for redundant manipulators, in *Proceedings of the 1985 IEEE International Conference on Robotics and Automation*, vol. 2 (1985), pp. 722–728

21. J. Baillieul, A constraint oriented approach to inverse problems for kinematically redundant manipulators, in *Proceedings of the 1987 IEEE International Conference on Robotics and Automation* (1987), pp. 1827–1833
22. J.W. Burdick, On the inverse kinematics of redundant manipulators: characterization of the self-motion manifolds, in *Advanced Robotics:1989* (Springer, Berlin, 1989), pp. 25–34
23. A.A. Almarkhi, A.A. Maciejewski, Maximizing the size of self-motion manifolds to improve robot fault tolerance. IEEE Robot. Autom. Lett. **4**(3), 2653–2660 (2019)
24. J.D. English, A.A. Maciejewski, Fault tolerance for kinematically redundant manipulators: anticipating free-swinging joint failures. IEEE Trans. Robot. Autom. **14**(4), 566–575 (1998)
25. Y. Yi, J.E. McInroy, Y. Chen, Fault tolerance of parallel manipulators using task space and kinematic redundancy. IEEE Trans. Robot. **22**(5), 1017–1021 (2006)
26. M.L. Visinsky, J.R. Cavallaro, I.D. Walker, A dynamic fault tolerance framework for remote robots. IEEE Trans. Robot. Autom. **11**(4), 477–490 (1995)
27. R.G. Roberts, A.A. Maciejewski, A local measure of fault tolerance for kinematically redundant manipulators. IEEE Trans. Robot. Autom. **12**(4), 543–552 (1996)
28. K.N. Groom, A.A. Maciejewski, V. Balakrishnan, Real-time failure-tolerant control of kinematically redundant manipulators. IEEE Trans. Robot. Autom. **15**(6), 1109–1115 (1999)
29. A.A. Maciejewski, R.G. Roberts, On the existence of an optimally failure tolerant 7R manipulator Jacobian. Appl. Math. Comput. Sci. **5**(2), 343–357 (1995)
30. K.M. Ben-Gharbia, A.A. Maciejewski, R.G. Roberts, Kinematic design of redundant robotic manipulators for spatial positioning that are optimally fault tolerant. IEEE Trans. Robot. **29**(5), 1300–1307 (2013)
31. C.J.J. Paredis, P.K. Khosla, Fault tolerant task execution through global trajectory planning. Reliab. Eng. Syst. Saf. **53**(3), 225–235 (1996)
32. C.L. Lewis, A.A. Maciejewski, Fault tolerant operation of kinematically redundant manipulators for locked joint failures. IEEE Trans. Robot. Autom. **13**(4), 622–629 (1997)
33. R.S. Jamisola, A.A. Maciejewski, R.G. Roberts, Failure-tolerant path planning for kinematically redundant manipulators anticipating locked-joint failures. IEEE Trans. Robot. **22**(4), 603–612 (2006)

Parallel Robots

Andreas Müller

Abstract Parallel Robots, also referred to as Parallel Kinematics Machines (PKM) are mechanisms possessing closed kinematic loops. This fundamental difference to classical serial robots offers significant advantages but necessitates novel approaches to the kinematic and dynamic modeling, synthesis, and control. The closed loop kinematics introduces singularities, which further complicates the motion planning and control. This chapter reviews research and development that was conducted in the last three decades, and presents a tailored approach to kinematics and dynamics modeling. Special attention is paid to the computational aspects and the modularity inherent to PKM. Therewith, model-based control schemes are presented that eventually allow exploiting the dynamic capabilities of PKM. Synthesis methods developed for PKM are summarized. The important issue of kinematic singularities is addressed.

1 Overview

A parallel robot is a closed loop mechanism where a moving platform is connected to a base by several kinematic chains (limbs). Parallel robots are synonymously referred to as parallel kinematic machines (PKM), which will be used throughout this chapter. The best known PKM is the Gough-Stewart platform, as in Fig. 1. The conceptual difference to serial robots is that the actuators share the load of the moving *platform*. This leads to remarkable advantages such as low moving mass, high payload and acceleration, as well as high maximal stiffness and dexterity. At the same time, these properties change significantly within the workspace. The special kinematics calls for tailored approaches to kinematic and dynamic modeling as well as synthesis and optimal design of PKM. Also model-based control schemes, as established for serial manipulators (SM), must be amended.

A. Müller (✉)
Institute of Robotics, Johannes Kepler University Linz, Linz, Austria
e-mail: a.mueller@jku.at

Fig. 1 Schematic drawing of a 6-DOF 6U$\underline{P}$S PKM with extensible struts. Such a PKM is usually called Gough-Stewart-platform or Hexapod. The underlined symbol indicates the actuated joint, in this case the prismatic joint

After a short overview of the historic development in Sect. 2, and a summary of relevant definitions and notations in Sects. 3 and 4, the topology representation is introduced in Sect. 5. A tailored kinematic modeling approach is described in Sect. 6. In this approach, the inverse kinematics solutions of each single limb (regarded as SM) is used to derive the inverse kinematics solution of the PKM (motions of all joints described by the platform motion). This is extended to PKM with complex limbs in Sect. 7 and to robots with general topology in Sect. 8. Therewith, the dynamic EOM are derived in Sect. 9 for non-redundant, in Sect. 10 for kinematically redundant, in Sect. 11 for PKM with complex limbs, and in Sect. 12 for robots with general topology. This makes use of the EOM of SM, i.e. of a tree-topology systems. The linearity of the EOM in the dynamic parameters is recalled in Sect. 13, and the significance for model identification is briefly discussed. Model-based control is then discussed in Sect. 14. In principle, any method for controlling SM can be applied to PKM. The difference is that motions of passive joints cannot be measured but must be computed by resolving the loop constraints,and different resolution schemes lead to different computational properties. Specific aspects of redundantly actuated PKM (RA-PKM), and their exploitation for secondary tasks are outlined in Sect. 14. Sections 15 and 16 address the existence of singularities and the variations of kinematic dexterity, which are central topics in the PKM design. The synthesis of PKM is addressed in Sect. 17.

2 Origin and History

Closed loop linkages have always been an elementary design element in mechanisms and machines. The design concept specific to PKM consists in connecting a moving *platform* to the ground (fixed platform) by several, usually structurally identical, actuated limbs, which leads to kinematic loops. The motion characteristics of these limbs collectively determine the motion of the platform. The best-known original 6-DOF PKM are the tire tester by Gough [54] and the flight simulator by Stewart [158]. Gough's PKM contains six extensible struts. The struts are often called legs, and PKM with six extensible struts are often referred to as Hexapods. The platform of Stewart's PKM is connected to the ground by three limbs each comprising two piston-type actuators acting in parallel at an intermediate link that is connected to the platform by a spherical joint. Gough's PKM is a special case of Stewart's invention where this intermediate link is the platform. Despite of this difference, PKM possessing six extensible struts are frequently called Gough-Stewart-platforms (GSP). A prototypical design of a 6U$\underline{P}$S GSP is shown in Fig. 1 (for notation see Sect. 3). It is to be mentioned that a patent for a PKM, which is essentially identical to Stewart's platform, was filed by Cappel [25] in 1964 (before publication of Stewart's paper) but did not receive due attention. A detailed discussion on the origin of hexapod PKM can be found in [17]. Extensive research in the 1990s has first focused on GSP type PKM, but then addressed the design of dedicated PKM. Today, PKM applications range from flight simulators and horse riding simulators for hippotherapy to high-precision and agile positioning devices.

2.1 Spatial and Lower-Mobility PKM

Aiming at high-speed applications, one major change from the original GSP design was to fix the actuators to the ground in order to reduce the moving mass. To this end, PKM with linear drives connected to the ground were developed, which also increases the workspace. One of the first concepts of spatial 6-DOF PKM with stationary linear drives was the Hexaglide [62], which is essentially a GSP with fixed leg lengths and movable foot points. Other 6-DOF PKM with stationary linear actuators, such as the HexaM [141] and the Linapod [181], were intended as 5-axes machine tools. An attempt to increase the workspace not only in one direction was the 6-DOF PKM [163, 166] and its TriPlanar prototype [91] equipped with planar drives, which can move on a plane. The latter used an earlier concept of the Tribias PKM [139], which aimed at high precision rather than large workspace.

Moving towards manipulation and handling applications, a number of lower mobility PKM with stationary drives were developed. A prominent example is the 4-DOF Delta robot proposed by Clavel [148] intended for highly dynamic applications. Figure 2 shows a commercial realization. It consists of a platform

Fig. 2 4-DOF Delta robot
(source: ABB)

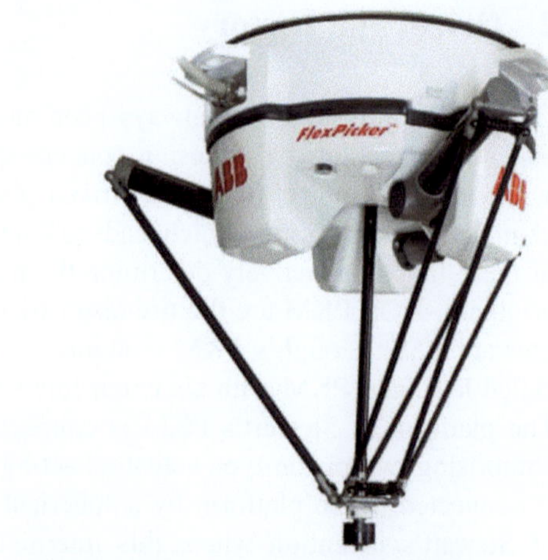

Fig. 3 H4, a 4-DOF PKM
whose moving platform
performs SCARA motion
(Courtesy of Sébastien Krut,
LIRMM, Montpellier, FR)

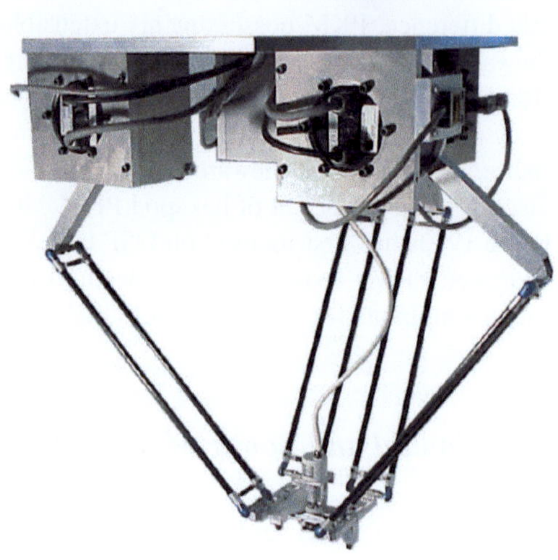

that can perform spatial translations only and is controlled by three arms. The end-effector (EE) is attached at the distal end of an additional telescope arm, which is positioned by the platform. The vertical rotation of the EE can be independently controlled by the telescope arm. Overall, the EE can perform Schönflies motions, also called SCARA motions. The 4-DOF Delta robot is, strictly speaking, not a PKM as its EE is not directly connected to all arms (Sect. 3).

The Delta has inspired creation of many other PKM that use lightweight (complex) arms with base-fixed motors. The H4 [140] in Fig. 3 was proposed as a modification of the 4-DOF Delta robot, where the EE rotation is controlled by an articulated platform. An optimized version, called Par4, was later presented in [142].

For positioning tasks, 3-DOF PKM were designed, where the platform can only translate in space, such as the 3P̲RRR mechanism reported in [26, 49, 52, 80, 81, 88]. The Tripteron in Fig. 4 is an implementation of this concept. Similar prototypes are the Isoglide3-T3 [49] and the CPM [80]. This PKM is fully isotropic and its input-output relation is linear, i.e. movement of one of the linear actuators leads to a translation of the platform along the direction of the actuator. This is exploited to achieve extremely high acceleration. The Orthoglide [30] is another 3-DOF positioning PKM. The 3-DOF Delta platform, without telescope arm, in Fig. 5 is

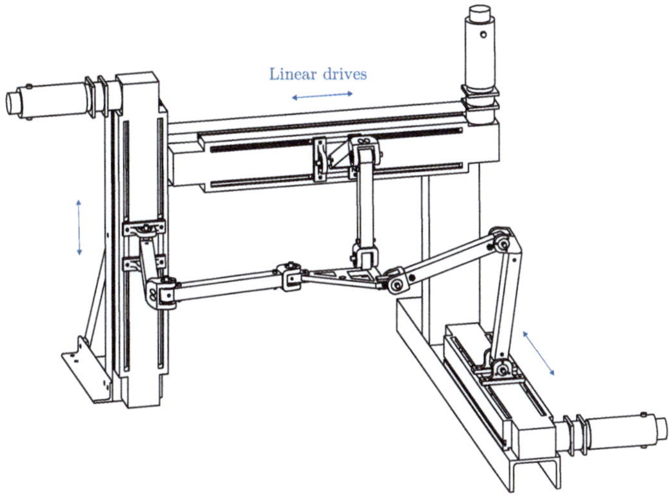

Fig. 4 Tripteron—a 3-DOF 3P̲RRR PKM (Courtesy of C. Gosselin, Laval University, Quebec)

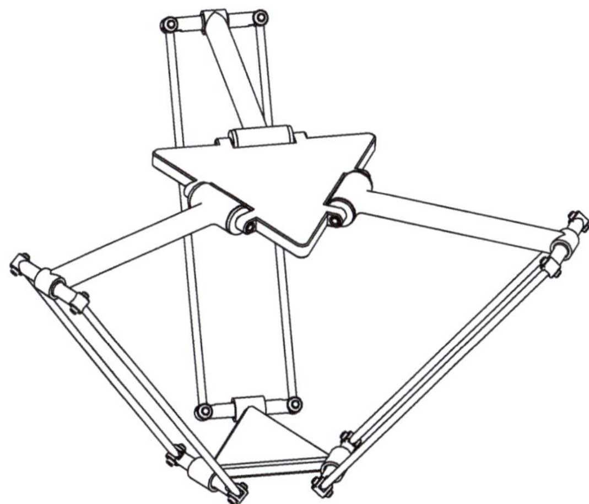

Fig. 5 3-DOF Delta robot proposed by Clavel [33]

Fig. 6 Agile Eye—a 3DOF
spherical 3RRR PKM
(Courtesy of C. Gosselin,
Laval University, Quebec)

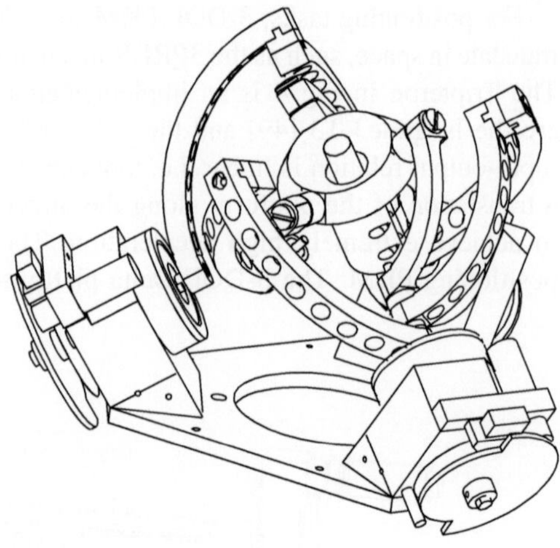

used as positioning device. Several 3-DOF PKM with spherical platform motions were proposed. Well-known is the Agile Eye in Fig. 6. It is tailored as a fast high-precision pointing device.

Planar 2-DOF PKM were proposed mainly for pick-and-place operations. Typical examples are the IRSBot-2 [45] in Fig. 8 and the Par2 [34, 143]. Notably, the Delta robot in Figs. 2 and 5, as well as the H4 in Fig. 4 have complex limbs, i.e. the limbs are not simple serial chain.

A summary of different PKM prototypes can be found in the text books [19, 110].

2.2 *Redundantly Actuated PKM*

The concept of actuation redundancy was introduced in the late 1990s [84–86, 92, 131, 185]. The motivation largely stemmed from the observation that non-redundant PKM exhibit significant variations of their kinetostatic properties.

Actuation redundancy refers to the situation that the number of 1-DOF actuators exceeds the DOF of the PKM—a peculiarity which does not exists for SM. It increases and homogenizes the kinematic manipulability and the elastic stiffness within the workspace, and allows to eliminate input singularities (Sect. 15), and thus to increase the usable workspace. From a dynamic point of view, actuation redundancy increases the dynamic capability (payload and acceleration) and homogenizes the load distribution among the actuators [44, 130]. This eventually reduces the load and thus the power consumption of the individual actuators allowing for the use of smaller drive units. Actuation redundancy also provides a certain level of robustness with respect to failures of actuators, and is thus also a means to design fault-tolerant manipulators. Actuation redundancy implies sensor redundancy, which can be exploited for geometric (self-)calibration and to increase the overall accuracy.

Fig. 7 2-DOF planar PKM
for machining applications
[174]

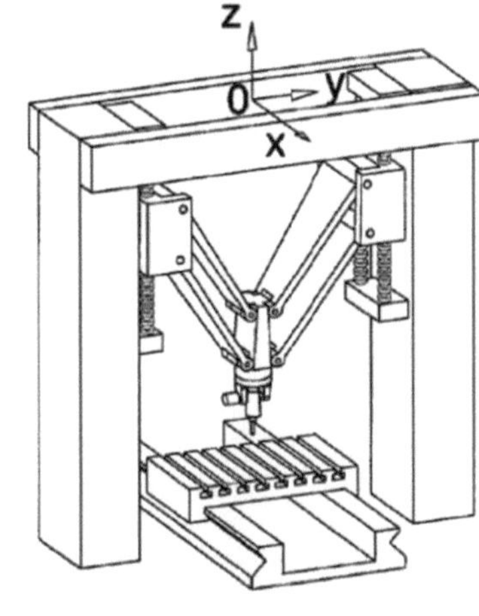

Fig. 8 IRSBot-2, a 2 DOF
fully parallel PKM with
complex limbs (courtesy of
Sébastien Briot, Laboratoire
des Sciences du Numérique
de Nantes, FR)

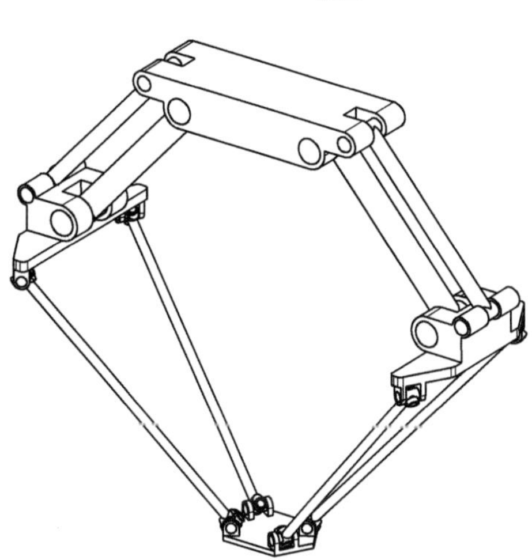

Redundantly actuated PKM (RA-PKM) are ideal candidates for use in high-precision applications, such as micro-manipulation, and robot-assisted surgery since they deliver the needed high accuracy, stiffness, dexterity, and reliability.

Actuation redundancy can be introduced in a non-redundantly actuated PKM by introducing additional limbs or/and by actuating passive joints. The former is deemed most relevant for practical applications, and is addressed in most publications. Various planar RA-PKM were proposed, such as Archi [1, 98], a 3-DOF planar RA-PKM actuated by four linear drives. Planar RA-PKM were proposed for use within machine tools exploiting their load capacity, e.g. [38] or the PKM [174] in Fig. 7. A planar RA-PKM intended for high-speed applications is the DualV [170] in Fig. 9. This RA-PKM is force balanced [169], i.e. it does not cause retroactive effects on the foundation. A spatial 6-DOF Octapod (a GSP with two additional struts) was reported in [167], which is a variant of the RA-PKM proposed in [164]. Another 6-DOF RA-PKM intended for rapid machining, is the

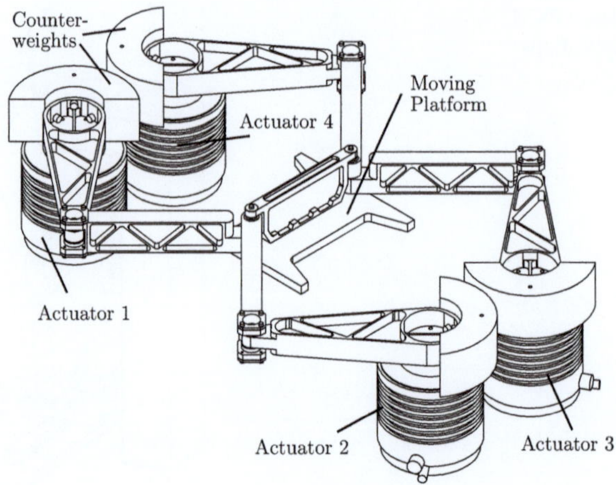

Fig. 9 DualV, a planar 3-DOF PKM (Courtesy of Volkert van der Wijk, TU Delft, NL)

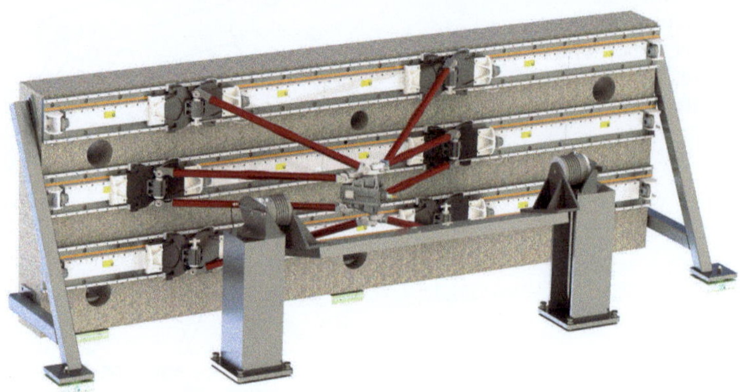

Fig. 10 ARROW, a 4-DOF PKM equipped with 6 linear drives whose EE performs a SCARA motion (courtesy of Ahmed Chemori, LIRM, Montpellier, FR)

Eclipse reported in [82, 83, 93], which is admits 360° spinning of the platform. A spatial RA-PKM used as test rig for structural testing was presented in [144]. The ARROW [15] (Fig. 10) is a 4-DOF RA-PKM that is intended to accommodate highly dynamic applications and as 5-axes machine tool. Current research focusses on lower mobility manipulators intended for agile handling and object manipulation. An example is the R4 designed for very high accelerations [152]. The SPIDER4 [39] is a serial-parallel robot with a 2-DOF wrist mechanism mounted at the platform of a redundantly actuated 3-DOF translational PKM. As for all PKM performing very fast motions in highly dynamic applications (e.g. Tripteron, ARROW, R4, DualV) the reduction of ground reaction forces is crucial.

Actuation redundancy enables generating antagonistic control forces, which can be exploited for secondary task such as the avoidance of backlash and joint clearance or the control of tangential stiffness (Sect. 14.4). It was already shown in [85] that backlash in gearboxes can be avoided by controlling the preload of actuators. The generation of desired antagonistic actuation can be incorporated in the model-based control [112].

2.3 Kinematically Redundant PKM

In analogy to SM, a PKM is kinematically redundant if the DOF of the manipulator is higher then the platform DOF. The first systematic approach to introduce kinematic redundancy to closed loop linkages led to so-called variable geometry trusses (VGT), which can be regarded as systems built by serially stacking up several PKM. The design of VGT was addressed in [111, 147]. Special attention must be paid to the elastic properties and net stiffness of VGT [75, 153]. The inverse and forward kinematics solution for VGT constructed from PKM with particular shape were addressed in [129, 134, 159]. Dynamics modeling of VGT was addressed in [41, 56, 151, 182]. If the number of PKM-modules making up the VGT becomes large, the manipulator can be regarded as hyper-redundant. Several design concepts with applications in large scale structures were presented in [160]. Introducing reconfigurability, VGT with changing topology were considered in [157].

Recent approaches to kinematic redundancy use additional actuated joints within the limbs. Adding an actuated prismatic joint in each limb of a planar 3$\underline{R}$RR PKM yields different planar 6-DOF PKM [28]. Kinematic redundancy of the planar 6-DOF 3$\underline{RP}$RR PKM in Fig. 11, which was proposed in [37], was exploited for singularity avoidance [27, 29]. The singularity avoiding motion planning of a planar 3$\underline{R}$RR PKM, where one limb was mounted on an actuated prismatic joint, was presented in [90]. There are only a few spatial kinematically redundant PKM. By mounting the limbs of a 6S$\underline{P}$S GSP on a circular slider joint, a 12 DOF PKM was constructed in [132]. Advancing a concept of [6, 13], a spatial 9-DOF PKM was investigated in [3], where the anchor points of a pair of limbs of a 6U$\underline{P}$S is mounted on a linear slider joint. Yet no wider application has been identified where the benefits justify the increased complexity accompanied with the kinematic redundancy. An interesting planar 4-DOF PKM was proposed in [53] whose platform can perform 3-DOF planar motions with unlimited rotation.

3 Definitions

A parallel kinematic *manipulator (PKM)* is a manipulator where the mobile platform, carrying the end-effector (EE), is connected to the fixed platform (base, ground) by at least two kinematic chains (limbs, legs) and there are no further kinematic chains. The name stems from the fact that the EE is connected to the base in parallel by several limbs. Various types of PKM can be distinguished regarding

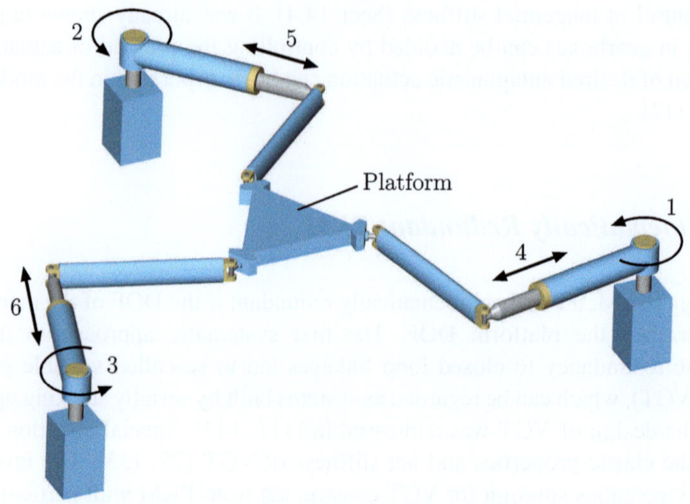

Fig. 11 A kinematically redundant planar 6-DOF 3$\underline{\text{R}}$PRR PKM [37] with six actuators

their kinematic topology, which describes the existence of joints connecting the bodies (links, members) of the PKM. A *fully parallel manipulator (FPKM)* is a PKM where each limb comprises exactly one drive unit (actuator). The terminology emphasizes that there are no passive limbs (e.g. in the Par2 PKM [143] and the Tricept), and all limbs contribute to the motion of the platform. A *limb* is an assembly of bodies and joints which links the platform to the ground but is not connected to other parts of the mechanism. A *simple limb* is a simple kinematic chain, i.e. contains no loops. A *complex limb* is a limb that comprises at least one kinematic loop itself. The DOF of the PKM is denoted with δ. The 3-DOF Delta manipulator (Fig. 5) and most Delta-like PKM, e.g. H4 (Fig. 3), are PKM with complex limbs. The DualV (Fig. 9) is another example. The 4-DOF Delta robot (Fig. 2) is not a FPKM as the EE is connected to the ground in parallel by the 3-DOF Delta platform and by the telescope arm. Also the H4 (Fig. 3) is not a FPKM.

The *platform DOF*, denoted δ_{p}, is the DOF of the platform relative to the base. In the terminology of mechanisms [2, 65], the platform DOF is determined by the connectivity of platform and base.

A *lower-mobility PKM* is one with a platform DOF less than 6. The Tripteron (Fig. 4) and the Delta platform (Fig. 5) are examples where the platform can only perform spatial translations. The Agile Eye (Fig. 6) is a spherical mechanism, whose EE can only perform spatial rotations. Many lower-mobility PKM are overconstrained mechanisms [105].

A PKM is *full-actuated* if it has at least as many 1-DOF actuators as its DOF ($\delta \leq N_{\mathrm{act}}$, with N_{act} denoting the number of 1-DOF actuators) [115]. A PKM is *redundantly actuated* (occasionally called overactuated) if the number of 1-DOF actuators is higher than its DOF ($\delta < N_{\mathrm{act}}$).

A PKM is *kinematically redundant* if its DOF is higher then the platform DOF ($\delta_P < \delta$). Kinematic redundancy allows strategic reconfiguration so to account for secondary tasks or to avoid obstacles, as it does for SM [Chapter 10: Redundant Robots]. The above definition of fully parallel manipulator, requiring that all actuators act in parallel at the platform, does hence not allow for kinematically redundant FPKM.

A *hybrid manipulator* is obtained by serially connecting parallel manipulators (modules). Such systems can be designed by serially stacking PKM where the moving platform of one PKM serves as the base platform for the succeeding PKM. A PKM is usually called *kinematically hyper-redundant* if its DOF is 'much higher' than the platform DOF. The Logabex [32] and the Orm [24] are early examples that can be regarded as hyper-redundant.

A *serial-parallel manipulator* is one where the EE is connected to the ground by a PKM followed/preceded by a simple (serial) kinematic chain. An example is the ABB IRB940 Tricept, where the EE is attached at a 2-DOF wrist mechanism (serial chain), which is mounted at the platform of a 3-DOF PKM.

It is remarked that kinematic loops can also be found in serial industrial manipulators (usually intended for load balancing). Examples are the ABB IRB4400, IRB6660, the Comau Smart NJ series, SR400, and the 4 DOF KUKA robots KR 40-PA, KR 50-PA, KR 700-PA.

4 Notation

The kinematic topology of FPKM with L structurally identical simple limbs is indicated by the notation $L\mathrm{J}_1\mathrm{J}_2 \ldots \mathrm{J}_{N_l}$, where J_i indicates the type of joint i in the limb. For instance, 6U$\underline{\mathrm{P}}$S refers to a GSP with $L = 6$ limbs, and each limb is a kinematic chain starting with a universal (U) joint at the ground followed by a prismatic (P) joint, succeeded by a spherical (S) joint at the platform. The Tripteron in Fig. 4 has a 3$\underline{\mathrm{P}}$RRR topology, and the Agile Eye in Fig. 6 is referred to as a 3$\underline{\mathrm{R}}$RR PKM. This notation does indeed not reveal the DOF nor the motion pattern of the PKM. Underlined symbols indicate the actuated joints. The notation is extended to FPKM with different limbs. The planar 3-DOF FPKM in Fig. 21 is referred to as a $\underline{\mathrm{R}}$R/2R$\underline{\mathrm{R}}$R PKM.

List of Symbols

δ	DOF of the PKM
δ_p	DOF of the platform
L	Number of limbs of PKM, limbs indexed with $l = 1, \ldots, L$
N	Number of joint variables in the PKM model
$\boldsymbol{\vartheta} \in \mathbb{V}^N$	Vector of joint variables in the PKM model
N_l	Number of joint variables associated to limb l
$\boldsymbol{\vartheta}_{(l)} \in \mathbb{V}^{N_l}$	Vector of joint variables of limb l
N_act	Number of variables assigned to actuated joints
$\boldsymbol{\vartheta}_\mathrm{act} \in \mathbb{V}^{N_\mathrm{act}}$	Vector of joint variable of actuated joints
n	Number of joint variables in tree-topology model when platform is removed
n_l	Number of joint variables of limb l when disconnected from platform
$\bar{\boldsymbol{\vartheta}} \in \mathbb{V}^n$	Vector of joint variables in PKM model when platform is removed
$\bar{\boldsymbol{\vartheta}}_{(l)} \in \mathbb{V}^{n_l}$	Vector of joint variables of limb l when disconnected from platform
$\mathbf{q}, \dot{\mathbf{q}} \in \mathbb{V}^\delta$	Generalized coordinate and velocity vector
$\mathscr{F}_\mathrm{p}$	Reference frame of the platform
$\mathscr{F}_0$	Inertial frame (IFR), also called global world frame
$\mathbf{C}_\mathrm{p} \in SE\,(3)$	Homogenous transformation matrix representing the platform pose
$\mathbf{r}_\mathrm{p} \in \mathbb{R}^3$	Position vector of origin of platform frame $\mathscr{F}_\mathrm{p}$ relative to global IFR $\mathscr{F}_0$
$\mathbf{R}_\mathrm{p} \in SO\,(3)$	Rotation matrix from $\mathscr{F}_\mathrm{p}$ to $\mathscr{F}_0$
$\mathbf{V}_\mathrm{p} = \left(\boldsymbol{\omega}_\mathrm{p}^T, \mathbf{v}_\mathrm{p}^T\right)^T \in \mathbb{R}^6$	Platform twist in ray coordinates, with angular velocity $\boldsymbol{\omega}_\mathrm{p}$ of platform frame $\mathscr{F}_\mathrm{p}$, and velocity $\mathbf{v}_\mathrm{p}$ of the origin of $\mathscr{F}_\mathrm{p}$, both expressed in the platform frame
$\mathbf{V}_\mathrm{t}$	Task space velocity vector composed of δ_p selected components of $\mathbf{V}_\mathrm{p}$
$\mathbf{J}_{\mathrm{p}(l)}$	Forward kinematics Jacobian of limb l so that $\mathbf{V}_\mathrm{p} = \mathbf{J}_{\mathrm{p}(l)}\dot{\boldsymbol{\vartheta}}_{(l)}$
$\mathbf{J}_{\mathrm{t}(l)}$	Task space Jacobian of limb l, so that $\mathbf{V}_\mathrm{t} = \mathbf{J}_{\mathrm{t}(l)}\dot{\boldsymbol{\vartheta}}_{(l)}$
$\mathbf{F}, \bar{\mathbf{F}}$	Inverse kinematics Jacobian of mechanism so that $\dot{\boldsymbol{\vartheta}} = \mathbf{F}\mathbf{V}_\mathrm{t}, \dot{\bar{\boldsymbol{\vartheta}}} = \bar{\mathbf{F}}\mathbf{V}_\mathrm{t}$
$\mathbf{J}_\mathrm{IK}, \mathbf{J}_\mathrm{FK}$	Inverse and forward kinematics Jacobian of PKM so that $\dot{\boldsymbol{\vartheta}}_\mathrm{act} = \mathbf{J}_\mathrm{IK}\mathbf{V}_\mathrm{t}, \mathbf{V}_\mathrm{t} = \mathbf{J}_\mathrm{FK}\dot{\boldsymbol{\vartheta}}_\mathrm{act}$
f_FK	forward kinematic mapping of a PKM, so that $\mathbf{C}_\mathrm{p} = f_\mathrm{FK}(\boldsymbol{\vartheta}_\mathrm{act})$
$f_{(l)}$	Forward kinematic mapping of limb l, so that $\mathbf{C}_\mathrm{p} = f_{(l)}(\boldsymbol{\vartheta}_{(l)})$
$\mathbf{M}, \mathbf{C}$	Generalized mass and Coriolis matrix of PKM model
$\mathbf{W}$	Vector of axis coordinates of a wrench
$\mathbf{Q}$	Vector of generalized forces
$\mathbf{u}$	Vector of actuator forces/torques
$\mathbf{A}^T$	Control matrix so that $\mathbf{A}^T\mathbf{u}$ are the effective control forces
$\mathbf{N}_\mathbf{J}$	Projector to the null space of matrix $\mathbf{J}$

List of Abbreviations

PKM	Parallel Kinematic Manipulator or Parallel Kinematics Machine (also called Parallel Manipulator)
FPKM	Fully parallel manipulator
RA-PKM	Redundantly actuated PKM
SM	Serial Manipulator
GSP	Gough-Stewart platform
EE	End-Effector
TCP	Tool Center Point
DOF	Degree Of Freedom
EOM	Equations of Motion
CTC	Computed Torque Control
c-space	Configuration Space

5 Topology of PKM

The *kinematic topology* of a mechanism refers to the existence and arrangement of bodies and joints. It is represented by a topological graph Γ [117, 179]. Figure 12a shows the topological graph of the 6U$\underline{P}$S Gough-Stewart platform, and Fig. 13a that of the 3-DOF Delta PKM in Fig. 5. Vertices represent bodies and edges represent joints. The type of joint is indicated by a symbol R (revolute), S (spherical), U (universal), P (prismatic), etc. The symbol referring to 'actuated joints' (joints at which actuators are located) are underlined. The topological graph gives a quick overview of the kinematic setup. It immediately reveals the 'parallel' nature of the manipulator with the platform being connected to the ground by L limbs. Limbs are indexed with $l = 1, \ldots, L$. Each limb corresponds to a subgraph $\Gamma_{(l)}$ including the base and platform vertex as shown in Figs. 12b and 13b. This shows that the GSP

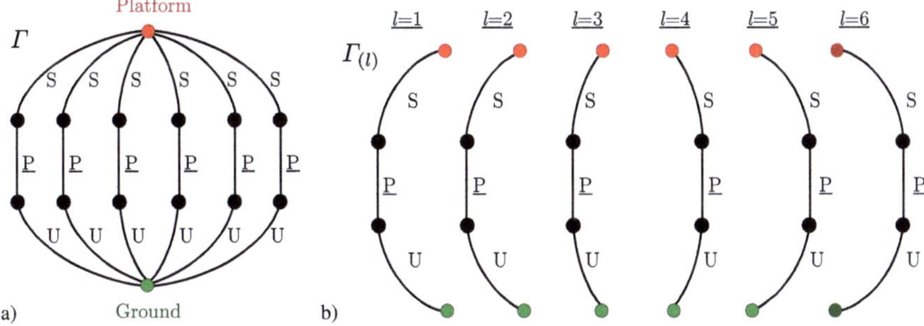

Fig. 12 (a) Topological graph Γ of the 6U$\underline{P}$S Gough-Stewart platform. (b) Subgraphs corresponding to the $L = 6$ simple limbs

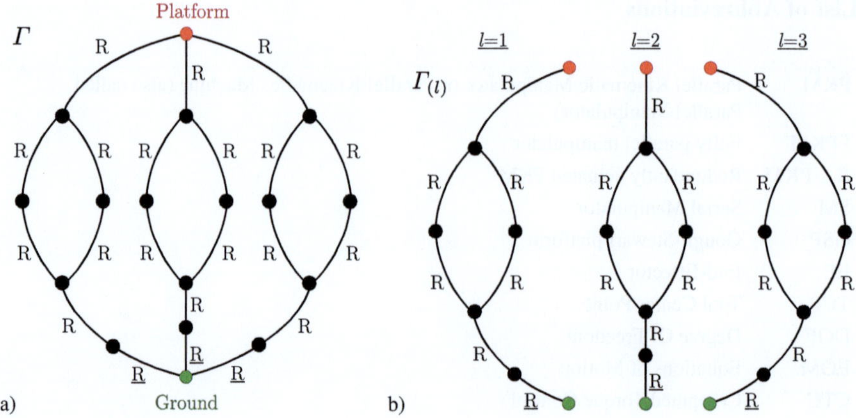

Fig. 13 (**a**) Topological graph Γ of the 3-DOF Delta platform. (**b**) Subgraphs corresponding to the $L = 3$ complex limbs

comprises simple limbs whereas a complex limb of the 3-DOF Delta possesses one kinematic loop.

Different labeling schemes of Γ were proposed in order to index joints and bodies [71, 179]. This will not be necessary in the following. Also there are similar topology representations such as the kinematic diagram and the layout graph [175].

6 Kinematics of Parallel Manipulators with Simple Limbs

Purpose of the kinematics modeling is to describe the PKM motion. Regarded as a kinetostatic transmission device, the platform motion (output) is expressed as a function of the actuator motion (input), or vice versa, which represent the forward respectively inverse kinematics relation of the manipulator that are used for dexterity assessment and singularity analysis. The dynamics modeling, on the other hand, relies on a kinematic model that admits describing the motion of all members of the PKM, either in terms of the platform motion or the actuator motion. Both tasks are addressed in this section.

The moving platform carries an end-effector (EE). A platform frame $\mathscr{F}_\mathrm{p}$ is located at the EE. The origin of this frame is also called the tool center point (TCP). The configuration of the EE-frame relative to the world frame $\mathscr{F}_0$ describes the *platform pose*. The transformation from platform to world frame is described by a homogenous transformation matrix $\mathbf{C}_\mathrm{p} \in SE(3)$ [9, 128, 154]. This is used to represent the EE-pose of a spatial manipulator. Different descriptions are used for lower-mobility PKM ($\delta_\mathrm{p} < 6$). The EE-pose of a Cartesian (translational, regional) PKM is described by the position vector $\mathbf{r}_\mathrm{p} \in \mathbb{R}^3$, that of a spherical PKM by the rotation matrix $\mathbf{R}_p \in SO(3)$, and that of a planar PKM by the corresponding homogenous transformation matrix $\mathbf{C}_p \in SE(2)$.

The platform pose is often described in terms of δ_p 'platform coordinates' summarized in $\mathbf{x} \in \mathbb{R}^{\delta_p}$ so that $\mathbf{C}_p = \mathbf{C}_p(\mathbf{x})$. Such explicit parameterizations in terms of δ_p parameters may lead to singularities when the platform performs spatial rotations. These are avoided by use of redundant parameters (Euler-parameter, unit quaternion, dual quaternions) [9]. With the platform coordinates, the platform twist is determined as

$$\mathbf{V}_p = \boldsymbol{\Omega}(\mathbf{x})\dot{\mathbf{x}} \tag{1}$$

where $\mathbf{V}_p = (\boldsymbol{\omega}_p^T, \mathbf{v}_p^T)^T$ consist of the angular and linear velocity of the platform frame represented in the world frame $\mathscr{F}_0$. The particular form of the matrix $\boldsymbol{\Omega}$ is given for Euler-angles in [155, Chapter 2], for instance.

6.1 Inverse Kinematics of the Manipulator

The inverse kinematics problem of the PKM is to determine the motion of the actuators for a given platform motion (known from motion planning in task space).

6.1.1 Position IKP

A solution of the position inverse kinematics problem has the form $\boldsymbol{\vartheta}_{act} = f_{IK}(\mathbf{C}_p)$, where $\boldsymbol{\vartheta}_{act} \in \mathbb{V}^{N_{act}}$ comprises the N_{act} joint variables of the actuated joints, and f_{IK} is the inverse kinematics mapping. For many PKM the inverse kinematics problem possesses a unique solution that is determined easily. The actuator coordinates of the GSP (elongation of the struts), for instance, are uniquely determined by the locations of the attachment points of the limbs at the moving platform and the ground. Also for the Delta-manipulator, the rotation angles of the actuated R-joints are readily determined by the platform position, when the solution is restricted to technically admissible values (no inside bending of limbs).

6.1.2 Velocity IKP

An elegant way to solve the velocity inverse kinematics problem for PKM with simple limbs is the reciprocal screw formulation originally introduced by Tsai [74, 163, 165]. Consider a FPKM so that one joint of each limb is actuated. This is usually a 1-DOF joint. There is a screw that is reciprocal to the screws of all joints but not to the actuated joint of the limb. The latter can be considered as the wrench that performs no work on any joint other than the actuated joint. Denote with $\mathbf{W}_{act(l)}$ the axis coordinate vector of the screw reciprocal to the screws of all joints but the actuated joint of limb l, and with $\mathbf{S}_{\mathrm{act}(l)}$ the screw coordinate vector of the actuated joint. Then the relation of platform twist and actuator velocity is

$$\mathbf{A}\mathbf{V}_{\mathrm{p}} = \mathbf{B}\dot{\boldsymbol{\vartheta}}_{\mathrm{act}} \tag{2}$$

with

$$\mathbf{A} = \begin{pmatrix} \mathbf{W}_{\mathrm{act}(1)}^{T} \\ \vdots \\ \mathbf{W}_{\mathrm{act}(L)}^{T} \end{pmatrix}, \quad \mathbf{B} = \begin{pmatrix} \mathbf{W}_{\mathrm{act}(1)}^{T}\mathbf{S}_{\mathrm{act}(1)} & & \mathbf{0} \\ & \ddots & \\ \mathbf{0} & & \mathbf{W}_{\mathrm{act}(L)}^{T}\mathbf{S}_{\mathrm{act}(L)} \end{pmatrix} \tag{3}$$

As $\mathbf{B}$ is a diagonal matrix, it is readily inverted to yield the $N_{\mathrm{act}} \times \delta_{\mathrm{p}}$ *inverse kinematics Jacobian*

$$\mathbf{J}_{\mathrm{IK}} = \begin{pmatrix} \mathbf{W}_{\mathrm{act}(1)}^{T}/(\mathbf{W}_{\mathrm{act}(1)}^{T}\mathbf{S}_{\mathrm{act}(1)}) \\ \vdots \\ \mathbf{W}_{\mathrm{act}(L)}^{T}/(\mathbf{W}_{\mathrm{act}(L)}^{T}\mathbf{S}_{\mathrm{act}(L)}) \end{pmatrix}. \tag{4}$$

The solution of the inverse velocity kinematics problem is thus

$$\dot{\boldsymbol{\vartheta}}_{\mathrm{act}} = \mathbf{J}_{\mathrm{IK}}\mathbf{V}_{\mathrm{p}}. \tag{5}$$

where $\boldsymbol{\vartheta}_{\mathrm{act}} = (\vartheta_{\mathrm{act}(1)}, \ldots, \vartheta_{\mathrm{act}(L)})^{T}$ consists of the joint variables associated to the actuated joints of the L limbs of the FPKM. This formulation is applicable to manipulators with more than one actuator per limb [74, 123, 163, 165]. For redundantly actuated PKM ($\delta_{\mathrm{p}} \leq \delta < N_{\mathrm{act}}$), $\mathbf{J}_{\mathrm{IK}}$ is not square.

Example 1 For the Gough-Stewart platform, $\mathbf{S}_{\mathrm{act}(l)} = (\mathbf{0}, \mathbf{e}_{l})^{T}$ are the ray coordinates of the infinite pitch screw corresponding to the pure translation of the actuated prismatic joint, and $\mathbf{W}_{\mathrm{act}(l)} = (\mathbf{e}_{l} \times \mathbf{p}_{l}, \mathbf{e}_{l})^{T}$ are the axis coordinates of a zero pitch screw passing through the attachment points of the limb at the base and the moving platform, respectively, corresponding to a pure force along the limb. $\mathbf{e}_{l}$ is a unit vector along the prismatic joint, and $\mathbf{p}_{l}$ is the vector to a point on the joint axis (e.g. the location of the spherical joint on the platform). The IK Jacobian is thus the 6×6 matrix

$$\mathbf{J}_{\mathrm{IK}} = \begin{pmatrix} \mathbf{e}_{1} \times \mathbf{p}_{1} & \mathbf{e}_{1} \\ \vdots & \vdots \\ \mathbf{e}_{6} \times \mathbf{p}_{6} & \mathbf{e}_{6} \end{pmatrix}. \tag{6}$$

Remark 1 The main advantage of the reciprocal screw formulation is that it only involves those vectors that define the screws of the actuated joints and the reciprocal screws. The vectors in (6) are readily known from the platform pose, for instance. That is, however, this formulation necessitates solving the geometric inverse kinematics problem of the mechanism (Sect. 6.4) since the screw coordinates $\mathbf{S}_{\mathrm{act}(l)}$ and $\mathbf{W}_{\mathrm{act}(l)}$ depend in general on the variables of the actuated joints but also of the passive joints, which must be expressed in terms of platform coordinates.

6.2 Forward Kinematics of the Manipulator

The forward (or direct) kinematics problem is to determine the platform motion for prescribed motion of the actuators.

6.2.1 Velocity FKP

The solution to the velocity forward kinematics problem is $\mathbf{V}_{\mathrm{p}} = \mathbf{J}_{\mathrm{FK}} \dot{\boldsymbol{\vartheta}}_{\mathrm{act}}$. The forward kinematics Jacobian $\mathbf{J}_{\mathrm{FK}}$ is the inverse of (4) if the manipulator is non-redundant. For a full-actuated ($\delta \leq N_{\mathrm{act}}$) kinematically redundant ($\delta_{\mathrm{p}} < N_{\mathrm{act}}$) PKM, $\mathbf{J}_{\mathrm{IK}}$ is not square, and the forward kinematics Jacobian is $\mathbf{J}_{\mathrm{FK}} = \mathbf{J}_{\mathrm{IK}}^{+}$, with the left pseudoinverse $\left(\mathbf{J}^T \mathbf{J}\right)^{-1} \mathbf{J}$.

6.2.2 Position FKP

A solution of the position forward kinematics problem has the form $\mathbf{C}_{\mathrm{p}} = f_{\mathrm{FK}}\left(\boldsymbol{\vartheta}_{\mathrm{act}}\right)$, where f_{FK} is the forward kinematics mapping. For most PKM, the forward kinematics problem does not have a unique solution. The solution branches are called *assembly modes*. Furthermore, f_{FK} cannot be determined explicitly. Solving the forward kinematics problem amounts to solve a system of non-linear equations. The Gough-Stewart platform with general geometry, for example, possesses 40 solutions [36, 146, 173], which reduces to a polynomial of degree 16 if two limbs are attached at the same point to the moving platform [107]. Specific solutions for various PKM were derived, e.g. for a 3UP̱U [162], and for a special class of 6-DOF PKM [22]. Special leg arrangements were identified that simplify the forward kinematics problem of GSP type PKM [110]. As general approach to the explicit solution of the forward kinematics problem, the use of algorithms from algebraic geometry were proposed [67, 102], which are applicable to general Gough-Stewart platforms, for instance.

The velocity forward kinematics solution gives rise to an iterative numerical algorithm for solving the position forward kinematics, which is generally applicable [108]. The transition from platform pose at time t_i, with rotation matrix $\mathbf{R}_{\mathrm{p}}\left(t_i\right)$ and position vector $\mathbf{r}_{\mathrm{p}}\left(t_i\right)$, to the pose at time step t_{i+1} is described by

$$\mathbf{r}_{\mathrm{p}}\left(t_{i+1}\right) = \varLambda \mathbf{r}_{\mathrm{p}} + \mathbf{r}_{\mathrm{p}}\left(t_i\right), \quad \mathbf{R}_{\mathrm{p}}\left(t_{i+1}\right) = \varDelta \mathbf{R}_{\mathrm{p}} \mathbf{R}_{\mathrm{p}}\left(t_i\right) \tag{7}$$

where $\Delta\mathbf{r}_\mathrm{p}$ is the position increment, and $\Delta\mathbf{R}_\mathrm{p} = \mathbf{R}_\mathrm{p}(t_{i+1})\mathbf{R}_\mathrm{p}^T(t_i)$ is the incremental rotation matrix. The latter is determined by an incremental rotation vector $\boldsymbol{\rho}_\mathrm{p}$, and $\Delta\mathbf{R}_\mathrm{p}(\boldsymbol{\rho}_\mathrm{p})$ is given by the Euler-Rodrigues formula [9, 105, 128, 154]. Assuming small increments, the elements of $\boldsymbol{\rho}_\mathrm{p}$ can also be regarded as Cardan angles (1-2-3 Euler angles), and $\Delta\mathbf{R}_\mathrm{p}$ be computed accordingly. The actuator coordinates at time step t_i and t_{i+1} are related by the joint increment vector $\Delta\boldsymbol{\vartheta}$ as $\boldsymbol{\vartheta}_\mathrm{act}(t_{i+1}) = \boldsymbol{\vartheta}_\mathrm{act}(t_i) + \Delta\boldsymbol{\vartheta}$. A first-order approximation of the incremental pose parameters is

$$\begin{pmatrix} \boldsymbol{\rho}_\mathrm{p} \\ \Delta\mathbf{r}_\mathrm{p} \end{pmatrix} = \mathbf{J}_\mathrm{FK}(\boldsymbol{\vartheta}_\mathrm{act}(t_i))\Delta\boldsymbol{\vartheta}_\mathrm{act} \tag{8}$$

assuming that the forward kinematics Jacobian $\mathbf{J}_\mathrm{FK}$ can be expressed in terms of the actuator coordinates $\boldsymbol{\vartheta}_\mathrm{act}$. The so obtain platform pose (7) will be a good approximation for small joint increments (except at singularities). Repeated application of this step yields a solution for the position forward kinematics problem with machine precision.

6.3 Limb Kinematics

The PKM comprises L limbs. Each limb is a simple kinematic chain resembling a serial manipulator with the platform as terminal body. This is clear from the topological graph, as shown in Fig. 12b for the GSP. Denote with $\boldsymbol{\vartheta}_{(l)} \in \mathbb{V}^{N_l}$ the vector of N_l joint variables of limb l. Each limb is a simple kinematic chain, and determines the platform pose as $\mathbf{C}_\mathrm{p} = f_{(l)}(\boldsymbol{\vartheta}_{(l)})$, where $f_{(l)}$ can be regarded as the forward kinematic mapping of separated limb l. The platform twist is determined by the joint velocities as

$$\mathbf{V}_\mathrm{p} = \mathbf{J}_{\mathrm{p}(l)}\dot{\boldsymbol{\vartheta}}_{(l)} \tag{9}$$

with the forward kinematics Jacobian $\mathbf{J}_{\mathrm{p}(l)}(\boldsymbol{\vartheta}_{(l)})$ of limb l [123, 125]. Denote with $r_l \leq N_l$ the maximal rank of $\mathbf{J}_{\mathrm{p}(l)}$. Now consider limb l, with the platform attached, when separated from the PKM, as in Fig. 14. The DOF of the separated limb with the platform attached is N_l. When connected to the separated limb l only, the DOF of the platform (relative to the base) is $r_l \leq N_l$.

If the platform DOF of the PKM is the same as the DOF of the platform when connected to a separated limb, i.e. $\delta_\mathrm{P} = r_l$, then the PKM is called *equimobile*. All 6 DOF PKM are equimobile, e.g. the GSP. An example for a non-equimobile FPKM is the Tripteron in Fig. 4. A limb of the Tripteron (Fig. 14) is a 4-DOF PRRR chain. The platform attached to such a limb can perform spatial translations and a rotation about an axis parallel to the P joint axis (Schönflies or SCARA motion). The platform of the Tripteron can only perform spatial translations, thus has a DOF $\delta_\mathrm{p} =$

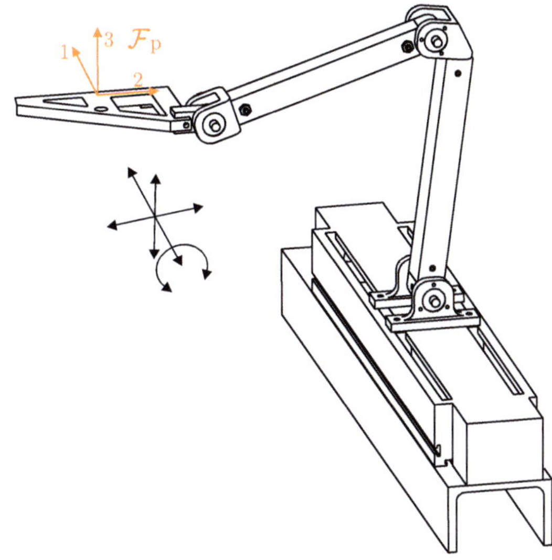

Fig. 14 Single $\underline{\mathrm{P}}$RRR limb of the 3-DOF Tripteron in Fig. 4 with the platform attached

3, however. Attaching a limb of a non-equimobile PKM to the platform apparently imposes further constraints.

6.4 Kinematics of the Mechanism for Kinematically Non-redundant PKM

The geometric inverse kinematics problem of the mechanism consists in determining the complete set of N joint variables $\boldsymbol{\vartheta} \in \mathbb{V}^N$ for given platform pose $\mathbf{C}_\mathrm{p}$. For many PKM a closed form solution $\boldsymbol{\vartheta} = f(\mathbf{C}_\mathrm{p})$, respectively $\boldsymbol{\vartheta} = f(\mathbf{x})$, of the position inverse kinematics can be derived. The derivation of such solutions is specific for the particular PKM. More generally, the geometric inverse kinematics problem can be solved numerically making use of the velocity inverse kinematics solution. The velocity inverse kinematics problem is to determine the velocity of all joints (active and passive) for given $\mathbf{V}_\mathrm{p}$. This proceeds by separately determining $\dot{\boldsymbol{\vartheta}}_{(l)}$ of all limbs. The platform DOF of the PKM is $\delta_\mathrm{P} \leq r_l, l = 1, \ldots, L$, and $\mathbf{V}_\mathrm{p}$ is in accordance to the feasible δ_P-dimensional platform motion. It is assumed that $\mathbf{J}_{\mathrm{p}(l)}$ has full rank r_l.

The geometric forward kinematics problem of the mechanism is to determine the values of all joint variables $\boldsymbol{\vartheta}$ for given vector of $\delta = n_{\mathrm{act}}$ actuator variables. In most cases, a closed form expression $\boldsymbol{\vartheta} = f(\boldsymbol{\vartheta}_{\mathrm{act}})$ cannot be derived. Therefore, a numerical solution is pursued employing the velocity forward kinematics, which is to determine $\dot{\boldsymbol{\vartheta}}$ for given $\dot{\boldsymbol{\vartheta}}_{\mathrm{act}}$.

6.4.1 General Solution to the Velocity Inverse Kinematics Problem

Since the PKM is kinematically non-redundant, the rank of $\mathbf{J}_{p(l)}$ is equal to the number of joint variables of the limb ($N_l = r_l$). The unique solution $\dot{\boldsymbol{\vartheta}}_{(l)} = \mathbf{J}_{p(l)}^{+} \mathbf{V}_p$ of (9) provides an inverse kinematics solution for limb l, with the left pseudoinverse $\mathbf{J}^{+} = \left(\mathbf{J}^{T}\mathbf{J}\right)^{-1}\mathbf{J}$. The overall inverse kinematics solution for the mechanism is

$$\dot{\boldsymbol{\vartheta}} = \mathbf{F}\mathbf{V}_p \tag{10}$$

with the $N \times 6$ matrix $\mathbf{F}(\boldsymbol{\vartheta})$ defined as

$$\mathbf{F} = \begin{pmatrix} \mathbf{J}_{p(1)}^{+} \\ \mathbf{J}_{p(2)}^{+} \\ \vdots \\ \mathbf{J}_{p(L)}^{+} \end{pmatrix} = \begin{pmatrix} \mathbf{F}_{(1)} \\ \mathbf{F}_{(2)} \\ \vdots \\ \mathbf{F}_{(L)} \end{pmatrix} \tag{11}$$

where $\mathbf{F}_{(l)} := \mathbf{J}_{p(l)}^{+}$. The solution (10) involves the pseudoinverse of the limb Jacobian. This can be avoided for the majority of PKM, as described in the following two sections.

6.4.2 Velocity Inverse Kinematics of Equimobile PKM

For an equimobile PKM, the rank of $\mathbf{J}_{p(l)}$ is equal to the task space dimension ($r_l = \delta_p$). Moreover, for most PKM, the EE frame can be defined so that exactly δ_p components of the platform twist $\mathbf{V}_p$ are prescribed. The δ_p non-zero components are summarized in the *task space velocity* vector $\mathbf{V}_t \in \mathbb{R}^{\delta_p}$. This is formalized as

$$\mathbf{V}_p = \mathbf{P}_p \mathbf{V}_t \tag{12}$$

with the unimodular $6 \times \delta_p$ velocity distribution matrix $\mathbf{P}_p$. With a platform frame $\mathscr{F}_p$ introduced accordingly, for spatial translations and rotations, planar motions, and Schönflies motions, for example, this matrix attains the form

$$\mathbf{P}_p^{\text{trans}} = \begin{pmatrix} \mathbf{0}_{3,3} \\ \mathbf{I}_3 \end{pmatrix}, \ \mathbf{P}_p^{\text{rot}} = \begin{pmatrix} \mathbf{I}_{3,3} \\ \mathbf{0}_3 \end{pmatrix}, \ \mathbf{P}_p^{\text{planar}} = \begin{pmatrix} 0 & 0 & 0 \\ 0 & 0 & 0 \\ 0 & 0 & 1 \\ 1 & 0 & 0 \\ 0 & 1 & 0 \\ 0 & 0 & 0 \end{pmatrix}, \ \mathbf{P}_p^{\text{SCARA}} = \begin{pmatrix} 0 & 0 & 0 \\ 0 & 0 & 0 \\ 0 & 0 & 1 \\ 1 & 0 & 0 \\ 0 & 1 & 0 \\ 0 & 0 & 1 \end{pmatrix}. \tag{13}$$

For instance: all $\delta_p = 6$ components of the platform twist of a 6-DOF PKM are prescribed, and $\mathbf{P}_p = \mathbf{I}_6$; the $\delta_p = 3$ components of the translation velocity of a Cartesian PKM (e.g. Tricept, Tripteron, 3-DOF Delta) with $\mathbf{P}_p^{\text{trans}}$; $\delta_p = 3$ components (two translations, one angular) of the platform twist of planar PKM (e.g. DualV) with $\mathbf{P}_p^{\text{planar}}$; $\delta_p = 4$ components of the platform twist (three components of the linear and one component of the angular velocity) for PKM generating Schönflies motions (e.g. H4, 4-DOF Delta) with $\mathbf{P}_p^{\text{SCARA}}$. Only for special lower-mobility PKM, which generate mixed platform motions, it is not possible to uniquely select δ_p independent velocity components. The parameterization of platform twist (1) yields the parameterization of the task velocity

$$\mathbf{V}_t = \boldsymbol{\Omega}\dot{\mathbf{x}}. \tag{14}$$

A $\delta_p \times N_l$ *task space Jacobian* $\mathbf{J}_{t(l)}$ of limb l is introduced so that

$$\mathbf{V}_t = \mathbf{J}_{t(l)}\dot{\boldsymbol{\vartheta}}_{(l)}, l = 1, \ldots, L. \tag{15}$$

That is, $\mathbf{J}_{t(l)}$ consists of the δ_p rows of $\mathbf{J}_{p(l)}$ corresponding to the relevant task space components. When the PKM is kinematically non-redundant ($\delta_p = N_l = r_l$), the relation (15) can be solved as $\dot{\boldsymbol{\vartheta}}_{(l)} = \mathbf{J}_{t(l)}^{-1}\mathbf{V}_t$, which provides an inverse kinematics solution for limb l. The overall inverse kinematics solution for the mechanism is

$$\dot{\boldsymbol{\vartheta}} = \mathbf{F}\mathbf{V}_t \tag{16}$$

with the $N \times \delta_p$ matrix $\mathbf{F}(\boldsymbol{\vartheta})$ defined as

$$\mathbf{F} = \begin{pmatrix} \mathbf{J}_{t(1)}^{-1} \\ \mathbf{J}_{t(2)}^{-1} \\ \vdots \\ \mathbf{J}_{t(L)}^{-1} \end{pmatrix} = \begin{pmatrix} \mathbf{F}_{(1)} \\ \mathbf{F}_{(2)} \\ \vdots \\ \mathbf{F}_{(L)} \end{pmatrix}. \tag{17}$$

6.4.3 Velocity Inverse Kinematics of Non-equimobile PKM

When the platform is connected to only one limb of a non-equimobile PKM, its DOF is higher then the platform DOF of the PKM, and the Jacobian $\mathbf{J}_{p(l)}$ of limb l in (9) has rank $r_l > \delta_p$. The following assumption is valid for most non-equimobile PKM used in practice.

Assumption 1 The EE frame is introduced so that $\mathbf{J}_{p(l)}$ contains exactly r_l non-zero rows.

The $r_l \times N_l$ task space Jacobian $\mathbf{J}_{t(l)}$ of limb l is now constructed from the r_l non-zero rows of $\mathbf{J}_{p(l)}$. The task space velocity vector $\mathbf{V}_t$ is again constructed from the δ_p non-zero components of the platform twist $\mathbf{V}_p$, according to (12). Since $r_l > \delta_p$, the relation $\mathbf{J}_{t(l)} \dot{\boldsymbol{\vartheta}}_{(l)}$ delivers the δ_p components of $\mathbf{V}_t$ but also $r_l - \delta_p$ components that must be zero. The latter correspond to constraints imposed on the motion of limb l. Introduce a $r_l \times \delta_p$ velocity distribution matrix $\mathbf{D}_{t(l)}$, which assigns the components of the task space velocity $\mathbf{V}_t$ to the relevant rows of the task space Jacobian of limb l. Then the forward kinematics of the limb and the imposed constraints are summarized in

$$\mathbf{D}_{t(l)} \mathbf{V}_t = \mathbf{J}_{t(l)} \dot{\boldsymbol{\vartheta}}_{(l)}, l = 1, \ldots, L. \tag{18}$$

If the PKM is equimobile, (15) holds true, and $\mathbf{D}_{t(l)}$ is the identity matrix.

For kinematically non-redundant PKM it is $N_l = r_l$, and the task space Jacobian $\mathbf{J}_{t(l)}$ is invertible. The overall solution for the mechanism is thus given by

$$\dot{\boldsymbol{\vartheta}} = \mathbf{F} \mathbf{V}_t \tag{19}$$

with

$$\mathbf{F} = \begin{pmatrix} \mathbf{J}_{t(1)}^{-1} \mathbf{D}_{t(1)} \\ \mathbf{J}_{t(2)}^{-1} \mathbf{D}_{t(2)} \\ \vdots \\ \mathbf{J}_{t(L)}^{-1} \mathbf{D}_{t(L)} \end{pmatrix} = \begin{pmatrix} \mathbf{F}_{(1)} \\ \mathbf{F}_{(2)} \\ \vdots \\ \mathbf{F}_{(L)} \end{pmatrix}. \tag{20}$$

Example 2 The geometric Jacobian of a limb of the 3-DOF Tripteron has rank $r_l = 4$. The PKM platform can only translate ($\delta_p = 3$), and the task space velocity vector $\mathbf{V}_t$ consists of the three components of the linear EE velocity $\mathbf{v}_p$. The selection matrix is thus

$$\mathbf{D}_{t(l)} = \begin{pmatrix} \mathbf{0}_{1,3} \\ \mathbf{I}_3 \end{pmatrix}, l = 1, 2, 3. \tag{21}$$

Remark 2 The EE-twist, and thus the taskspace velocity, is expressed in the platform frame. The explicit use of matrix $\mathbf{D}_{t(l)}$ is avoided when implementing (19). It merely eliminates those columns of $\mathbf{J}_{t(l)}^{-1}$ that do not correspond to a component of $\mathbf{V}_t$. These columns are simply omitted in the multiplication of $\mathbf{J}_{t(l)}^{-1}$ and $\mathbf{D}_{t(l)}$.

It may occasionally be advantageous to represent the EE twists and the geometric Jacobian $\mathbf{J}_{p(l)}$ in different frames. Then the matrix $\mathbf{D}_{t(l)}$ is not constant, and must be used explicitly in (19).

Remark 3 In situations where it is not possible to introduce an EE frame so that $\mathbf{J}_{p(l)}$ contains $r_l - \delta_p$ non-zero components, it is not allowed to just arbitrarily select

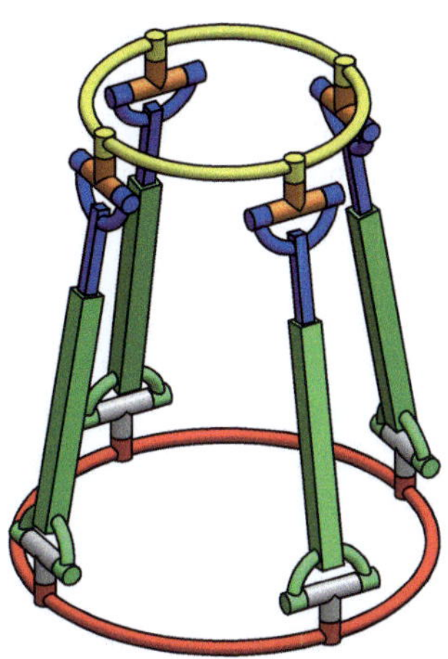

Fig. 15 A 4-DOF 4$U\underline{P}U$ non-equimobile PKM

r_l rows for constructing $\mathbf{J}_{t(l)}$, and the formulation presented in this section cannot be used without further amendment. For instance, the geometric Jacobian of a limb, including platform, of the 4-DOF 4U$\underline{P}$U FPKM in Fig. 15 has rank $r_l = 5$. Also the platform attached to this open chain has 5 DOF. Due to the special geometric arrangement (the vertical axes of the U joints are parallel), the platform of the PKM can translate in all three directions and rotate about the vertical axis (Schönflies motions). Hence, the platform DOF is $\delta_p = 4$, and the PKM is thus non-equimobile. There is no choice of EE frame that would render any of the rows of $\mathbf{J}_{p(l)}$ non-zero. While $\mathbf{J}_{p(l)}$ always has rank 5 (unless the prismatic joint is fully contracted), any choice of $r_l = 5$ rows to construct $\mathbf{J}_{t(l)}$ introduces (artificial) singularities. Yet it is possible to switch between different selections of 5 independent rows, to be used in the presented formulation.

6.4.4 Velocity Inverse Kinematics of the Manipulator

The inverse kinematics solution (10), (16), and (19), respectively, yields the actuator velocities. Thus, if the inverse kinematics of the mechanism is evaluated (e.g. for computing the dynamic EOM), the reciprocal screw formulation (5) is dispensable. The inverse kinematics Jacobian $\mathbf{J}_{\mathrm{IK}}$ in (5) is then readily available as the submatrix of $\mathbf{F}$.

6.4.5 Geometric Inverse Kinematics

If no explicit solution of the position inverse kinematics is available, or if a case-by-case treatment is not desirable, the motion of all joints can be computed for given platform motion by time integration of the velocity relation (10), respectively

(16) or (19). Alternatively, the joint variables $\boldsymbol{\vartheta}$ can be determined at discrete sampling times. The relative translation and orientation of the EE-frame between two consecutive time instants, represented in IFR, are $\Delta \mathbf{r}_{\mathrm{p}}$ and $\Delta \mathbf{R}_{\mathrm{p}}(\boldsymbol{\rho}_{\mathrm{p}})$ in (7). The corresponding increments of the joint variables are

$$\Delta \boldsymbol{\vartheta} = \mathbf{F} \mathbf{P}_{\mathrm{p}}^{T} \begin{pmatrix} \boldsymbol{\rho}_{\mathrm{p}} \\ \Delta \mathbf{r}_{\mathrm{p}} \end{pmatrix} \tag{22}$$

The joint variable vector at time step i is then approximated as $\boldsymbol{\vartheta}\,(t_i) \approx \boldsymbol{\vartheta}\,(t_{i-1}) + \Delta \boldsymbol{\vartheta}$. This approximation is repeated until $\boldsymbol{\vartheta}\,(t_i)$ satisfies the geometric constraints $\mathbf{C}_{\mathrm{p}} = f_{(l)}(\boldsymbol{\vartheta}_{(l)}), l = 1, \ldots, L$.

6.4.6 Forward Kinematics

Combining the solution of the velocity inverse kinematics of the mechanism with the solution of the velocity forward kinematics problem of the manipulator yields the joint velocities in terms of actuator velocities

$$\dot{\boldsymbol{\vartheta}} = \mathbf{F} \mathbf{J}_{\mathrm{FK}} \dot{\boldsymbol{\vartheta}}_{\mathrm{act}}. \tag{23}$$

The forward kinematics Jacobian is the inverse of $\mathbf{J}_{\mathrm{IK}}$, which is either determined with the reciprocal screw approach as (4) or as the submatrix of $\mathbf{F}$ (Sect. 6.4.4). If $\mathbf{J}_{\mathrm{FK}}$ is constructed as submatrix of $\mathbf{F}$, the FKP can thus be solved using $\mathbf{F}$ only.

The relation (23) gives rise to an iterative solution method for the geometric forward kinematics problem. Starting from a configuration $\boldsymbol{\vartheta}$ with actuator variables $\boldsymbol{\vartheta}_{\mathrm{act}}$, the joint increment $\Delta \boldsymbol{\vartheta}$ so that $\boldsymbol{\vartheta} + \Delta \boldsymbol{\vartheta}$ is in correspondence to the actuator variables $\boldsymbol{\vartheta}_{\mathrm{act}} + \Delta \boldsymbol{\vartheta}_{\mathrm{act}}$ is determined by (repeated application of) the step increment

$$\Delta \boldsymbol{\vartheta} = \mathbf{F} \mathbf{J}_{\mathrm{FK}} \Delta \boldsymbol{\vartheta}_{\mathrm{act}}. \tag{24}$$

An alternative to solving the geometric forward kinematic problem of the mechanism is to use a closed form solution $\boldsymbol{\vartheta} = f\left(\mathbf{C}_{\mathrm{p}}\right)$ or $\boldsymbol{\vartheta} = f\left(\mathbf{x}\right)$ of the inverse kinematics problem of the mechanism. First, a solution $\mathbf{x} = f\left(\boldsymbol{\vartheta}_{\mathrm{act}}\right)$ to the forward kinematics problem is determined, and then $\boldsymbol{\vartheta} = f\left(\mathbf{x}\right)$ is evaluated. This is usually computationally simpler than the incremental solution (24).

6.5 Inverse Kinematics of the Mechanism for Kinematically Redundant PKM

A PKM is kinematically redundant if one of its limbs is kinematically redundant, which implies that $r_l < N_l$. The limb can perform $N_l - r_l$-dimensional 'self-motions' even if the platform is locked. The general solution (10) does not apply to a kinematically redundant limb. An explicit inverse kinematics solution can be derived if Assumption 1 holds true. This solution is constructed by adopting the joint space decomposition approach, introduced in [172] for serial manipulators, to the individual limbs.

6.5.1 Velocity Inverse Kinematics

The N_l joint variables of limb l are partitioned into a vector $\boldsymbol{\vartheta}_{\mathrm{r}(l)}$ of $N_l - r_l$ 'redundant' variables and a vector $\boldsymbol{\vartheta}_{\mathrm{nr}(l)}$ of r_l 'non-redundant' joint variables. The vector of 'redundant' variables $\boldsymbol{\vartheta}_{\mathrm{r}(l)}$ consists of selected variables of actuated joints. The notation stems from the fact that they allow for self-motion of the limb, which does not affect the platform motion. The partitioning of the joint coordinate vector is formalized by means of unimodular matrices $\mathbf{P}_{\mathrm{nr}(l)} \in \mathbb{R}^{N_l, r_l}$ and $\mathbf{P}_{\mathrm{r}(l)} \in \mathbb{R}^{N_l, N_l - r_l}$ so that $\boldsymbol{\vartheta}_{(l)} = \mathbf{P}_{\mathrm{nr}(l)} \boldsymbol{\vartheta}_{\mathrm{nr}(l)} + \mathbf{P}_{\mathrm{r}(l)} \boldsymbol{\vartheta}_{\mathrm{r}(l)}$, where $\mathbf{P}_{\mathrm{nr}(l)}^T \mathbf{P}_{\mathrm{r}(l)} = \mathbf{0}_{r_l, N_l - r_l}$. Accordingly, (18) splits as

$$\mathbf{D}_{\mathrm{t}(l)} \mathbf{V}_{\mathrm{t}} = \mathbf{J}_{\mathrm{t}(l),\mathrm{nr}} \dot{\boldsymbol{\vartheta}}_{\mathrm{nr}(l)} + \mathbf{J}_{\mathrm{t}(l),\mathrm{r}} \dot{\boldsymbol{\vartheta}}_{\mathrm{r}(l)} \tag{25}$$

with $\mathbf{J}_{\mathrm{t}(l),\mathrm{nr}} := \mathbf{J}_{\mathrm{t}(l)} \mathbf{P}_{\mathrm{nr}(l)}$ and $\mathbf{J}_{\mathrm{t}(l),\mathrm{r}} := \mathbf{J}_{\mathrm{t}(l)} \mathbf{P}_{\mathrm{r}(l)}$. The $r_l \times r_l$ matrix $\mathbf{J}_{\mathrm{t}(l),\mathrm{nr}}$ is invertible, and (25) can be solved as

$$\dot{\boldsymbol{\vartheta}}_{(l)} = \mathbf{P}_{\mathrm{nr}(l)} \mathbf{J}_{\mathrm{t}(l),\mathrm{nr}}^{-1} \mathbf{D}_{\mathrm{t}(l)} \mathbf{V}_{\mathrm{t}} + (\mathbf{P}_{\mathrm{r}(l)} - \mathbf{P}_{\mathrm{nr}(l)} \mathbf{J}_{\mathrm{t}(l),\mathrm{nr}}^{-1} \mathbf{J}_{\mathrm{t}(l),\mathrm{r}}) \dot{\boldsymbol{\vartheta}}_{\mathrm{r}(l)}. \tag{26}$$

This is a solution for the inverse velocity kinematics problem of a limb.

The vector of *generalized velocities* $\mathbf{s} \in \mathbb{R}^\delta$ of the kinematically redundant PKM is introduced as

$$\mathbf{s} := \begin{pmatrix} \mathbf{V}_{\mathrm{t}} \\ \dot{\boldsymbol{\vartheta}}_{\mathrm{r}(1)} \\ \vdots \\ \dot{\boldsymbol{\vartheta}}_{\mathrm{r}(L)} \end{pmatrix} = \begin{pmatrix} \mathbf{V}_{\mathrm{t}} \\ \dot{\boldsymbol{\vartheta}}_{\mathrm{r}} \end{pmatrix} \tag{27}$$

where $\delta = \delta_{\mathrm{p}} + N - r_1 - r_2 \ldots - r_L$ is the DOF of the PKM, and $N - r_1 - r_2 - \ldots - r_L$ is the total number of 'redundant' joint variables summarized in $\boldsymbol{\vartheta}_{\mathrm{r}}$. The inverse kinematics solution for the mechanism of a kinematically redundant PKM is

$$\dot{\boldsymbol{\vartheta}} = \mathbf{F}\mathbf{s} \tag{28}$$

with

$$
\mathbf{F} = \begin{pmatrix}
\mathbf{P}_{\mathrm{nr}(1)}\mathbf{J}_{\mathrm{t}(1),\mathrm{nr}}^{-1}\mathbf{D}_{\mathrm{t}(1)} & \mathbf{P}_{\mathrm{r}(1)} - \mathbf{P}_{\mathrm{nr}(1)}\mathbf{J}_{\mathrm{t}(1),\mathrm{nr}}^{-1}\mathbf{J}_{\mathrm{t}(1),\mathrm{r}} & \mathbf{0} \\
\mathbf{P}_{\mathrm{nr}(2)}\mathbf{J}_{\mathrm{t}(2),\mathrm{nr}}^{-1}\mathbf{D}_{\mathrm{t}(2)} & \mathbf{0} & \mathbf{P}_{\mathrm{r}(2)} - \mathbf{P}_{\mathrm{nr}(2)}\mathbf{J}_{\mathrm{t}(2),\mathrm{nr}}^{-1}\mathbf{J}_{\mathrm{t}(2),\mathrm{r}} \\
\vdots & \vdots & \vdots \\
\mathbf{P}_{\mathrm{nr}(L)}\mathbf{J}_{\mathrm{t}(L),\mathrm{nr}}^{-1}\mathbf{D}_{\mathrm{t}(L)} & \mathbf{0} & \mathbf{0}
\end{pmatrix}
$$

$$
\begin{pmatrix}
\cdots & \mathbf{0} & \\
\cdots & \mathbf{0} & \\
\ddots & \vdots & \\
\cdots & \mathbf{P}_{\mathrm{r}(L)} - \mathbf{P}_{\mathrm{nr}(L)}\mathbf{J}_{\mathrm{t}(L),\mathrm{nr}}^{-1}\mathbf{J}_{\mathrm{t}(L),\mathrm{r}}
\end{pmatrix}
$$

$$
= \begin{pmatrix} \mathbf{F}_{(1)} \\ \mathbf{F}_{(2)} \\ \vdots \\ \mathbf{F}_{(L)} \end{pmatrix}. \tag{29}
$$

The velocity of all joints is thus expressed in terms of the task space velocity and velocities of selected actuators, which explicitly determine the self-motion. Equimobile PKM are included with $\mathbf{D}_{\mathrm{t}(l)} = \mathbf{I}$. The expression (29) provides a systematic formulation, with the matrices $\mathbf{P}_{\mathrm{nr}(l)}$ and $\mathbf{P}_{\mathrm{r}(l)}$ simply allocating the respective 'non-redundant' and 'redundant' joint coordinates within the joint coordinate vector.

Example 3 The planar 6-DOF $3\underline{RP}RR$ PKM in Fig. 11 can be regarded as an amended version of the 3-DOF $3\underline{R}RR$ PKM by introducing a controllable length of the first link in each limb. The platform of this PKM has 3-DOF. The degree of kinematic redundancy is $\delta - \delta_{\mathrm{p}} = 3$, so that 3 'redundant' joint variables must be selected, one per limb. The coordinate vector $\boldsymbol{\vartheta}_{(l)} = (\vartheta_1, \vartheta_2, \vartheta_3, \vartheta_4)_{(l)}^T$ of limb l contains the rotation angle ϑ_1 of the actuated revolute joint at the base, the displacement coordinate ϑ_2 of the actuated prismatic joint, and the angles ϑ_3, ϑ_4 of the remaining two passive revolute joints. The displacement coordinates $\vartheta_{2(l)}$ of the prismatic joints (actuators 4,5,6) are used as the 'redundant' variables. The corresponding selection matrices are

$$
\mathbf{P}_{\mathrm{r}(l)} = \begin{pmatrix} 0 \\ 1 \\ 0 \\ 0 \end{pmatrix}, \quad \mathbf{P}_{\mathrm{nr}(l)} = \begin{pmatrix} 1 & 0 & 0 \\ 0 & 0 & 0 \\ 0 & 1 & 0 \\ 0 & 0 & 1 \end{pmatrix}. \tag{30}
$$

Remark 4 Notice that (28) contains $\delta - \delta_{\mathrm{p}}$ trivial equations of the form $\dot{\boldsymbol{\vartheta}}_{\mathrm{r}(l)} = \dot{\boldsymbol{\vartheta}}_{\mathrm{r}(l)}$. In the above $3\underline{RP}RR$ example, these are $\dot{\vartheta}_{2(l)} = \dot{\vartheta}_{2(l)}$.

6.5.2 Velocity Inverse Kinematics of the Manipulator

The relation (28) also represents a solution to the velocity inverse kinematics problem of the manipulator. Collecting the N_{act} relevant rows of (28), this attains the form

$$\dot{\boldsymbol{\vartheta}}_{act} = \mathbf{J}_{IK}\mathbf{V}_t + \mathbf{S}\dot{\boldsymbol{\vartheta}}_r \tag{31}$$

where $\mathbf{J}_{IK}$ and $\mathbf{S}$ are the corresponding submatrices of $\mathbf{F}$ in (28). This is a particular solution where the null-space velocity is parameterized by 'redundant' joint velocities $\dot{\boldsymbol{\vartheta}}_r$, i.e. by the velocities of selected actuators.

6.5.3 Geometric Inverse Kinematics of the Mechanism

The position inverse kinematics problem is to compute all joint variables $\boldsymbol{\vartheta}$ for given platform pose $\mathbf{C}_p$ and 'redundant' joint variables $\boldsymbol{\vartheta}_r$. Relation (28) gives rise to the following Newton-iteration step

$$\Delta\boldsymbol{\vartheta} = \mathbf{F}\begin{pmatrix} \Delta\mathbf{x} \\ \Delta\boldsymbol{\vartheta}_r \end{pmatrix} \tag{32}$$

with $\Delta\mathbf{x} = \mathbf{P}_p^T \begin{pmatrix} \boldsymbol{\rho}_p \\ \Delta\mathbf{r}_p \end{pmatrix}$, which yield $\boldsymbol{\vartheta}\,(t_i) \approx \boldsymbol{\vartheta}\,(t_{i-1}) + \Delta\boldsymbol{\vartheta}$. This is repeated until $\boldsymbol{\vartheta}\,(t_i)$ satisfies the geometric constraints $\mathbf{C}_p = f_{(l)}(\boldsymbol{\vartheta}_{(l)}), l = 1, \ldots, L$.

6.5.4 Forward Kinematics

Actuator coordinates are used as 'redundant' coordinates, and (31) yields the forward kinematics solution for the manipulator

$$\mathbf{V}_t = \mathbf{J}_{IK}^{+}\dot{\boldsymbol{\vartheta}}_{act}. \tag{33}$$

A forward kinematics solution for the mechanism is then obtained with (28) as

$$\dot{\boldsymbol{\vartheta}} = \mathbf{F}\begin{pmatrix} \mathbf{J}_{IK}^{+}\dot{\boldsymbol{\vartheta}}_{act} \\ \dot{\boldsymbol{\vartheta}}_r \end{pmatrix} \tag{34}$$

Since $\boldsymbol{\vartheta}_r$ is contained in $\boldsymbol{\vartheta}_{act}$, this can be reduced to $\dot{\boldsymbol{\vartheta}} = \mathbf{A}\dot{\boldsymbol{\vartheta}}_{act}$, with $N \times \delta$ matrix $\mathbf{A}$. The solution to the geometric forward kinematics is obtained by integrating (34) or with an iterative Newton-scheme.

7 Kinematics of PKM with Complex Limbs

The approach to model PKM with simple limbs is carried over to PKM with complex limbs by resolving the loop constraints within the limbs, so that each limb resembles a simple kinematic chain. In the following, non-redundant PKM will be considered only.

7.1 Modeling Complex Limbs by Functionally Equivalent Joints

PKM with complex limbs were often modeled by replacing the complex limbs by an equivalent simple limb, i.e. a simple kinematic chain. The basic principle of this modeling method consists in replacing each loop within a limb by a 'virtual joint', i.e. an equivalent kinetostatic transmission element. This allows modeling the overall transmission characteristics of the respective loop. It does, however, not directly determine the motion of all members of the loop (and may thus not capture internal singularities), which is necessary for dynamics modeling.

While the geometric loop constraints are difficult to solve in general, many of the proposed PKM possess closed form solutions. As such, several PKM were proposed that comprise planar 4-bar parallelogram linkages, which function as so-called Π-joints, e.g. the Orthoglide [30]. The IRSBot-2 [45] in Fig. 8 comprises a planar 4R parallelogram as well as a loop with four U joints, which can be solved explicitly.

7.2 Modeling Complex Limbs by Resolution of Velocity Loop Constraints

A complex limb l possesses γ_l kinematic loops. A limb of the 3-DOF Delta robot in Fig. 5 has $\gamma_l = 1$ loop (formed by the rods of the parallelogram), while $\gamma_l = 2$ loops can be identified in a limb of the 2-DOF IRSBot-2 in Fig. 8 (one formed by the planar 4-bar and another by the four U joints). This is reflected by the existence of γ_l independent cycles, denoted $\Lambda_{k(l)}, k = 1, \ldots, \gamma_l$, in the topological graph $\Gamma_{(l)}$ of a separated limb, as shown in Figs. 16a and 17a. A spanning tree $G_{(l)}$ is introduced on this subgraph, which is obtained by removing one joint (the *cut-joint*) from each loop (Figs. 16b and 17b).

Denote with $\boldsymbol{\eta}_{(l)} \in \mathbb{V}^{\nu_l}$ the vector of ν_l joint variables of the tree-joints (vector $\boldsymbol{\vartheta}_{(l)}$ with cut-joint variables removed). The geometric constraints due to the γ_l loops of limb l give rise to the system of m_l non-linear equations $g_{(l)}(\boldsymbol{\eta}_{(l)}) = \mathbf{0}$, with corresponding velocity constraints

$$\mathbf{G}_{(l)}\dot{\boldsymbol{\eta}}_{(l)} = \mathbf{0} \tag{35}$$

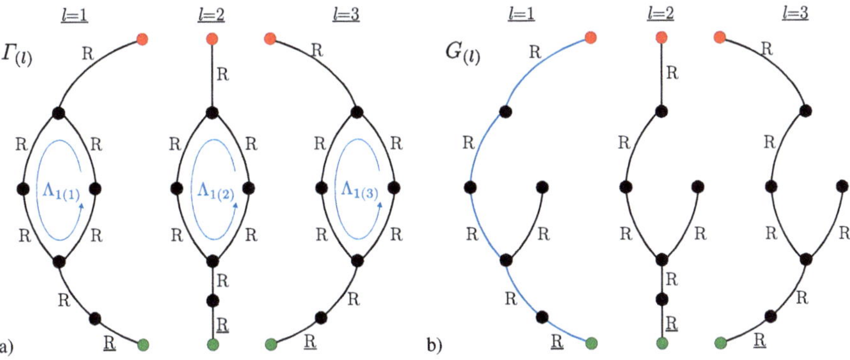

Fig. 16 (**a**) Subgraphs $\Gamma_{(l)}$ for the $L = 3$ limbs of the Delta robot. (**b**) Spanning trees $G_{(l)}$ obtained by removing one edge from $\Gamma_{(l)}$

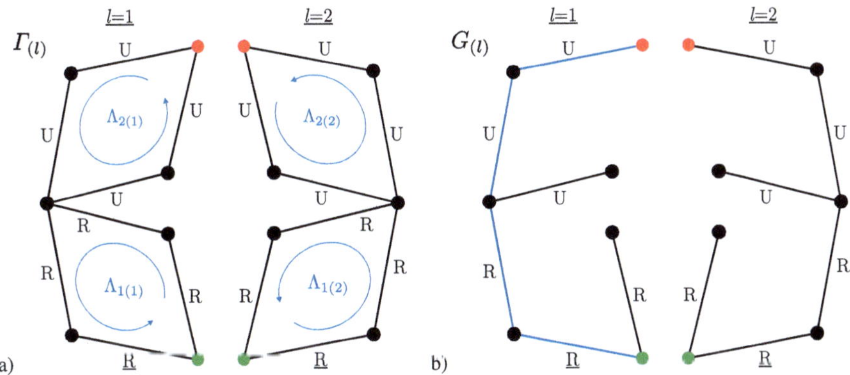

Fig. 17 (**a**) Subgraphs $\Gamma_{(l)}$ for the $L = 2$ limbs of the IRSBot-2 in Fig. 8. (**b**) Spanning trees $G_{(l)}$ obtained by removing one edge from $\Gamma_{(l)}$

where $\mathbf{G}_{(l)}(\boldsymbol{\eta}_{(l)})$ is the $m_l \times \nu_l$ constraint Jacobian. Assuming independence of the m_l constraints, the separated limb has $\delta_l := \nu_l - m_l$ DOF. The velocity constraints (35) can then be solved as

$$\dot{\boldsymbol{\eta}}_{(l)} = \mathbf{H}_{(l)}\dot{\mathbf{y}}_{(l)} \tag{36}$$

where $\dot{\mathbf{y}}_{(l)}$ is a vector of $\nu_l - m_l$ independent velocities and $\mathbf{H}_{(l)}$ is an orthogonal complement to $\mathbf{G}_{(l)}$, i.e. the $\nu_l - m_l$ columns of $\mathbf{H}_{(l)}$ form a basis for the null-space of $\mathbf{G}_{(l)}$. $\mathbf{H}_{(l)}$ can be determined after block partitioning of $\mathbf{G}_{(l)}$ or using the reciprocal screw approach.

The geometric Jacobian of the platform, $\mathbf{J}_{\mathrm{p}(l)}$ in (9), is now the Jacobian of the kinematic chain connecting platform to the base in the tree-topology (indicated in blue color for limb 1 in Figs. 16b and 17b, such that $\mathbf{V}_\mathrm{p} = \mathbf{J}_{\mathrm{p}(l)}\dot{\boldsymbol{\eta}}_{(l)}$. Inserting (36) yields the platform twist

$$\mathbf{V}_\mathrm{p} = \bar{\mathbf{J}}_{\mathrm{p}(l)}\dot{\mathbf{y}}_{(l)} \tag{37}$$

where $\bar{\mathbf{J}}_{\mathrm{p}(l)}(\boldsymbol{\eta}_{(l)})$ is the *composite Jacobian* of complex limb l defined as

$$\bar{\mathbf{J}}_{\mathrm{p}(l)} := \mathbf{J}_{\mathrm{p}(l)}\mathbf{H}_{(l)}. \tag{38}$$

The *composite task space Jacobian of the complex limb l* is the $\delta_\mathrm{p} \times \delta_l$ matrix $\bar{\mathbf{J}}_{\mathrm{t}(l)}$ comprising the δ_p relevant rows of $\bar{\mathbf{J}}_{\mathrm{p}(l)}$.

Remark 5 The number of constraints depends on the DOF of the cut-joint, i.e. on the number of constraints the cut-joint introduces. The constraints (35) do not involve the cut-joint variables. Cut-joints can be selected so to reduce the number of remaining variables involved in the constraints as well as the number of constraint equations. This is an advantage of the cut-joint formulation.

Remark 6 Resolving the loop constraints within a complex limb independently from the overall loop constraints of the PKM is also called *constraint embedding*. This approach was introduced in [72, 73] for general multibody systems aiming to obtain a tree-topology representation for systems with general topology. A tailored constraint embedding for PKM with complex loops is presented in [124].

Example 4 A limb of the IRSBot-2 in Fig. 8 contains $\gamma_l = 2$ loops. The proximal loop consists of 4 revolute joints forming a planar parallelogram. Removing one R joint, the remaining branch with three R joints is subjected to 2 independent constraints. The solution to the geometric loop constraints is obvious. The loop comprising the four U joints is opened by cutting one of the U joints, which leaves a branch with three U joints, and 6 joint variables. A U joint introduces $m_l = 4$ constraints (35) to which the 6 U joint variables are subjected. The $m_l = 4$ constraints are independent so that the loop itself has 2-DOF, and a solution (36) is easily obtained. The constraints for both loops yields a system of $m_l = 6$ constraints for the remaining $\nu_l = 9$ tree joint variables, so that the limb (and the platform) has $\nu_l - m_l = 9 - 6 = 3$ DOF. When both limbs are connected, the platform DOF is $\delta_\mathrm{p} = 2$ DOF (no rotation possible when the two limbs are connected), so that the IRSBot-2 is non-equimobile.

7.3 Inverse Kinematics

7.3.1 Velocity Inverse Kinematics of the Mechanism

After opening the kinematic loops of all L limbs, the model comprises $\nu = \nu_1 + \ldots + \nu_L$ joint variables. The vector of independent joint velocities of limb (l) of a kinematically non-redundant PKM is uniquely determined by the platform twist as $\dot{\mathbf{y}}_{(l)} = \bar{\mathbf{J}}_{\mathrm{p}(l)}^{+} \mathbf{V}_{\mathrm{p}}$. The overall vector of v joint velocities $\dot{\boldsymbol{\eta}}$ is then found with (36). In terms of the taskspace velocity, the solution to the inverse kinematics problem of the mechanism is

$$\dot{\boldsymbol{\eta}} = \mathbf{F} \mathbf{V}_{\mathrm{t}} \tag{39}$$

with the $v \times \delta_{\mathrm{p}}$ matrix $\mathbf{F}(\boldsymbol{\vartheta})$ defined as

$$\mathbf{F} = \begin{pmatrix} \mathbf{H}_{(1)} \bar{\mathbf{J}}_{\mathrm{t}(1)}^{-1} \mathbf{D}_{\mathrm{t}(1)} \\ \mathbf{H}_{(2)} \bar{\mathbf{J}}_{\mathrm{t}(2)}^{-1} \mathbf{D}_{\mathrm{t}(2)} \\ \vdots \\ \mathbf{H}_{(L)} \bar{\mathbf{J}}_{\mathrm{t}(L)}^{-1} \mathbf{D}_{\mathrm{t}(L)} \end{pmatrix} = \begin{pmatrix} \mathbf{F}_{(1)} \\ \mathbf{F}_{(2)} \\ \vdots \\ \mathbf{F}_{(L)} \end{pmatrix}. \tag{40}$$

Equimobile PKM with complex limbs are included with $\mathbf{D}_{\mathrm{t}(l)} = \mathbf{I}$.

7.3.2 Velocity Inverse Kinematics of the Manipulator

The actuator variables are included in the vector $\boldsymbol{\eta}$. Thus, (39) provides a solution $\dot{\boldsymbol{\vartheta}}_{\mathrm{act}} = \mathbf{J}_{\mathrm{IK}} \mathbf{V}_{\mathrm{t}}$ to the velocity inverse kinematics problem of the manipulator, where $\mathbf{J}_{\mathrm{IK}}$ is the relevant submatrix of $\mathbf{F}$ in (40).

7.3.3 Geometric Inverse Kinematics of the Mechanism

If no closed form solution is available, time integration of the kinematic relation (39), for given $\mathbf{V}_{\mathrm{t}}(t)$, yields the motion $\boldsymbol{\eta}(t)$ of all tree-joints. For implementation in the robot controller, the solution can again be determined with a Newton-iteration analogous (22). The update step follows directly from (39) as

$$\Delta \boldsymbol{\eta} = \mathbf{F} \mathbf{P}_{\mathrm{p}}^{T} \begin{pmatrix} \boldsymbol{\rho}_{\mathrm{p}} \\ \Delta \mathbf{r}_{\mathrm{p}} \end{pmatrix} \tag{41}$$

approximating the joint variable vector at time step i as $\boldsymbol{\vartheta}(t_i) \approx \boldsymbol{\vartheta}(t_{i-1}) + \Delta \boldsymbol{\vartheta}$. The iteration step (41) is repeated until $\boldsymbol{\eta}(t_i)$ satisfies the geometric platform constraints $\mathbf{C}_{\mathrm{p}} = f_{(l)}(\boldsymbol{\vartheta}_{(l)}), l = 1, \ldots, L$ and the loop constraints $g_{(l)}(\boldsymbol{\eta}_{(l)}) = \mathbf{0}, l = 1, \ldots, L$.

The solution of loop constraints can be performed independently for each of the L limbs. Therewith, the limb can be regarded as a serial chain, i.e. a simple limb. This approach is referred to as (local) constraint embedding. A solution scheme for numerical constraint embedding was reported in [124].

8 Kinematics of Manipulators with General Topology

The modeling of general multibody system (MBS) is adopted to treat manipulators with general topology (e.g. hybrid or serial-parallel manipulators) [178, 179]. To this end, a spanning tree G on the topological graph Γ is introduced. The $\bar{N}$ joint variables of the remaining tree-joints are denoted with ζ. The joint variables that are not contained in ζ are called cut-joint variables. The γ kinematic loops give rise to geometric loop constraints that can be expressed as

$$\mathbf{g}(\vartheta) = \mathbf{0} \quad \text{or} \quad \bar{\mathbf{g}}(\zeta) = \mathbf{0} \tag{42}$$

which is a system of m respectively $\bar{m}$ non-linear equations. The corresponding velocity constraints attain the form

$$\mathbf{G}(\vartheta)\,\dot{\vartheta} = \mathbf{0} \quad \text{or} \quad \bar{\mathbf{G}}(\zeta)\,\dot{\zeta} = \mathbf{0}. \tag{43}$$

A set of $\delta = N - m = \bar{N} - \bar{m}$ independent joint variables is selected that serve as generalized coordinates, denoted $\mathbf{q} \in \mathbb{V}^\delta$. A solution of (43) is then

$$\dot{\vartheta} = \mathbf{F}\dot{\mathbf{q}} \quad \text{or} \quad \dot{\zeta} = \bar{\mathbf{F}}\dot{\mathbf{q}} \tag{44}$$

where $\mathbf{F}$ and $\bar{\mathbf{F}}$ is an orthogonal complement to $\mathbf{G}$ respectively $\bar{\mathbf{G}}$. Actuator velocities are determined by $\dot{\vartheta}_{\text{act}} = \mathbf{A}\dot{\mathbf{q}}$ or $\dot{\zeta}_{\text{act}} = \bar{\mathbf{A}}\dot{\mathbf{q}}$, where $\mathbf{A}$ and $\bar{\mathbf{A}}$ is the relevant submatrix of $\mathbf{G}$ and $\bar{\mathbf{G}}$, respectively. This formulation follows the classical approach to modeling MBS described in terms of relative coordinates (joint variables), and is applicable to any manipulator.

9 Dynamics Modeling of Kinematically Non-redundant PKM with Simple Limbs

Early approaches to the dynamics modeling of PKM (specifically of the GSP) considered the platform as the member whose inertia dominates the dynamics, and the effect of other parts were neglected. The development of model-based non-linear control schemes (Sect. 14) necessitates dynamic models with higher fidelity, however.

In the following, an approach to derive the EOM is presented that is tailored to PKM. It was presented in different forms in [4, 9, 19, 123]. In order to derive the EOM, the dynamic equations of the platform and the L limbs are derived separately. The EOM are then obtained by imposing the constraints. To this end, the joint velocities are expressed in terms of the task space velocity $\mathbf{V}_t$, which serves as generalized velocity for a kinematically non-redundant PKM. This approach allows for modular modeling as well as parallel computation. The fact that the inverse kinematics problem can be solved explicitly for many PKM suggests expressing the EOM in terms of platform coordinates. However, the inverse kinematics problem (of the mechanism) may not have a unique solution, and deriving the explicit relation of all joint variables and the platform coordinates is difficult and rather involved in general. Moreover, when using EOM for model-based control, the platform coordinates are not accessible since the EE-pose cannot be measured, and they must by computed from the actuator coordinates. Therefore, the EOM are derived in terms of $\mathbf{V}_t$ as generalized velocity while the dependency on all joint variables $\boldsymbol{\vartheta}$ is kept. The latter are expressed either in terms of platform coordinates or actuator coordinates, using the inverse and forward kinematics solution, respectively.

9.1 *Kinematics of Limbs Without Platform*

Eliminating the platform and all joints connecting it to the limbs of a PKM yields a tree-topology system as shown in Fig. 18 for the GSP. Each limb gives rise to a simple kinematic chain. Denote with $\bar{\boldsymbol{\vartheta}}_{(l)} \in \mathbb{V}^{n_l}$ the vector of $n_l < N_l$ joint variables of the kinematic chain obtained from limb l after eliminating the last joint connecting the limb to the platform. The joint velocities are determined by the inverse kinematics solution (10), respectively (16) or (19). Denote with $\bar{\mathbf{F}}_{(l)}$ the submatrix of $\mathbf{F}_{(l)}$ corresponding to the remaining n_l joint variables, then

$$\dot{\bar{\boldsymbol{\vartheta}}} = \bar{\mathbf{F}}\mathbf{V}_t \tag{45}$$

with the $n \times \delta_p$ matrix

$$\bar{\mathbf{F}} = \begin{pmatrix} \bar{\mathbf{F}}_{(1)} \\ \bar{\mathbf{F}}_{(2)} \\ \vdots \\ \bar{\mathbf{F}}_{(L)} \end{pmatrix}. \tag{46}$$

Fig. 18 Tree graph obtained by removing the platform vertex from the topological graph Γ of the Gough-Stewart platform in Fig. 12

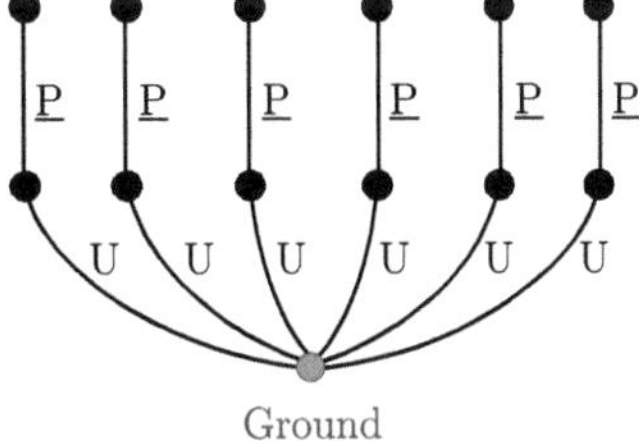

9.2 Motion Equations of a Single Limb

The EOM of limb l can be written in the form

$$\bar{\mathbf{M}}_{(l)}\ddot{\bar{\boldsymbol{\vartheta}}}_{(l)} + \bar{\mathbf{C}}_{(l)}\dot{\bar{\boldsymbol{\vartheta}}}_{(l)} + \bar{\mathbf{Q}}_{(l)} = \bar{\mathbf{Q}}_{(l)}^{\text{act}} \tag{47}$$

which resembles exactly the EOM of a serial robot [9, 128]. Here $\bar{\mathbf{M}}_{(l)}(\bar{\boldsymbol{\vartheta}}_{(l)})$ is the generalized mass matrix, $\bar{\mathbf{C}}_{(l)}(\bar{\boldsymbol{\vartheta}}_{(l)}, \dot{\bar{\boldsymbol{\vartheta}}}_{(l)}, t)$ is the Coriolis matrix, and $\bar{\mathbf{Q}}_{(l)}(\dot{\bar{\boldsymbol{\vartheta}}}_{(l)}, \bar{\boldsymbol{\vartheta}}_{(l)})$ accounts for the generalized forces due to gravity, joint friction, elastic elements, etc. The non-zero entries of $\bar{\mathbf{Q}}_{(l)}^{\text{act}}(t)$ are the drive forces/torques of the actuated joints.

The EOM can be derived in closed form by means of various approaches. As such Lagrange equations and the natural orthogonal complement (NOC) approach [9, 10] was used [186]. Several conceptually equivalent formulations were employed that (implicitly) exploit the recursive kinematic and dynamic relations inherent to open chain systems. Approaches exploiting geometric nature of the Lie group SE (3) of rigid body motions [96, 118, 123] are a systematic generalization of these formulations. Besides their compactness, a main advantage is the use of (coordinate-invariant) geometric model data, instead of using Denavit-Hartenberg convention, for instance. In particular, the joint kinematics is described by means of screw coordinates defined by joint position and direction vectors. The EOM (47) can also be evaluated with recursive O (n) algorithms [40, 122, 138].

9.3 Equations of Motion of the Platform

The platform is regarded as a rigid body whose dynamics is governed by the Newton-Euler equations. The platform twist $\mathbf{V}_p = \left(\boldsymbol{\omega}_p^T, \mathbf{v}_p^T\right)^T$ consists of the linear velocity $\mathbf{v}_p$ and angular velocity $\boldsymbol{\omega}_p$ of the platform frame $\mathscr{F}_p$, relative to the world frame $\mathscr{F}_0$. A wrench $\mathbf{W}_p = \left(\mathbf{m}_p^T, \mathbf{f}_p^T\right)^T$ acting at the platform, represented in $\mathscr{F}_p$, is defined by a torque $\mathbf{m}_p$ and a force $\mathbf{f}_p$. The wrenches acting at the platform are the wrench $\mathbf{W}_p^{\text{grav}}$ due to gravity and the EE-wrench $\mathbf{W}_p^{\text{EE}}$ (due to interaction of the PKM). The Newton-Euler equations of the platform are

$$\mathbf{M}_p\dot{\mathbf{V}}_p + \mathbf{G}_p\mathbf{M}_p\mathbf{V}_p + \mathbf{W}_p^{\text{grav}} = \mathbf{W}_p^{\text{EE}} \tag{48}$$

with the constant 6×6 inertia matrix $\mathbf{M}_p$ and gyroscopic matrix $\mathbf{G}_p$, respectively,

$$\mathbf{M}_p = \begin{pmatrix} \boldsymbol{\Theta} & m\widetilde{\mathbf{d}} \\ -m\widetilde{\mathbf{d}} & m\mathbf{I} \end{pmatrix}, \quad \mathbf{G}_p\left(\mathbf{V}_p\right) = \begin{pmatrix} \widetilde{\boldsymbol{\omega}}_p & \widetilde{\mathbf{v}}_p \\ \mathbf{0} & \widetilde{\boldsymbol{\omega}}_p \end{pmatrix} \tag{49}$$

where $\boldsymbol{\Theta}$ is the body-fixed inertia tensor w.r.t. $\mathscr{F}_p$, and $\mathbf{d}$ is the position vector to the COM represented in $\mathscr{F}_p$. The NE equations (49) holds true in an arbitrary body-fixed frame $\mathscr{F}_p$. In terms of the adjoint action on se (3) (the algebra of screws), the gyroscopic matrix is expressed as $\mathbf{G}_p\left(\mathbf{V}_p\right) = -\mathbf{ad}_{\mathbf{V}_p}^T$ [128, 154].

9.4 Task Space Formulation of the Equations of Motion

With the inverse kinematics solution (45) and with (12), the velocity of the PKM is determined by the task space velocity. The principle of virtual power yields

$$\left(\begin{array}{c} \mathbf{P}_p \\ \bar{\mathbf{F}} \end{array}\right)^T \left(\begin{array}{c} \mathbf{M}_p\dot{\mathbf{V}}_p + \mathbf{G}_p\mathbf{M}_p\mathbf{V}_p + \mathbf{W}_p^{\mathrm{grav}} \\ \bar{\mathbf{M}}_{(1)}\ddot{\boldsymbol{\vartheta}}_{(1)} + \bar{\mathbf{C}}_{(1)}\dot{\boldsymbol{\vartheta}}_{(1)} + \bar{\mathbf{Q}}_{(1)} \\ \vdots \\ \bar{\mathbf{M}}_{(L)}\ddot{\boldsymbol{\vartheta}}_{(L)} + \bar{\mathbf{C}}_{(L)}\dot{\boldsymbol{\vartheta}}_{(L)} + \bar{\mathbf{Q}}_{(L)} \end{array}\right) = \left(\begin{array}{c} \mathbf{P}_p \\ \bar{\mathbf{F}} \end{array}\right)^T \left(\begin{array}{c} \mathbf{W}_p^{\mathrm{EE}} \\ \bar{\mathbf{Q}}_{(1)}^{\mathrm{act}} \\ \vdots \\ \bar{\mathbf{Q}}_{(L)}^{\mathrm{act}} \end{array}\right). \tag{50}$$

Introducing (45) and its derivative yields the *task space formulation* of the EOM

$$\mathbf{M}_t\dot{\mathbf{V}}_t + \mathbf{C}_t\mathbf{V}_t + \mathbf{W}_t = \mathbf{W}_t^{\mathrm{EE}} + \mathbf{J}_{\mathrm{IK}}^T\mathbf{u}\left(t\right) \tag{51}$$

with the inverse kinematics Jacobian $\mathbf{J}_{\mathrm{IK}}$, and the vector $\mathbf{u} \in \mathbb{R}^{N_{\mathrm{act}}}$ of $N_{\mathrm{act}} \geq \delta$ actuator forces/torques, and the $\delta \times \delta$ generalized mass and Coriolis matrix

$$\mathbf{M}_t(\boldsymbol{\vartheta}):= \sum_{l=1}^{L} \bar{\mathbf{F}}_{(l)}^T\bar{\mathbf{M}}_{(l)}\bar{\mathbf{F}}_{(l)} + \mathbf{P}_p^T\mathbf{M}_p\mathbf{P}_p \tag{52}$$

$$\mathbf{C}_t(\boldsymbol{\vartheta}, \dot{\boldsymbol{\vartheta}}):= \sum_{l=1}^{L} \bar{\mathbf{F}}_{(l)}^T(\bar{\mathbf{C}}_{(l)}\bar{\mathbf{F}}_{(l)} + \bar{\mathbf{M}}_{(l)}\dot{\bar{\mathbf{F}}}_{(l)}) - \mathbf{P}_p^T\mathbf{G}_p\mathbf{M}_p\mathbf{P}_p \tag{53}$$

where $\mathbf{G}_p = \mathbf{G}_p\left(\mathbf{P}_p\mathbf{V}_t\right)$. The vector of generalized forces accounting for EE-loads, actuation, and all other loads are

$$\mathbf{W}_t^{\mathrm{EE}}\left(t\right) := \mathbf{P}_p^T\mathbf{W}_p^{\mathrm{EE}}\left(t\right) \tag{54}$$

$$\mathbf{W}_t(\boldsymbol{\vartheta}, \dot{\boldsymbol{\vartheta}}, t) := \sum_{l=1}^{L} \bar{\mathbf{F}}_{(l)}^T\bar{\mathbf{Q}}_{(l)} + \mathbf{P}_p^T\mathbf{W}_p^{\mathrm{grav}}\left(\boldsymbol{\vartheta}\right). \tag{55}$$

Notice that $\mathbf{W}_p^{\mathrm{grav}}$ depends on the pose of the platform, which is determined by the joint variables of the limbs (Sect. 6.3). The equations (51)–(55) are functions of all N (dependent) joint variables.

A task space formulation in terms of the $\delta_p = \delta$ platform coordinates $\mathbf{x}$ is obtained by inserting the inverse kinematics solution $\boldsymbol{\vartheta} = f(\mathbf{x})$ and (10) in the EOM (51) so that all expressions depend solely on $\mathbf{x}$ and $\mathbf{V}_t$

$$\mathbf{M}_t(\mathbf{x})\dot{\mathbf{V}}_t + \mathbf{C}_t(\mathbf{x}, \mathbf{V}_t)\mathbf{V}_t + \mathbf{W}_t(\mathbf{x}, \mathbf{V}_t, t) = \mathbf{W}_t^{\mathrm{EE}}(t) + \mathbf{J}_{\mathrm{IK}}^T(\mathbf{x})\mathbf{u}(t). \tag{56}$$

Finally the task space velocity is eliminated using the kinematic relation (14), which yields the EOM

$$\bar{\mathbf{M}}_t(\mathbf{x})\ddot{\mathbf{x}} + \bar{\mathbf{C}}_t(\mathbf{x}, \dot{\mathbf{x}})\dot{\mathbf{x}} + \bar{\mathbf{W}}_t(\mathbf{x}, \dot{\mathbf{x}}, t) = \bar{\mathbf{W}}_t^{\mathrm{EE}}(t) + \bar{\mathbf{J}}_{\mathrm{IK}}^T(\mathbf{x})\mathbf{u}(t) \tag{57}$$

with

$$\bar{\mathbf{M}}_t(\mathbf{x}) := \boldsymbol{\Omega}^T\mathbf{M}_t\boldsymbol{\Omega}, \ \ \bar{\mathbf{C}}_t(\mathbf{x}) := \boldsymbol{\Omega}^T(\mathbf{C}_t\boldsymbol{\Omega} + \mathbf{M}_t\dot{\boldsymbol{\Omega}}) \tag{58}$$

and $\bar{\mathbf{W}}_t := \boldsymbol{\Omega}^T\mathbf{W}_t, \bar{\mathbf{J}}_{\mathrm{IK}} := \mathbf{J}_{\mathrm{IK}}\boldsymbol{\Omega}$.

Remark 7 The task space formulation uses platform coordinates to parameterize the PKM motion and to express the EOM. This parameterization fails in type 1 singularities, where different inverse kinematics solutions meet (Sect. 15). It is not affected by type 2 singularities, where different forward kinematics solutions meet. Since for most PKM the inverse kinematics is unique, the task space formulation is deemed robust w.r.t. singularities.

9.5 Equations of Motion in Terms of Actuator Coordinates

9.5.1 Non-redundantly Actuated PKM

The $N_{\mathrm{act}} = \delta$ actuator coordinates serve as generalized coordinates. The task space velocity is determined by these generalized velocities as

$$\mathbf{V}_t = \mathbf{J}_{\mathrm{IK}}^{-1}\dot{\boldsymbol{\vartheta}}_{\mathrm{act}} = \mathbf{J}_{\mathrm{FK}}\dot{\boldsymbol{\vartheta}}_{\mathrm{act}}. \tag{59}$$

Introducing this into (51) yields the EOM

$$\mathbf{M}_a\ddot{\boldsymbol{\vartheta}}_{\mathrm{act}} + \mathbf{C}_a\dot{\boldsymbol{\vartheta}}_{\mathrm{act}} + \mathbf{Q}_a = \mathbf{Q}_a^{\mathrm{EE}} + \mathbf{u} \tag{60}$$

with the $\delta \times \delta$ generalized mass and Coriolis matrix

$$\mathbf{M}_a(\boldsymbol{\vartheta}) := \mathbf{J}_{\mathrm{FK}}^T\mathbf{M}_t\mathbf{J}_{\mathrm{FK}} \tag{61}$$

$$\mathbf{C}_a(\boldsymbol{\vartheta}, \dot{\boldsymbol{\vartheta}}) := \mathbf{J}_{\mathrm{FK}}^T(\mathbf{C}_t\mathbf{J}_{\mathrm{FK}} + \mathbf{M}_t\dot{\mathbf{J}}_{\mathrm{FK}}) = \mathbf{J}_{\mathrm{FK}}^T(\mathbf{C}_t - \mathbf{M}_t\mathbf{J}_{\mathrm{FK}}\dot{\mathbf{J}}_{\mathrm{IK}})\mathbf{J}_{\mathrm{FK}}. \tag{62}$$

The vector $\mathbf{u}$ of actuator forces/torques appears explicitly, while generalized EE forces and the vector of all remaining forces are

$$\mathbf{Q}_a^{EE}(\boldsymbol{\vartheta}, t) := \mathbf{J}_{FK}^T \mathbf{W}_t^{EE} \tag{63}$$

$$\mathbf{Q}_a(\boldsymbol{\vartheta}, \dot{\boldsymbol{\vartheta}}, t) := \mathbf{J}_{FK}^T \mathbf{W}_t. \tag{64}$$

The EOM (60) depend on the complete vector $\boldsymbol{\vartheta}$ of all joint variables. If a closed form solution $\boldsymbol{\vartheta} = f(\boldsymbol{\vartheta}_{act})$ of the geometric forward kinematics exists, it can be inserted in (60) along with (23) to yield the formulation in *actuator coordinates*

$$\mathbf{M}_a(\boldsymbol{\vartheta}_{act})\ddot{\boldsymbol{\vartheta}}_{act} + \mathbf{C}_a(\boldsymbol{\vartheta}_{act}, \dot{\boldsymbol{\vartheta}}_{act})\dot{\boldsymbol{\vartheta}}_{act} + \mathbf{Q}_a(\boldsymbol{\vartheta}_{act}, \dot{\boldsymbol{\vartheta}}_{act}) = \mathbf{Q}_a^{EE}(\boldsymbol{\vartheta}_{act}) + \mathbf{u}. \tag{65}$$

The forward kinematics is generally solved numerically with (24) as explicit closed form relations are only available for special PKM.

A formulation that is advantageous for control purposes, is obtained after inserting the inverse kinematics solution $\boldsymbol{\vartheta} = f(\mathbf{x})$ of the mechanism

$$\mathbf{M}_a(\mathbf{x})\ddot{\boldsymbol{\vartheta}}_{act} + \mathbf{C}_a(\mathbf{x}, \dot{\mathbf{x}})\dot{\boldsymbol{\vartheta}}_{act} + \mathbf{Q}_a(\mathbf{x}, \dot{\mathbf{x}}) = \mathbf{Q}_a^{EE}(\mathbf{x}) + \mathbf{u}. \tag{66}$$

This can be regarded as a *hybrid formulation* as it involves actuators coordinates as well as task space coordinates.

Remark 8 The formulations (60) suffers from type 2 singularities, where different forward kinematics solutions meet (Sect. 15). It is thus vital to identify all singularities as the forward kinematics of many PKM does not posses a unique solution.

9.5.2 Redundantly Actuated PKM

A redundantly full-actuated PKM possesses more actuators than its DOF, $\delta < N_{act}$. The EOM (60) can be adopted by selecting δ of the N_{act} actuator coordinates. They are summarized in the vector $\mathbf{q} \in \mathbb{V}^\delta$, and serve as generalized coordinates. The corresponding generalized velocities are related to the task space velocity by

$$\dot{\mathbf{q}} = \mathbf{B}\mathbf{V}_t \tag{67}$$

where $\mathbf{B}$ is the relevant $\delta \times \delta$ submatrix of the IK Jacobian $\mathbf{J}_{IK}$. As above, introducing the inverse relation $\mathbf{V}_t = \mathbf{B}^{-1}\dot{\mathbf{q}}$ into (51) yields the EOM of the form

$$\mathbf{M}_q\ddot{\mathbf{q}} + \mathbf{C}_q\dot{\mathbf{q}} + \mathbf{Q}_q = \mathbf{Q}_q^{EE} + \mathbf{A}^T\mathbf{u} \tag{68}$$

where the expressions (61)–(64) remain valid except that $\mathbf{J}_{FK}$ is replaced by $\mathbf{B}^{-1}$ and $\mathbf{J}_{IK}$ by $\mathbf{B}$. The actuator velocities are related to the δ generalized velocities by $\dot{\boldsymbol{\vartheta}}_{act} = \mathbf{A}\dot{\mathbf{q}}$, with $\mathbf{A} := \mathbf{J}_{IK}\mathbf{B}^{-1}$. The $\delta \times N_{act}$ matrix $\mathbf{A}^T$ is the *force distribution matrix* or *control matrix*. For simple redundancy ($N_{act} - \delta = 1$), the matrix possesses a simple form [112].

Remark 9 The EOM (68) are exactly the EOM (60) of a non-redundantly actuated PKM with the δ actuator coordinates as $\mathbf{q}$. It hence suffers from type 2 singularities (Sect. 15) of the corresponding non-redundant PKM, even if the motion of the RA-PKM is well-defined by the N_{act} actuator coordinates. For instance, selecting the coordinates of joint 1 and 2 as generalized coordinates $\mathbf{q} = (\vartheta_1, \vartheta_2)^T$ of the RA-PKM in Fig. 21b, the mass matrix $\mathbf{M}_q$ in (68) becomes singular at the type 2 singularity of the non-redundant counterpart in Fig. 21a.

9.6 Forward Dynamics

The forward dynamics problem consists in computing the PKM motion for given actuator forces $\mathbf{u}$, which amounts to numerically solving the dynamic EOM. The different formulations of EOM give rise to different ODE systems.

9.6.1 Task Space Formulation

The EOM (57) represents a system of $\delta = \delta_p$ second-order ODEs in the state $(\mathbf{x}, \dot{\mathbf{x}})$, which can be solved for given initial platform state. One drawback of expressing the EOM in terms of platform coordinates according to (58) is that it leads to complex expressions (basically caused by expressing the angular velocity and acceleration with the time derivatives $\dot{\mathbf{x}}$ and $\ddot{\mathbf{x}}$). This is avoided by using the Eq. (56) complemented by the kinematic relation (14)

$$\mathbf{M}_t(\mathbf{x})\dot{\mathbf{V}}_t + \mathbf{C}_t(\mathbf{x}, \mathbf{V}_t)\mathbf{V}_t + \mathbf{W}_t(\mathbf{x}, \mathbf{V}_t, t) = \mathbf{W}_t^{EE} + \mathbf{J}_{IK}^T(\mathbf{x})\mathbf{u}\,(t) \tag{69}$$

$$\mathbf{V}_t = \boldsymbol{\Omega}\,(\mathbf{x})\dot{\mathbf{x}} \tag{70}$$

which is a system of 2δ first-order ODEs in platform coordinates $\mathbf{x}$ and task space velocity $\mathbf{V}_t$.

If no closed form solution of the geometric inverse kinematics problem of the mechanism is available, or if this is not desirable, Eq. (51) are complemented with the respective inverse kinematics solution (16), (10) or (19), which gives rise to the system of $N + \delta$ first-order ODEs in terms of $\boldsymbol{\vartheta}$ and $\mathbf{V}_t$

$$\mathbf{M}_t(\boldsymbol{\vartheta})\dot{\mathbf{V}}_t + \mathbf{C}_t(\boldsymbol{\vartheta}, \dot{\boldsymbol{\vartheta}})\mathbf{V}_t + \mathbf{W}_t(\boldsymbol{\vartheta}, \dot{\boldsymbol{\vartheta}}, t) = \mathbf{W}_t^{EE}\,(\boldsymbol{\vartheta}, t) + \mathbf{J}_{IK}^T(\boldsymbol{\vartheta})\mathbf{u} \tag{71}$$

$$\dot{\boldsymbol{\vartheta}} = \mathbf{F}(\boldsymbol{\vartheta})\mathbf{V}_t. \tag{72}$$

The dependency of $\mathbf{C}_t$ on $\dot{\boldsymbol{\vartheta}}$ can be avoided by inserting (72) so that $\mathbf{C}_t = \mathbf{C}_t(\boldsymbol{\vartheta}, \mathbf{V}_t)$. Then (71) is linear in $\dot{\mathbf{V}}_t$ and $\dot{\boldsymbol{\vartheta}}$, which is beneficial for numerical solution. The formulation (71), (72) is generally applicable, and does not involve solving the geometric inverse kinematics of the mechanism. Its solution is the time evolution of $\mathbf{V}_t$ as well as $\boldsymbol{\vartheta}$ for given actuation $\mathbf{u}$ and EE-loads $\mathbf{W}_t^{\mathrm{EE}}$. The platform pose is computed with the forward kinematic mapping $\mathbf{C}_p = f_{(l)}(\boldsymbol{\vartheta}_{(l)})$ of one limb l. Although the dimension $(N + \delta)$ of (71), (72) is larger than that of (69), (70) it is computationally comparable when no closed form inverse solution is used since the latter is determined as the solution of (72).

9.6.2 Formulation in Actuator Coordinates

The equations (65) form a system of second-order ODEs in the state $(\boldsymbol{\vartheta}_{\mathrm{act}}, \dot{\boldsymbol{\vartheta}}_{\mathrm{act}})$. However, it relies on an explicit solution $\boldsymbol{\vartheta} = f(\boldsymbol{\vartheta}_{\mathrm{act}})$ of the geometric forward kinematics, which is not available in general. An alternative formulation is obtained by complementing (60) with the solution (23) of the forward velocity kinematics problem of the mechanism. This yields the overall system of N EOM

$$\mathbf{M}_{\mathrm{a}}(\boldsymbol{\vartheta})\ddot{\boldsymbol{\vartheta}}_{\mathrm{act}} + \mathbf{C}_{\mathrm{a}}(\boldsymbol{\vartheta}, \dot{\boldsymbol{\vartheta}})\dot{\boldsymbol{\vartheta}}_{\mathrm{act}} + \mathbf{Q}_{\mathrm{a}}(\boldsymbol{\vartheta}, \dot{\boldsymbol{\vartheta}}, t) = \mathbf{Q}_{\mathrm{a}}^{\mathrm{EE}}(\boldsymbol{\vartheta}) + \mathbf{u} \tag{73}$$

$$\dot{\boldsymbol{\vartheta}} = \mathbf{F}(\boldsymbol{\vartheta})\mathbf{J}_{\mathrm{FK}}(\boldsymbol{\vartheta})\dot{\boldsymbol{\vartheta}}_{\mathrm{act}}. \tag{74}$$

The dynamic equations (73) form a second-order ODE system in $\boldsymbol{\vartheta}_{\mathrm{act}}$, whereas the kinematic equations (74) form a first-order system. Usually, (74) is replaced by its time derivative so that (73), (74) becomes a second-order ODE system in $\boldsymbol{\vartheta}$.

10 Dynamics Modeling of Kinematically Redundant PKM with Simple Limbs

10.1 Kinematics of Limbs Without Platform

The joint velocities of the tree-topology system, obtained by removing the platform, are known from the inverse kinematics solution (28). Extracting the relevant parts of (28) yields

$$\dot{\boldsymbol{\vartheta}} = \bar{\mathbf{F}}\mathbf{s} \tag{75}$$

where the $n \times \delta$ matrix

$$\bar{\mathbf{F}} = \begin{pmatrix} \bar{\mathbf{F}}_{(1),\mathrm{t}} & \bar{\mathbf{F}}_{(1),\mathrm{r}} & \mathbf{0} & \cdots & \mathbf{0} \\ \bar{\mathbf{F}}_{(2),\mathrm{t}} & \mathbf{0} & \bar{\mathbf{F}}_{(2),\mathrm{r}} & \cdots & \mathbf{0} \\ \vdots & \vdots & \vdots & \ddots & \vdots \\ \bar{\mathbf{F}}_{(L),\mathrm{t}} & \mathbf{0} & \mathbf{0} & \cdots & \bar{\mathbf{F}}_{(L),\mathrm{r}} \end{pmatrix} \tag{76}$$

comprises the $n_l \times \delta$ matrices $\bar{\mathbf{F}}_{(l)}$, which consists of the first n_l rows of $\mathbf{F}_{(l)}$ in (29).

10.2 Task Space Formulation of the Equations of Motion

The EOM are formulated in terms of $\mathbf{s}$ that serve as generalized velocities. It determines the platform and the joint velocity of the tree-system by

$$\begin{pmatrix} \mathbf{V}_p \\ \dot{\vartheta} \end{pmatrix} = \begin{pmatrix} \mathbf{P}_p \; \mathbf{0}_{6,(\delta-\delta_p)} \\ \bar{\mathbf{F}} \end{pmatrix} \mathbf{s}. \tag{77}$$

This kinematic relation and Jourdain's principle of virtual power, along with the EOM of the platform (49) and the tree-system (47), gives rise to the *task space formulation* of the EOM

$$\mathbf{M}\dot{\mathbf{s}} + \mathbf{C}\mathbf{s} + \mathbf{Q} = \mathbf{Q}^{\text{EE}} + \mathbf{Q}^{\text{act}} \tag{78}$$

with generalized mass and Coriolis matrix

$$\mathbf{M}(\vartheta) = \begin{pmatrix} \mathbf{M}_{\text{tt}} & \mathbf{M}_{\text{tr}(1)} & \mathbf{M}_{\text{tr}(2)} & \cdots & \mathbf{M}_{\text{tr}(L)} \\ \mathbf{M}_{\text{tr}(1)}^T & \mathbf{M}_{(1)} & \mathbf{0} & \cdots & \mathbf{0} \\ \mathbf{M}_{\text{tr}(2)}^T & \mathbf{0} & \mathbf{M}_{(2)} & \cdots & \mathbf{0} \\ \vdots & \vdots & \vdots & \ddots & \vdots \\ \mathbf{M}_{\text{tr}(L)}^T & \mathbf{0} & \mathbf{0} & \cdots & \mathbf{M}_{(L)} \end{pmatrix} \tag{79}$$

$$\mathbf{C}(\vartheta, \dot{\vartheta}) = \begin{pmatrix} \mathbf{C}_{\text{tt}} & \mathbf{C}_{\text{tr}(1)} & \mathbf{C}_{\text{tr}(2)} & \cdots & \mathbf{C}_{\text{tr}(L)} \\ \mathbf{C}_{\text{rt}(1)} & \mathbf{C}_{(1)} & \mathbf{0} & \cdots & \mathbf{0} \\ \mathbf{C}_{\text{rt}(2)} & \mathbf{0} & \mathbf{C}_{(2)} & \cdots & \mathbf{0} \\ \vdots & \vdots & \vdots & \ddots & \vdots \\ \mathbf{C}_{\text{rt}(L)} & \mathbf{0} & \mathbf{0} & \cdots & \mathbf{C}_{(L)} \end{pmatrix}. \tag{80}$$

The block elements are given explicitly as follows

$$\mathbf{M}_{\text{tt}} := \sum_{l=1}^{L} \bar{\mathbf{F}}_{(l),\text{t}}^T \bar{\mathbf{M}}_{(l)} \bar{\mathbf{F}}_{(l),\text{t}} + \mathbf{P}_p^T \mathbf{M}_p \mathbf{P}_p,$$

$$\mathbf{M}_{(l)} := \bar{\mathbf{F}}_{(l),\text{r}}^T \bar{\mathbf{M}}_{(l)} \bar{\mathbf{F}}_{\eta(l),\text{r}}, \quad \mathbf{M}_{\text{tr}(l)} := \bar{\mathbf{F}}_{(l),\text{t}}^T \bar{\mathbf{M}}_{(l)} \bar{\mathbf{F}}_{\eta(l),\text{r}}$$

$$\mathbf{C}_{\text{tt}} := \sum_{l=1}^{L} \bar{\mathbf{F}}_{(l),\text{t}}^T (\bar{\mathbf{M}}_{(l)} \dot{\bar{\mathbf{F}}}_{(l),\text{t}} + \bar{\mathbf{C}}_{(l)} \bar{\mathbf{F}}_{(l),\text{t}}) + \mathbf{P}_p^T \mathbf{G}_p \mathbf{M}_p \mathbf{P}_p$$

$$\mathbf{C}_{(l)} := \bar{\mathbf{F}}_{(l),\mathrm{r}}^{T}(\bar{\mathbf{M}}_{(l)}\dot{\bar{\mathbf{F}}}_{(l),\mathrm{r}} + \bar{\mathbf{C}}_{(l)}\bar{\mathbf{F}}_{(l),\mathrm{r}}),$$

$$\mathbf{C}_{\mathrm{tr}(l)} := \bar{\mathbf{F}}_{(l),\mathrm{t}}^{T}(\bar{\mathbf{M}}_{(l)}\dot{\bar{\mathbf{F}}}_{(l),\mathrm{r}} + \bar{\mathbf{C}}_{(l)}\bar{\mathbf{F}}_{(l),\mathrm{r}})$$

$$\mathbf{C}_{\mathrm{rt}(l)} := \bar{\mathbf{F}}_{(l),\mathrm{r}}^{T}(\bar{\mathbf{M}}_{(l)}\dot{\bar{\mathbf{F}}}_{(l),\mathrm{t}} + \bar{\mathbf{C}}_{(l)}\bar{\mathbf{F}}_{(l),\mathrm{t}}) \tag{81}$$

where $\bar{\mathbf{M}}_{(l)}$ and $\bar{\mathbf{C}}_{(l)}$ are the mass and Coriolis matrix in the EOM (47) of limb l, and $\mathbf{G}_{\mathrm{p}}\left(\mathbf{P}_{\mathrm{p}}\mathbf{V}_{\mathrm{t}}\right)$ is the gyroscopic term in (49). The generalized forces are

$$\mathbf{Q}^{\mathrm{EE}}(\boldsymbol{\vartheta}, t) := \begin{pmatrix} \mathbf{P}_{\mathrm{p}}^{T}\mathbf{W}_{\mathrm{p}}^{\mathrm{EE}} \\ \mathbf{0}_{\delta-\delta_{\mathrm{p}}} \end{pmatrix} \tag{82}$$

$$\mathbf{Q}(\boldsymbol{\vartheta}, \dot{\boldsymbol{\vartheta}}, t) := \begin{pmatrix} \sum\limits_{l=1}^{L} \bar{\mathbf{F}}_{(l),\mathrm{t}}^{T}\bar{\mathbf{Q}}_{(l)} + \mathbf{P}_{\mathrm{p}}^{T}\mathbf{W}_{\mathrm{p}}^{\mathrm{grav}} \\ \bar{\mathbf{F}}_{(1),\mathrm{r}}^{T}\bar{\mathbf{Q}}_{(1)} \\ \vdots \\ \bar{\mathbf{F}}_{(L),\mathrm{r}}^{T}\bar{\mathbf{Q}}_{(L)} \end{pmatrix} \tag{83}$$

$$\mathbf{Q}^{\mathrm{act}}(\boldsymbol{\vartheta}, t) := \begin{pmatrix} \sum\limits_{l=1}^{L} \bar{\mathbf{F}}_{(l),\mathrm{t}}^{T}\bar{\mathbf{Q}}_{(l)}^{\mathrm{act}} \\ \bar{\mathbf{F}}_{(1),\mathrm{r}}^{T}\bar{\mathbf{Q}}_{(1)}^{\mathrm{act}} \\ \vdots \\ \bar{\mathbf{F}}_{(L),\mathrm{t}}^{T}\bar{\mathbf{Q}}_{(L)}^{\mathrm{act}} \end{pmatrix} = \begin{pmatrix} \mathbf{J}_{\mathrm{IK}}^{T} \\ \mathbf{S}^{T} \end{pmatrix} \mathbf{u} \tag{84}$$

with matrix $\mathbf{S}$ in (31), and $\mathbf{u}$ is the vector of actuator forces/torques. In this formulation the kinematic redundancy is resolved using the coordinates of the $\delta - \delta_{\mathrm{p}}$ 'redundant' actuated joints, according to the joint space decomposition method [172].

Replacing the joint velocities $\dot{\boldsymbol{\vartheta}}$ by the generalized velocity $\mathbf{s}$ using (28), and the joint variables $\boldsymbol{\vartheta}$ by a solution of the geometric inverse kinematics (if available) yields the task space formulation (78) with all terms now solely depending on the state $(\mathbf{x}, \boldsymbol{\vartheta}_{\mathrm{r}}, \mathbf{V}_{\mathrm{t}}, \dot{\boldsymbol{\vartheta}}_{\mathrm{r}})$.

10.3 Equations of Motion in Terms of Actuator Coordinates

For a non-redundantly actuated PKM, the inverse kinematics solution (31) can be inverted to yield the generalized velocities in terms of actuator velocities

$$\mathbf{s} = (\mathbf{J}_{\mathrm{IK}} \ \mathbf{S})^{-1} \dot{\boldsymbol{\vartheta}}_{\mathrm{act}}. \tag{85}$$

Inserting this expression, and its time derivative, into (78) leads to the EOM in terms of actuator coordinates of the form (60). In case of redundantly actuated PKM, the EOM of the form (68) are obtained after partitioning the actuator coordinates accordingly. A formulation (65), with all terms depending on the state $(\boldsymbol{\vartheta}_{\mathrm{act}}, \dot{\boldsymbol{\vartheta}}_{\mathrm{act}})$, is obtained by inserting a closed form solution $\boldsymbol{\vartheta} = f(\boldsymbol{\vartheta}_{\mathrm{act}})$ of the geometric forward kinematics, if such is available.

10.4   Forward Dynamics

10.4.1   Task Space Formulation

The ODE formulations suited for numerical integration are adopted from non-redundant PKM. The task space formulation (69), (70) becomes

$$\mathbf{M}(\mathbf{x}, \boldsymbol{\vartheta}_{\mathrm{r}})\dot{\mathbf{s}} + \mathbf{C}(\mathbf{x}, \boldsymbol{\vartheta}_{\mathrm{r}}, \mathbf{s})\mathbf{s} + \mathbf{Q}(\mathbf{x}, \boldsymbol{\vartheta}_{\mathrm{r}}, \mathbf{s}, t) = \mathbf{Q}^{\mathrm{EE}} + \mathbf{Q}^{\mathrm{act}}(\mathbf{x}, \boldsymbol{\vartheta}_{\mathrm{r}}, t) \qquad (86)$$

$$\mathbf{V}_{\mathrm{t}} = \boldsymbol{\Omega}(\mathbf{x})\dot{\mathbf{x}}. \qquad (87)$$

Also the formulation (71), (72) remains valid with (72) replaced by (28).

10.4.2   Formulation in Actuator Coordinates

If the EOM can be expressed solely in terms of $\boldsymbol{\vartheta}_{\mathrm{act}}$ as in (65), they form a system of δ ODEs that can be numerically solved. To account for the more general situation, the EOM (78) are combined with (28) to form a system of $N + \delta$ ODEs in the $N + \delta$ velocities $\mathbf{s}$ and $\dot{\boldsymbol{\vartheta}}$. Since the 'redundant' coordinates in $\boldsymbol{\vartheta}_{\mathrm{r}}$ are contained in $\boldsymbol{\vartheta}$, $\delta - \delta_{\mathrm{p}}$ equations in (28) can be eliminated (Remark 4). The remaining $N + \delta_{\mathrm{p}}$ equations form the final ODE system to be solved.

11   Dynamics Modeling of Kinematically Non-redundant PKM with Complex Limbs

11.1   Kinematics of Complex Limbs Without Platform

Elimination of the cut-joints from a separated limb already leads to a tree-topology with corresponding ν_l joint variables $\boldsymbol{\eta}_{(l)}$. Now the platform is removed along with the joint connecting it to the limb. Overall, this yields a tree-topology system, as shown in Fig. 19 for the 3-DOF Delta robot. The vector of the remaining $\bar{\nu}_l$ joint

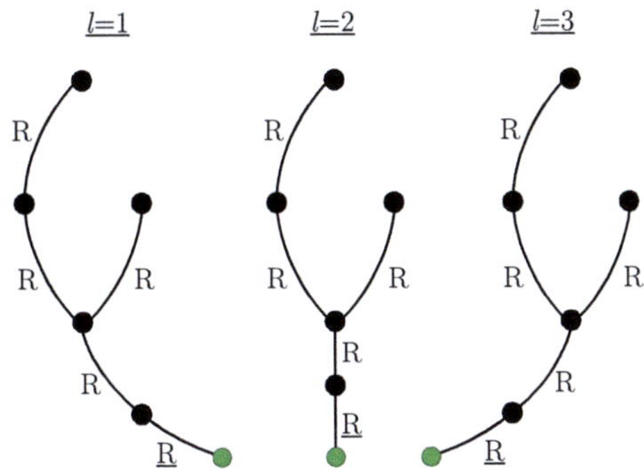

Fig. 19 Tree-graphs obtained by removing the platform vertex and one (cut-joint) edge for each loop of the topological graphs Γ_l of the complex limbs of the 3-DOF Delta robot in Fig. 13a

variables of limb l is denoted with $\bar{\boldsymbol{\eta}}_{(l)}$. The total number of joint variables is $\bar{\nu} = \bar{\nu}_1 + \ldots + \bar{\nu}_L$. Denote with $\bar{\mathbf{F}}_{(l)}$ the $\bar{\nu}_l \times \delta_{\mathrm{p}}$ submatrix of $\mathbf{F}_{(l)}$ in (39) so that

$$\dot{\bar{\boldsymbol{\eta}}} = \bar{\mathbf{F}} \mathbf{V}_{\mathrm{t}} \tag{88}$$

with the $\bar{\nu} \times \delta_{\mathrm{p}}$ matrix constructed as in (46). Notice that $\bar{\mathbf{F}}$ depends on $\boldsymbol{\eta}$.

11.2 Motion Equations of a Single Complex Limb

The EOM of the tree-topology system of limb l (without platform) have the form (47) with $\bar{\boldsymbol{\vartheta}}_{(l)}$ replaced by $\bar{\boldsymbol{\eta}}_{(l)}$. Expressing $\bar{\boldsymbol{\eta}}_{(l)}$ in terms of $\dot{\mathbf{y}}_{(l)}$, using (36), the EOM of limb l are

$$\tilde{\mathbf{M}}_{(l)}\ddot{\mathbf{y}}_{(l)} + \tilde{\mathbf{C}}_{(l)}\dot{\mathbf{y}}_{(l)} + \tilde{\mathbf{Q}}_{(l)} = \tilde{\mathbf{Q}}_{(l)}^{\mathrm{act}} \tag{89}$$

with

$$\tilde{\mathbf{M}}_{(l)}(\boldsymbol{\eta}_{(l)}) := \mathbf{H}_{(l)}^T \bar{\mathbf{M}}_{(l)} \mathbf{H}_{(l)}$$

$$\tilde{\mathbf{C}}_{(l)}(\boldsymbol{\eta}_{(l)}, \dot{\boldsymbol{\eta}}_{(l)}) := \mathbf{H}_{(l)}^T (\bar{\mathbf{M}}_{(l)} \dot{\mathbf{H}}_{(l)} + \bar{\mathbf{C}}_{(l)} \mathbf{H}_{(l)})$$

$$\tilde{\mathbf{Q}}_{(l)}(\boldsymbol{\eta}_{(l)}) := \mathbf{H}_{(l)}^T \bar{\mathbf{Q}}_{(l)}.$$

These are the EOM of a limb with the velocity and acceleration loop constraints resolved.

11.3 Task Space Formulation of the Equations of Motion

Once the EOM of a complex limb is formulated as in (89), the approach for PKM with simple limbs can be adopted. The task space formulation of the EOM of the form (51) is obtained by using the EOM (89) in the formulation (50) along with $\bar{\mathbf{F}}$ in (88). The mass and Coriolis matrix, $\mathbf{M}_t(\eta)$ and $\mathbf{C}_t(\eta, \dot{\eta})$, and the generalized forces $\mathbf{W}_t^{\text{act}}(\eta, t)$ and $\mathbf{W}_t(\eta, \dot{\eta}, t)$, depend on the joint variables η of the tree-topology system. This formulation captures the complete dynamics of all members, in contrast to tailored formulations that replace the kinematic loops by functionally equivalent joints.

11.4 Equations of Motion in Terms of Actuator Coordinates

A solution to the velocity forward kinematics of the form (59) is obtained with the inverse kinematics Jacobian $\mathbf{J}_{\text{IK}}$ in Sect. 7.3.2. Introducing this in the derivation of (51) yields the joint space formulation of EOM (60), which now depend on η and $\dot{\eta}$.

11.5 Forward Dynamics

For numerical integration of the EOM, solving the forward dynamics problem, the system (89) is complemented by the inverse kinematics solution (39). This yields a system of $\nu + \delta$ ODEs in terms of the velocity $\mathbf{V}_t$ and the joint variables η of the tree-topology system

$$\mathbf{M}_t(\eta)\dot{\mathbf{V}}_t + \mathbf{C}_t(\eta, \dot{\eta})\mathbf{V}_t + \mathbf{W}_t(\eta, \dot{\eta}, t) = \mathbf{W}_t^{\text{EE}}(\eta, t) + \mathbf{J}_{\text{IK}}^T(\eta)\mathbf{u} \tag{90}$$

$$\dot{\eta} = \mathbf{F}(\eta)\mathbf{V}_t. \tag{91}$$

The ODE system in terms of actuator coordinates is formally the same as (73), (74), but with all terms depending on η and $\dot{\eta}$.

12 Dynamics of Manipulators with General Topology

The kinematics of a manipulator with general topology is described in terms of δ general coordinates $\mathbf{q}$ (Sect. 8). The number of bodies is identical to the number of tree-joints of the tree-topology system obtained after removing a cut-joint for each loop. The EOM are derived from the NE equations (48) of the bodies along with

Jourdain's principle of virtual power. With the kinematic relation (44), and $\bar{\mathbf{F}} = \frac{\partial \dot{\zeta}}{\partial \dot{\mathbf{q}}}$, this yields

$$\sum_{i=1}^{\bar{N}} \frac{\partial \dot{\zeta}}{\partial \dot{\mathbf{q}}} \frac{\partial \mathbf{V}_i}{\partial \dot{\zeta}} \left(\mathbf{M}_i \dot{\mathbf{V}}_i + \mathbf{G}_i \mathbf{M}_i \mathbf{V}_i + \mathbf{W}_i \right) = \sum_{i=1}^{\bar{N}} \frac{\partial \dot{\zeta}}{\partial \dot{\mathbf{q}}} \frac{\partial \mathbf{V}_i}{\partial \dot{\zeta}} \mathbf{W}_i^{\mathrm{act}} \tag{92}$$

$$\bar{\mathbf{F}}^T \left(\bar{\mathbf{M}} \ddot{\zeta} + \bar{\mathbf{C}} \dot{\zeta} + \bar{\mathbf{Q}} \right) = \bar{\mathbf{F}}^T \bar{\mathbf{Q}}^{\mathrm{act}} \tag{93}$$

and finally

$$\mathbf{M}_{\mathrm{q}} \ddot{\mathbf{q}} + \mathbf{C}_{\mathrm{q}} \dot{\mathbf{q}} + \mathbf{Q}_{\mathrm{q}} = \mathbf{u} \tag{94}$$

with $\mathbf{M}_{\mathrm{q}} := \bar{\mathbf{F}}^T \bar{\mathbf{M}} \bar{\mathbf{F}}$, $\mathbf{C}_{\mathrm{q}} := \bar{\mathbf{F}}^T (\bar{\mathbf{C}} \bar{\mathbf{F}} + \bar{\mathbf{M}} \dot{\bar{\mathbf{F}}})$, $\mathbf{Q}_{\mathrm{q}} = \bar{\mathbf{F}}^T \bar{\mathbf{Q}}$, and $\mathbf{u}$ are the generalized actuator forces. The term in brackets in (93) are the EOM of the tree-topology system. They could be derived by any other method, e.g. Lagrange equations, rather than from the NE equations (the term in brackets in (92)).

13 EOM Linear in Dynamic Parameters

The EOM of serial kinematic chains can be written in a form that is linear in the dynamic parameters [12, 79]. The same holds true for PKM. Assuming a parameterization in terms of platform coordinates, the parameter-linear form of the EOM is

$$\boldsymbol{\Theta} \left(\mathbf{x}, \dot{\mathbf{x}}, \ddot{\mathbf{x}}, t \right) \mathbf{p} = \mathbf{A}^T \mathbf{u} \tag{95}$$

where $\mathbf{p}$ consists of the dynamic parameters of all bodies expressed in body-fixed frames located at the joint axes. The linearity w.r.t. dynamic parameters is important for model-based adaptive control [63, 133]. The most important usage of (95), however, is the dynamic parameter identification. The parameter vector $\mathbf{p}$ comprises the mass, the first moments (position vector of COM multiplied with the respective mass), and the inertia moments of all bodies. It further contains friction parameters (assuming friction models linear in joint velocities). Not all of these parameters are independent, and only a subset of the parameters can be identified, which are referred to as *base parameters* [12, 79]. The parameter vector $\mathbf{p}$ is expressed in terms of these base parameters by means of a numerical decomposition of the matrix $\boldsymbol{\Theta}$ (with given geometry).

The EOM of PKM attain the same form as (95). Thus the identification methods developed for serial manipulators are applicable in principle. However, the kinematic loops introduce further dependences of the dynamic parameters leading to a further reduction of the set of independent base parameters. Identification of non-redundant PKM was described in [79, 103], and amended identification schemes were reported in [20, 21]. For PKM also the identification of joint friction is more involved [5]. This is so because the measured control forces/torques arise from a combination friction forces/torques of passive as well as actuated joints.

14 Model-Bases Control Schemes

A model-based control scheme uses the non-linear PKM model to provide a feedforward control command to steer the PKM along the desired trajectory. A (usually linear) feedback regulator term then compensates for deviations from the desired trajectory due to external deviations and model inaccuracies. In this section the combination of model-based feedforward with linear feedback and the computed torques controller are presented as they are two widely used control schemes [?][128]. The presentation focuses on PKM with simple limbs. The extension to kinematically redundant PKM and to such with complex limbs is straightforward. The desired trajectory is described by the platform pose $\mathbf{C}_p^d$, respectively platform coordinates $\mathbf{x}^d$, and the taskspace velocity $\mathbf{V}_t^d$ and acceleration $\dot{\mathbf{V}}_t^d$.

14.1 Inverse Dynamics

The inverse dynamics problem is to compute the actuation forces/torques for a given motion of the PKM. The latter is determined by a prescribed EE-motion or by a prescribed motion of the actuators. Which motion is prescribed usually depends on how the trajectory was planned. As such, it involves solving the forward or inverse kinematics followed by evaluation of the EOM. The inverse dynamics is used to deliver the feedforward control command in model-based control schemes. It is evaluated at discrete time instances, according to the control cycle time.

14.1.1 Kinematically Non-redundant PKM

If the PKM motion is described in terms of platform coordinates $\mathbf{x}\,(t)$, the task space formulation (57) yields the inverse dynamics solution (EE-loads are omitted as they are usually unknown)

$$\begin{aligned}
\bar{\mathbf{u}} &= \bar{\mathbf{J}}_{\mathrm{IK}}^{-T}\,(\bar{\mathbf{M}}_t(\mathbf{x})\ddot{\mathbf{x}} + \bar{\mathbf{C}}_t(\mathbf{x},\dot{\mathbf{x}})\dot{\mathbf{x}} + \bar{\mathbf{W}}_t(\mathbf{x},\dot{\mathbf{x}})) \\
&= (\bar{\mathbf{J}}_{\mathrm{IK}}^{T})^{+}\,(\bar{\mathbf{M}}_t(\mathbf{x})\ddot{\mathbf{x}} + \bar{\mathbf{C}}_t(\mathbf{x},\dot{\mathbf{x}})\dot{\mathbf{x}} + \bar{\mathbf{W}}_t(\mathbf{x},\dot{\mathbf{x}})) + \mathbf{N}_{\bar{\mathbf{J}}_{\mathrm{IK}}^{T}}\mathbf{u}_0 \\
&=: \bar{\mathbf{g}}_t\,(\mathbf{x},\dot{\mathbf{x}},\ddot{\mathbf{x}})\,.
\end{aligned} \tag{96}$$

The second line applies to redundantly actuated PKM, where $(\mathbf{J}_{\mathrm{IK}}^{T})^{+}$ is the pseudoinverse of $\mathbf{J}_{\mathrm{IK}}^{T}$, and $\mathbf{N}_{\mathbf{J}_{\mathrm{IK}}^{T}}$ is a projector to its null-space, and $\mathbf{u}_0$ is the vector of desired control forces/torques, which do not affect the PKM motion but generate internal prestress (Sect. 14.4.1). When the taskspace velocity is used, instead of $\dot{\mathbf{x}}$, the inverse dynamics in taskspace coordinates is given with (56) as

$$
\begin{aligned}
\mathbf{u} &= \mathbf{J}_{\mathrm{IK}}^{-T}\left(\mathbf{M}_{\mathrm{t}}(\mathbf{x})\dot{\mathbf{V}}_{\mathrm{t}} + \mathbf{C}_{\mathrm{t}}(\mathbf{x}, \mathbf{V}_{\mathrm{t}})\mathbf{V}_{\mathrm{t}} + \mathbf{W}_{\mathrm{t}}(\mathbf{x}, \mathbf{V}_{\mathrm{t}})\right) \\
&= (\mathbf{J}_{\mathrm{IK}}^{T})^{+}\left(\mathbf{M}_{\mathrm{t}}(\mathbf{x})\dot{\mathbf{V}}_{\mathrm{t}} + \mathbf{C}_{\mathrm{t}}(\mathbf{x}, \mathbf{V}_{\mathrm{t}})\mathbf{V}_{\mathrm{t}} + \mathbf{W}_{\mathrm{t}}(\mathbf{x}, \mathbf{V}_{\mathrm{t}})\right) + \mathbf{N}_{\mathbf{J}_{\mathrm{IK}}^{T}}\mathbf{u}_0 \\
&=: \mathbf{g}_{\mathrm{t}}(\mathbf{x}, \mathbf{V}_{\mathrm{t}}, \dot{\mathbf{V}}_{\mathrm{t}})
\end{aligned}
\tag{97}
$$

where the pseudoinverse applies to redundantly actuated PKM. A computationally advantageous variation of (97) for general PKM, which separates the inverse dynamics from the inverse kinematics, is obtain from (51) as

$$
\begin{aligned}
\mathbf{u} &= \mathbf{J}_{\mathrm{IK}}^{-T}\left(\mathbf{M}_{\mathrm{t}}(\boldsymbol{\vartheta})\dot{\mathbf{V}}_{\mathrm{t}} + \mathbf{C}_{\mathrm{t}}(\boldsymbol{\vartheta}, \dot{\boldsymbol{\vartheta}})\mathbf{V}_{\mathrm{t}} + \mathbf{W}_{\mathrm{t}}(\boldsymbol{\vartheta}, \dot{\boldsymbol{\vartheta}})\right) \\
&=: \mathbf{g}_{\mathrm{t}}(\boldsymbol{\vartheta}, \dot{\boldsymbol{\vartheta}}, \mathbf{V}_{\mathrm{t}}, \dot{\mathbf{V}}_{\mathrm{t}}).
\end{aligned}
\tag{98}
$$

The inverse kinematics is solved before evaluating (98). For given $\mathbf{V}_{\mathrm{t}}$, joint rates $\dot{\boldsymbol{\vartheta}}$ are computed with (10), and $\boldsymbol{\vartheta}$ is computed from $\mathbf{C}_{\mathrm{p}}$ with (22), or with $\boldsymbol{\vartheta} = f(\mathbf{x})$ if possible.

If the actuator motion $\boldsymbol{\vartheta}_{\mathrm{act}}(t)$ is given, the EOM (65) yield an inverse dynamics solution. However, this formulation is only applicable to special PKM where an explicit solution to the forward kinematics is available. A generally applicable inverse dynamics solution is obtained from (60)

$$
\begin{aligned}
\mathbf{u} &= \mathbf{M}_{\mathrm{a}}(\boldsymbol{\vartheta})\ddot{\boldsymbol{\vartheta}}_{\mathrm{act}} + \mathbf{C}_{\mathrm{a}}(\boldsymbol{\vartheta}, \dot{\boldsymbol{\vartheta}})\dot{\boldsymbol{\vartheta}}_{\mathrm{act}} + \mathbf{Q}_{\mathrm{a}}(\boldsymbol{\vartheta}, \dot{\boldsymbol{\vartheta}}) \\
&=: \mathbf{g}_{\mathrm{a}}(\boldsymbol{\vartheta}, \dot{\boldsymbol{\vartheta}}, \ddot{\boldsymbol{\vartheta}}_{\mathrm{act}})
\end{aligned}
\tag{99}
$$

and for redundantly actuated PKM from (68)

$$
\begin{aligned}
\mathbf{u} &= (\mathbf{A}^{T})^{+}\left(\mathbf{M}_{\mathrm{a}}(\boldsymbol{\vartheta})\ddot{\mathbf{q}} + \mathbf{C}_{\mathrm{a}}(\boldsymbol{\vartheta}, \dot{\boldsymbol{\vartheta}})\dot{\mathbf{q}} + \mathbf{Q}_{\mathrm{a}}(\boldsymbol{\vartheta}, \dot{\boldsymbol{\vartheta}})\right) + \mathbf{N}_{\mathbf{A}^{T}}\mathbf{u}_0 \\
&=: \mathbf{g}_{\mathrm{a}}(\boldsymbol{\vartheta}, \dot{\boldsymbol{\vartheta}}, \ddot{\mathbf{q}}) + \mathbf{N}_{\mathbf{A}^{T}}\mathbf{u}_0
\end{aligned}
\tag{100}
$$

The velocity $\dot{\boldsymbol{\vartheta}}$ is computed via forward kinematics solution (74), and the joint variables $\boldsymbol{\vartheta}$ are obtained with a Newton-iteration, which is performed on the controller, with step increment

$$
\Delta\boldsymbol{\vartheta} = \mathbf{F}\mathbf{J}_{\mathrm{FK}}\Delta\boldsymbol{\vartheta}_{\mathrm{act}} \quad \text{or} \quad \Delta\boldsymbol{\vartheta} = \mathbf{F}\mathbf{B}^{-1}\Delta\mathbf{q}
\tag{101}
$$

or by numerically integrating (74).

If the inverse kinematics can be solved in closed form, then (66) yields a mixed form of inverse dynamics solution involving task space and actuator coordinates

$$\mathbf{u} = \mathbf{M}_a(\mathbf{x})\ddot{\boldsymbol{\vartheta}}_{\text{act}} + \mathbf{C}_a(\mathbf{x}, \dot{\mathbf{x}})\dot{\boldsymbol{\vartheta}}_{\text{act}} + \mathbf{Q}_a(\mathbf{x}, \dot{\mathbf{x}})$$

$$=: \mathbf{g}_{a,t}(\mathbf{x}, \dot{\mathbf{x}}, \dot{\boldsymbol{\vartheta}}_{\text{act}}, \ddot{\boldsymbol{\vartheta}}_{\text{act}}) \tag{102}$$

and for redundantly actuated PKM, with (68),

$$\mathbf{u} = (\mathbf{A}^T)^+ \left(\mathbf{M}_q(\mathbf{x})\ddot{\mathbf{q}} + \mathbf{C}_q(\mathbf{x}, \dot{\mathbf{x}})\dot{\mathbf{q}} + \mathbf{Q}_q(\mathbf{x}, \dot{\mathbf{x}})\right) + \mathbf{N}_{\mathbf{A}^T}\mathbf{u}_0$$

$$=: \mathbf{g}_{a,t}(\mathbf{x}, \dot{\mathbf{x}}, \dot{\mathbf{q}}, \ddot{\mathbf{q}}) + \mathbf{N}_{\mathbf{A}^T}\mathbf{u}_0. \tag{103}$$

The above inverse dynamics solutions are immediately adopted to PKM with complex limbs.

14.1.2 Kinematically Redundant PKM

If the platform motion is specified, then in addition the motion of the $N_{\text{act}} - \delta$ 'redundant' actuators is prescribed. The EOM (78) yield the inverse dynamics solution for a non-redundantly actuated PKM

$$\mathbf{u} = (\mathbf{J}_{\text{IK}}\ \mathbf{S})^{-T} (\mathbf{M}\dot{\mathbf{s}} + \mathbf{C}\mathbf{s} + \mathbf{Q}) \tag{104}$$

and for a redundantly actuated kinematically redundant PKM

$$\mathbf{u} = \begin{pmatrix} \mathbf{J}_{\text{IK}}^T \\ \mathbf{S}^T \end{pmatrix}^+ (\mathbf{M}\dot{\mathbf{s}} + \mathbf{C}\mathbf{s} + \mathbf{Q}) + \mathbf{N}_{(\mathbf{J}_{\text{IK}}\ \mathbf{S})^T}\mathbf{u}_0. \tag{105}$$

The (104) and (105) are expressed in terms of the state $(\mathbf{x}, \boldsymbol{\vartheta}_r, \mathbf{s})$, as in (86). If this is not possible, then $\boldsymbol{\vartheta}$ and $\dot{\boldsymbol{\vartheta}}$ are determined by the inverse kinematic solutions (32) and (31), respectively. If actuator motions are prescribed, the inverse dynamics solution proceeds as for non-redundant PKM.

14.1.3 Modularity and Parallel Computation

The kinematic and dynamic formulation reflect the modularity of a PKM (L limb modules, one platform module). This modular approach admits a distributed parallel computation, which is advantageous when numerically evaluating the EOM at a given state.

The mass matrix (52) is determined from the mass matrices of the L limbs by means of the L matrices in (46). The matrices $\bar{\mathbf{M}}_{(l)}, \bar{\mathbf{F}}_{(l)}$ and the vector $\bar{\mathbf{Q}}_{(l)}$ depend on the joint variables $\boldsymbol{\vartheta}_{(l)}$ of limb l only, and can thus be evaluated in parallel for each limb. The joint velocities $\dot{\boldsymbol{\vartheta}}_{(l)}$ and accelerations $\ddot{\boldsymbol{\vartheta}}_{(l)}$ are also independently computed from the tasks space velocity $\mathbf{V}_t$ by (10), (16), (19) and their derivatives, which allows evaluating all $\bar{\mathbf{C}}_{(l)}$ in parallel. The last step consists in evaluating (52)–(55), and finally (51). The described parallel computation steps apply equally to PKM with complex limbs. The inverse kinematics solutions (39), the EOM of the complex limb (89), and thus the overall EOM (51) can be computed in parallel.

The computational complexity of the parallel evaluation may be reduced further by evaluating the EOM (47) of the individual limbs in parallel using an $O(n)$ algorithm [40, 122, 138]. Then the matrices $\bar{\mathbf{F}}_{(l)}$ and subsequently $\dot{\boldsymbol{\vartheta}}_{(l)}, \ddot{\boldsymbol{\vartheta}}_{(l)}$ are computed from the given configuration $\boldsymbol{\vartheta}$ and task velocity $\mathbf{V}_t$, and (47) along with the NE equations (49) are evaluated. This is then used to evaluate (50).

The modularity of the EOM also allows for limbs with different kinematics as this is accounted for by the particular EOM (47). Moreover, the limbs as well as the platform may possess complex kinematics (e.g. articulated platforms as in case of the H4 [140] in Fig. 3), which require dedicated modeling approaches.

14.2 PID Control with Non-linear Feedforward

14.2.1 Control in Taskspace

If the platform motion is described with local coordinates $\mathbf{x}$, the position tracking error is defined as $\bar{\mathbf{e}}_t := \mathbf{x}^d - \mathbf{x}$, and the velocity error as $\dot{\bar{\mathbf{e}}}_t := \dot{\mathbf{x}}^d - \dot{\mathbf{x}}$. The PID control law with feedforward compensation computed with (96) is

$$\bar{\mathbf{u}}_{\text{FF+PID}} = \bar{\mathbf{g}}_t(\mathbf{x}^d, \dot{\mathbf{x}}^d, \ddot{\mathbf{x}}^d) + \mathbf{K}_P\bar{\mathbf{e}}_t + \mathbf{K}_D\dot{\bar{\mathbf{e}}}_t + \text{sat}\left(\mathbf{K}_I \int \bar{\mathbf{e}}_t dt\right) \tag{106}$$

where a saturation function is used to limit the value of the integral term (anti-windup).

To avoid parameterization singularities, and generally to eliminate the 'artificial dynamics' due to the parameterization, the tracking error is computed directly from the platform pose $\mathbf{C}_p$ as $\mathbf{e}_t := \mathbf{P}_p^T \log(\mathbf{C}_p^{-1}\mathbf{C}_p^d)$ using the log function on $SE(3)$ [96, 128]. The taskspace velocity tracking error is then introduced as $\dot{\mathbf{e}}_t := \mathbf{V}_t^d - \mathbf{V}_t$. With the inverse dynamics solution in taskspace coordinates (97), the control scheme is

$$\mathbf{u}_{\text{FF+PID}} = \mathbf{g}_t(\mathbf{x}^d, \mathbf{V}_t^d, \dot{\mathbf{V}}_t^d) + \mathbf{K}_P\mathbf{e}_t + \mathbf{K}_D\dot{\mathbf{e}}_t + \text{sat}\left(\mathbf{K}_I \int \mathbf{e}_t dt\right). \tag{107}$$

For small tracking errors the log function is approximated as

$$\mathbf{e}_t \approx \mathbf{P}_p^T \begin{pmatrix} \xi \\ \eta \end{pmatrix}, \text{ with } \eta = \mathbf{r}_p^d - \mathbf{r}_p, \ \tilde{\xi} = \frac{1}{2}\left(\Delta\mathbf{R} - \Delta\mathbf{R}^T\right) \tag{108}$$

with $\Delta \mathbf{R} := \mathbf{R}_{\mathrm{p}}^{T} \mathbf{R}_{\mathrm{p}}^{\mathrm{d}}$, and $\widetilde{\boldsymbol{\xi}} \in so\,(3)$ is the skew symmetric matrix associated to vector $\boldsymbol{\xi} \in \mathbb{R}^3$. Notice that the scaled rotation vector $\boldsymbol{\xi}$, describing the rotation error, is the same as that obtained when using unit quaternions to describe rotations.

14.2.2 Control in Joint Space

The tracking error in actuator coordinates is defined as $\mathbf{e}_{\mathrm{a}} := \boldsymbol{\vartheta}_{\mathrm{act}}^{\mathrm{d}} - \boldsymbol{\vartheta}_{\mathrm{act}}$. The control law in actuator coordinates is, with (99),

$$\mathbf{u}_{\mathrm{FF+PID}} = \mathbf{g}_{\mathrm{a}}(\boldsymbol{\vartheta}^{\mathrm{d}}, \dot{\boldsymbol{\vartheta}}^{\mathrm{d}}, \ddot{\boldsymbol{\vartheta}}_{\mathrm{act}}^{\mathrm{d}}) + \mathbf{K}_{\mathrm{P}}\mathbf{e}_{\mathrm{a}} + \mathbf{K}_{\mathrm{D}}\dot{\mathbf{e}}_{\mathrm{a}} + \mathrm{sat}\,(\mathbf{K}_{\mathrm{I}} \int \mathbf{e}_{\mathrm{a}} dt) \tag{109}$$

or using the mixed formulation (102)

$$\mathbf{u}_{\mathrm{FF+PID}} = \mathbf{g}_{\mathrm{a,t}}(\mathbf{x}^{\mathrm{d}}, \dot{\mathbf{x}}^{\mathrm{d}}, \dot{\boldsymbol{\vartheta}}_{\mathrm{act}}^{\mathrm{d}}, \ddot{\boldsymbol{\vartheta}}_{\mathrm{act}}^{\mathrm{d}}) + \mathbf{K}_{\mathrm{P}}\mathbf{e}_{\mathrm{a}} + \mathbf{K}_{\mathrm{D}}\dot{\mathbf{e}}_{\mathrm{a}} + \mathrm{sat}\,(\mathbf{K}_{\mathrm{I}} \int \mathbf{e}_{\mathrm{a}} dt) \tag{110}$$

where $\boldsymbol{\vartheta}_{\mathrm{act}}$ is computed from $\mathbf{x}$ by solving the inverse kinematics problem. The control schemes for redundantly actuated PKM are obtained with (100) and (103).

14.2.3 Features of the Control Schemes

Common properties of the control schemes in taskspace and joint space:

- The model-based feedforward according to the *desired* trajectory is used frequently, and already leads to a significant performance improvement.
- Computation of the feedforward term is robust against measurement noise since it is computed from the desired trajectory. It can be precomputed for the planned trajectory and stored on the controller.
- The feedforward term can be computed with a recursive algorithm and admits parallel computing, as discussed in Sect. 14.1.3.
- Tuning of the controller gains $\mathbf{K}_{\mathrm{P}}, \mathbf{K}_{\mathrm{D}}, \mathbf{K}_{\mathrm{I}}$ is difficult.
- Instead of the PID feedback loop any other linear control method can be used.

Features of the taskspace controller:

- The controller regulates directly in task space.
- The tracking error is computed in task space. The platform motion cannot be measured, however, and must be computed from actuator motion (encoder readings). This necessitates solving the geometric forward kinematics problem, which is a drawback of the task space control of PKM. To mitigate this issue, the desired actuator coordinates are computed with the inverse kinematics solution $\boldsymbol{\vartheta}_{\mathrm{act}}^{\mathrm{d}} = f(\mathbf{x}^{\mathrm{d}})$, and the actuator tracking error $\mathbf{e}_{\mathrm{a}} := \boldsymbol{\vartheta}_{\mathrm{act}}^{\mathrm{d}} - \boldsymbol{\vartheta}_{\mathrm{act}}$ is computed with the current actuator coordinates $\boldsymbol{\vartheta}_{\mathrm{act}}$. Assuming small tracking errors,

the task space error is estimated as $\mathbf{e}_t = \mathbf{J}_{FK}\mathbf{e}_a$. The error rate $\dot{\mathbf{e}}_t$ is known from the current (estimated) task space velocity computed as $\mathbf{V}_t = \mathbf{J}_{FK}\dot{\boldsymbol{\vartheta}}_{act}$. The correctness of the so-computed tracking error significantly depends on the accuracy of the geometric model. Unless the PKM is geometrically calibrated the taskspace tracking will exhibit a systematic error. Furthermore this method relies on a regular absolute referencing since the platform pose is only determined implicitly via the joint tracking error.

- If a closed form solution $\boldsymbol{\vartheta} = f(\mathbf{x})$ of the inverse kinematics is to be avoided (e.g. in a general control framework, i.e. without deriving the geometric inverse kinematics solution of the particular PKM), the EOM (56), and thus the inverse dynamics (97), are not applicable. Then, (98) can be used to deliver the feedforward term. The system state is computed with inverse kinematics solutions (10) and (22).

Features of the joint space controller:

- The control scheme (109) requires a solution $\boldsymbol{\vartheta}^d = f(\boldsymbol{\vartheta}^d_{act})$ of the forward kinematics of the mechanism.
- The control scheme (110) is preferable as it involves the prescribed platform motion and the measured actuator motion. The desired joint motion is computed with the inverse kinematics solution $\boldsymbol{\vartheta} = f(\mathbf{x})$.

14.3 Computed Torque Control (CTC)

While the feedforward term computed with the desired trajectory significantly improves tracking performance, its two main issues are that it does not guaranty stability and that it is difficult to select the controller gains. The so-called computed torque control (CTC) scheme [11] overcomes these limitations. Denote with $\mathbf{x}$, $\boldsymbol{\vartheta}$, and $\mathbf{V}_t$ the current platform and joint coordinates, and the task space velocity of the PKM. The CTC schemes for the different formulations of EOM are

- Taskspace control in terms of platform coordinates:

$$\begin{aligned}
\bar{\mathbf{u}}_{CT} &= \bar{\mathbf{g}}_t\,(\mathbf{x}, \dot{\mathbf{x}}, \bar{\mathbf{v}}_t) \\
&= \bar{\mathbf{J}}_{IK}^{-T}\left(\bar{\mathbf{M}}_t(\mathbf{x})\bar{\mathbf{v}}_t + \bar{\mathbf{C}}_t(\mathbf{x}, \dot{\mathbf{x}})\dot{\mathbf{x}} + \bar{\mathbf{W}}_t(\mathbf{x}, \dot{\mathbf{x}})\right) \quad (111) \\
&= (\bar{\mathbf{J}}_{IK}^T)^+\left(\bar{\mathbf{M}}_t(\mathbf{x})\bar{\mathbf{v}}_t + \bar{\mathbf{C}}_t(\mathbf{x}, \dot{\mathbf{x}})\dot{\mathbf{x}} + \bar{\mathbf{W}}_t(\mathbf{x}, \dot{\mathbf{x}})\right) + \mathbf{N}_{\bar{\mathbf{J}}_{IK}^T}\mathbf{u}_0 \quad (112)
\end{aligned}$$

with $\bar{\mathbf{v}}_t := \ddot{\mathbf{x}}^d + \mathbf{K}_P\bar{\mathbf{e}}_t + \mathbf{K}_D\dot{\bar{\mathbf{e}}}_t + \mathrm{sat}\,(\mathbf{K}_I \int \bar{\mathbf{e}}_t dt)$.

- Taskspace control in terms of task space coordinates:

$$\mathbf{u}_{CT} = \mathbf{g}_t(\mathbf{x}, \mathbf{V}_t, \mathbf{v}_t)$$

$$= \mathbf{J}_{IK}^{-T}\left(\mathbf{M}_t(\mathbf{x})\mathbf{v}_t + \mathbf{C}_t(\mathbf{x}, \mathbf{V}_t)\mathbf{V}_t + \mathbf{W}_t(\mathbf{x}, \mathbf{V}_t)\right) \tag{113}$$

$$= (\mathbf{J}_{IK}^T)^+\left(\mathbf{M}_t(\mathbf{x})\mathbf{v}_t + \mathbf{C}_t(\mathbf{x}, \mathbf{V}_t)\mathbf{V}_t + \mathbf{W}_t(\mathbf{x}, \mathbf{V}_t)\right) + \mathbf{N}_{\mathbf{J}_{IK}^T}\mathbf{u}_0 \tag{114}$$

with $\mathbf{v}_t := \dot{\mathbf{V}}_t^d + \mathbf{K}_{P}\mathbf{e}_t + \mathbf{K}_{D}\dot{\mathbf{e}}_t + \mathrm{sat}\,(\mathbf{K}_I \int \mathbf{e}_t dt)$.

- Joint space control using the EOM in terms of all joint variables and actuator coordinates:

$$\mathbf{u}_{CT} = \mathbf{g}_a(\boldsymbol{\vartheta}, \dot{\boldsymbol{\vartheta}}, \mathbf{v}_a)$$

$$= \mathbf{M}_a(\boldsymbol{\vartheta})\mathbf{v}_a + \mathbf{C}_a(\boldsymbol{\vartheta}, \dot{\boldsymbol{\vartheta}})\dot{\boldsymbol{\vartheta}}_{act} + \mathbf{Q}_a(\boldsymbol{\vartheta}, \dot{\boldsymbol{\vartheta}}) \tag{115}$$

$$= (\mathbf{A}^T)^+\left(\mathbf{M}_q(\boldsymbol{\vartheta})\mathbf{v}_q + \mathbf{C}_q(\boldsymbol{\vartheta}, \dot{\boldsymbol{\vartheta}})\dot{\mathbf{q}} + \mathbf{Q}_q(\boldsymbol{\vartheta}, \dot{\boldsymbol{\vartheta}})\right) + \mathbf{N}_{\mathbf{A}^T}\mathbf{u}_0 \tag{116}$$

with $\mathbf{v}_a := \ddot{\boldsymbol{\vartheta}}_{act}^d + \mathbf{K}_{P}\mathbf{e}_a + \mathbf{K}_{D}\dot{\mathbf{e}}_a + \mathrm{sat}\,(\mathbf{K}_I \int \mathbf{e}_a dt)$ and $\mathbf{v}_q := \ddot{\mathbf{q}}^d + \mathbf{K}_{P}\mathbf{e}_q + \mathbf{K}_{D}\dot{\mathbf{e}}_q + \mathrm{sat}\,(\mathbf{K}_I \int \mathbf{e}_q dt)$, where $\mathbf{e}_q := \mathbf{q}^d - \mathbf{q}$ is the tracking error for the δ actuator variables used as generalized coordinates.

- Joint space control using the mixed formulation of EOM in terms of platform and actuator coordinates:

$$\mathbf{u}_{CT} = \mathbf{g}_{a,t}(\mathbf{x}, \dot{\mathbf{x}}, \dot{\boldsymbol{\vartheta}}_{act}, \mathbf{v}_a)$$

$$= \mathbf{M}_a(\mathbf{x})\mathbf{v}_a + \mathbf{C}_a(\mathbf{x}, \dot{\mathbf{x}})\dot{\boldsymbol{\vartheta}}_{act} + \mathbf{Q}_a(\mathbf{x}, \dot{\mathbf{x}}) \tag{117}$$

$$= (\mathbf{A}^T)^+\left(\mathbf{M}_q(\mathbf{x})\mathbf{v}_q + \mathbf{C}_q(\mathbf{x}, \dot{\mathbf{x}})\dot{\mathbf{q}} + \mathbf{Q}_q(\mathbf{x}, \dot{\mathbf{x}})\right) + \mathbf{N}_{\mathbf{A}^T}\mathbf{u}_0. \tag{118}$$

Features of the CTC schemes:

- The CT controller leads to a feedback linearization of the closed loop error dynamics, which becomes

$$\mathbf{K}_{D}\ddot{\mathbf{e}} + \mathbf{K}_{D}\dot{\mathbf{e}} + \mathbf{K}_{P}\mathbf{e} + \mathrm{sat}\,\left(\mathbf{K}_I \int \mathbf{e}dt\right) = \mathbf{0}. \tag{119}$$

The gain matrices $\mathbf{K}_P, \mathbf{K}_D, \mathbf{K}_I$ can be tuned so to achieve exponentially stable trajectory tracking.

- The dependence of all terms on the current state makes the formulation sensitive to measurement noise. The state is deduced from the measurements of the actuator positions $\boldsymbol{\vartheta}_{act}$ from which $\dot{\boldsymbol{\vartheta}}_{act}$ and $\ddot{\boldsymbol{\vartheta}}_{act}$ are obtained with a linear filter.
- All terms in the non-linear feedforward are evaluated with the current state of the PKM. Consequently, all CTC formulations require a forward kinematics solution: (111)–(114), (117), (118) need the forward kinematics solution $\mathbf{x} =$

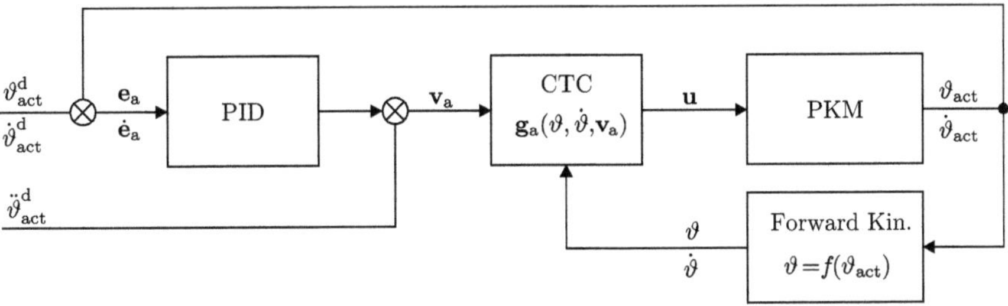

Fig. 20 Block diagram of the CTC controller (115) in joint space

$f(\boldsymbol{\vartheta}_{\mathrm{act}})$, while (115) and (116) need the forward kinematics solution $\boldsymbol{\vartheta} = f(\boldsymbol{\vartheta}_{\mathrm{act}})$.

- The CTC formulation (115) is usually computationally most efficient. The control scheme is shown in Fig. 20. It involves the measured joint variables $\boldsymbol{\vartheta}_{\mathrm{act}}$ and (estimated) $\dot{\boldsymbol{\vartheta}}_{\mathrm{act}}, \ddot{\boldsymbol{\vartheta}}_{\mathrm{act}}$. It also involves solving the forward kinematics $\boldsymbol{\vartheta} = f(\boldsymbol{\vartheta}_{\mathrm{act}})$ but does not need the inverse kinematics solution $\boldsymbol{\vartheta} = f(\mathbf{x})$ to express the EOM in terms of $\mathbf{x}$.
- The control forces are computed as the inverse dynamics with acceleration replaced by the control input $\mathbf{v}$. This admits using computationally efficient evaluation of the inverse dynamics (Sect. 14.1.3).

14.4 Particular Issues of Redundantly Actuated PKM

14.4.1 Exploiting Actuation Redundancy for Secondary Tasks

Redundancy Resolution for Optimal Load Distribution

The $m > \delta$ control forces/torques $\mathbf{u}$ (computed with any control scheme) act upon a RA-PKM via the $\delta \times N_{\mathrm{act}}$ control matrix $\mathbf{A}^{T}$. The control forces/torques are thus not unique, which admits solving for actuator forces/torques so that the overall load is distributed optimally between the m actuators. A particular solution that minimizes the $\mathbf{G}$-weighted 2-norm of $\mathbf{u}$, defined by $\|\mathbf{u}\|_{\mathbf{G}}^{2} = \mathbf{u}^{T}\mathbf{G}\mathbf{u}$, is

$$\mathbf{u} = (\mathbf{A}^{T})_{\mathbf{G}}^{+}(\mathbf{M}_{\mathrm{a}}\ddot{\mathbf{q}} + \mathbf{C}_{\mathrm{a}}\dot{\mathbf{q}} + \mathbf{Q}_{\mathrm{a}}) \tag{120}$$

where $(\mathbf{A}^{T})_{\mathbf{G}}^{+} = \mathbf{G}^{-1}\mathbf{A}(\mathbf{A}^{T}\mathbf{G}^{-1}\mathbf{A})^{-1}$ is the $\mathbf{G}$-weighted pseudoinverse, with $\mathbf{G}$ being a positive definite symmetric matrix. The weighting matrix is usually a diagonal matrix with entries according to the load capacity of the drives. A solution that minimizes the maximal actuator force/torques can be determined using the ∞-norm. There is, however, no closed form solution.

Avoidance of Backlash in Joints and Drives

The general inverse dynamics solution solving (68) is

$$\mathbf{u} = (\mathbf{A}^T)_{\mathbf{G}}^{+}(\mathbf{M}_a\ddot{\mathbf{q}} + \mathbf{C}_a\dot{\mathbf{q}} + \mathbf{Q}_a) + \mathbf{N}_{\mathbf{A}^T,\mathbf{G}}\mathbf{u}_0 \tag{121}$$

with $\mathbf{N}_{\mathbf{A}^T,\mathbf{G}} = \mathbf{I}_m - (\mathbf{A}^T)_{\mathbf{G}}^{+}\mathbf{A}^T$ being a projector to the nullspace of $\mathbf{A}^T$. Control commands belonging to the null space of $\mathbf{A}^T$ (second term in (121)) do not affect the PKM motion, but rather generate internal prestress. This leads to antagonistic drive forces/torques that can be applied to avoid joint backlash [85, 86, 168]. A backlash-avoiding control scheme was presented in [112]. The backlash-avoiding control must determine the vector $\mathbf{u}_0$ of desired control forces so that the preload of the actuators remains above a specified level.

Stiffness Modulation

Antagonistic controls computed from the null-space solution in (121) in a given configuration do not cause apparent EE forces/torques. Due to the non-linear kinematics, small perturbations from this configuration change the transmission from actuator to EE forces/torques, and the null-space solution leads to apparent EE forces. This relation of change of EE pose and resulting wrench contributes to the tangential EE stiffness. The controlled prestress generates a passive compliance, in contrast to active admittance control [95].

Several approaches to control the tangential EE stiffness (also called open-loop stiffness control) by means of actuation redundancy were reported [23, 31, 35, 55, 84, 94, 113, 161, 183, 184]. The degree of redundancy determines how many components of the EE stiffness can be controlled. The principal challenge remains to ensure stability so that the internal prestress does not destabilize the RA-PKM when disturbed.

Decentralized Control

As any robotic manipulator, PKM can be controlled with standard decentralized control schemes (e.g. PID control with additional gravity compensation). The dominating approach to control PKM is the standard decentralized control, where the desired actuator motion is sent to the individual drives, which are controlled independently. The decentralized PD control scheme, for instance, computes the control command $\mathbf{u} \in \mathbb{R}^m$ as

$$\mathbf{u} = \mathbf{K}_P\mathbf{e}_a + \mathbf{K}_D\dot{\mathbf{e}}_a \tag{122}$$

with positive definite diagonal gain matrices. The crucial point, however, is that the uncoordinated PD control (or any other type) of the individual drives leads to control commands in the nullspace of $\mathbf{A}^T$, which causes antagonistic control forces/torques and thus (uncontrolled) internal prestress [126]. This is inherent to any decentralized control of redundantly actuated PKM as each actuator is controlled independently within the control time step. Antagonistic control commands are also due to measurement errors and finite encoder resolutions.

A simple yet efficient way to reduce the effect of antagonistic control commands is to project the control command $\mathbf{u}$ in (122) to the range of $\mathbf{A}^T$, which yields the effective control commands as

$$\mathbf{u}_{\text{eff}} = \mathbf{R}_{\mathbf{A}^T}\mathbf{u} \tag{123}$$

with the projector $\mathbf{R}_{\mathbf{A}^T} := \mathbf{I} - \mathbf{N}_{\mathbf{A}^T} = (\mathbf{A}^T)^+\mathbf{A}^T$ to the range of $\mathbf{A}^T$. In this context $\mathbf{R}_{\mathbf{A}^T}$ is called *antagonism filter* [16, 126]. Input to the decentralized controllers of the actuators are the desired joint motions. Therefore, the control commands (123) are used to correct the desired joint motions as

$$\mathbf{q}_{\text{eff}}^{\text{d}} = \mathbf{q}_{\text{a}} - \mathbf{e}_{\text{eff}}, \quad \dot{\mathbf{q}}_{\text{eff}}^{\text{d}} = \dot{\mathbf{q}}_{\text{a}} - \dot{\mathbf{e}}_{\text{eff}} \tag{124}$$

where

$$\mathbf{e}_{\text{eff}} = \mathbf{R}_{\mathbf{A}^T}\mathbf{K}_{\text{P}}\mathbf{e}_{\text{a}}, \quad \dot{\mathbf{e}}_{\text{eff}} = \mathbf{R}_{\mathbf{A}^T}\mathbf{K}_{\text{D}}\dot{\mathbf{e}}_{\text{a}}. \tag{125}$$

The corrected commands $\mathbf{q}_{\text{eff}}^{\text{d}}$ and $\dot{\mathbf{q}}_{\text{eff}}^{\text{d}}$ are sent to the decentralized PD controllers. If the PD controllers of all drives have identical gains, i.e. $\mathbf{K}_P = k_\text{P}\mathbf{I}$ and $\mathbf{K}_D = k_\text{D}\mathbf{I}$, then the projection matrices simplify to $\mathbf{N}_{\mathbf{A}^T\mathbf{K}_\text{P}} = \mathbf{N}_{\mathbf{A}^T\mathbf{K}_\text{D}} = \mathbf{N}_{\mathbf{A}^T}$, and the antagonism filters are all identical.

14.4.2 Antagonistic Control Forces Due to Geometric Imperfections

In any model-based control scheme, geometric uncertainties of the model lead to systematic control errors. In case of redundantly actuated PKM, they additionally cause inadvertent internal prestresses, which increase the energy consumption but can also severely affect the controller stability [114]. An ad hoc method to alleviate this problem is to use the error $\mathbf{e}_{\text{q}} = \mathbf{q}^{\text{d}} - \mathbf{q}$ of δ independent actuators in the linear feedback of the controller (106) and (107), and use δ components of $\bar{\mathbf{v}}_{\text{t}}$ and $\mathbf{v}_{\text{t}}$ in (111) and (113).

15 Singularities

15.1 Singularities of the Input-Output Formulation

Singularities of the manipulator are related to the singularities of the matrices $\mathbf{A}$ and $\mathbf{B}$ [51] in the relation (2) of taskspace velocity (output) and actuator velocities (input). An established terminology for singularities of PKM is the following: A configuration ϑ is a *type 1* singularity when rank of $\mathbf{B}$ drops at ϑ, a *type 2* singularity when rank of $\mathbf{A}$ drops, and a *type 3* singularity if rank of both, $\mathbf{A}$ and $\mathbf{B}$, drops. At type 2 singularities, usually different branches of the geometric forward kinematics meet. They are therefore called *parallel singularities* as this can only happen for PKM. At type 1 singularities, usually different branches of the inverse kinematics meet, which is why they are also called *serial singularities* as this applies to PKM as well as to SM (which are included with $\mathbf{A} = \mathbf{I}$ and $\mathbf{B} = \mathbf{J}_{\mathrm{FK}}$). Type 3 singularities are also called architecture) singularities.

It is important to notice that this classification is solely based on the input-output relation and does not account for the 'internal' kinematics of the mechanism. Both matrices may have full rank, although the instantaneous DOF of certain members of the PKM changes. This is apparent from (3) as all passive joint rates are eliminated by means of the reciprocal screws. Moreover, (3) does not represent the complete system of constraints of the mechanism, and rank deficiency of $\mathbf{A}$ and $\mathbf{B}$ is not sufficient to concluded a constraint singularity. The relation (3) is merely a 'projection' of the overall configuration space to the input and output space. Nevertheless, for many PKM (3) provides a simple approach to check for singularities.

15.2 Singularity Identification Using Line Geometry

The formulation (3) gives rise to geometric conditions on $\mathbf{A}$ and $\mathbf{B}$ to have a certain rank. For the GSP, for instance, they, and the inverse kinematics Jacobian (6), are completely defined by the screw coordinates of the actuated prismatic screws (Plücker coordinates of lines at infinity). The rank of $\mathbf{A}$ and $\mathbf{B}$ is thus related to the linear dependence of the screws.

Geometric conditions for a set of lines to form a system of certain dimension is provided by the Grassmann line geometry. This fact was used in [106] to derive geometric conditions for the singularities of Gough-Stewart platforms, which was later extended in [57]. By allowing for general types of joints the reciprocal screw approach has become a standard method to determine type 1 and 2 singularities. The crucial point is that the conditions for a set of screws to form a k-system are known [46, 47, 99, 100]. A complete enumeration of singularities of planar 3-DOF PKM was derived in [18]. As a further formalization, the Grassmann-Cayley algebra was used [14]. This provides a coordinate-free algebraic framework to handle screws by means of abstract geometric operations. It was used in [76] to analyze singularities of lower mobility PKM with linear and rotational actuators.

In [77] the special geometries of GSP were determined so that the rank of matrix **A** is permanently less than 6. These situations were called 'permanent singularities'. This term is an oxymoron, however, since a configuration is a kinematic singularity if the rank of **A** changes. When **A** is permanently rank deficient, then the PKM is merely not controllable with the allocated actuator inputs. Such an example is the 4 DOF 4U$\underline{P}$U PKM in Fig. 15 whose mobility is well-defined, but it cannot be controlled by the four linear actuators.

15.3 Singularity-Free Assembly Mode Changing

The solution branches of the geometric forward kinematics problem are called assembly modes (for fixed actuator coordinates), in analogy to the assembly modes of SM, which refer to the solution branches of the inverse kinematics problem (for fixed EE-pose). A change of assembly mode is usually accomplished by passage through a type 2 singularity. An obvious question is whether this is necessary. It is know that 3R positioning SM can change assembly modes without encountering a singularity if the image of the singular set in workspace exhibits a cusp. The latter means that three inverse kinematic solutions meet. Therefore, such SM are called *cuspidal*. This property was first reported in [135, 136]. An exhaustive classification of all 3R positioning SM was given in [176]. The cuspidality argument can be adopted to PKM by swapping joint space and taskspace. It was shown for 3-DOF PKM [70, 177] that, when the set of forward kinematic singularities in joint space possesses a cusp, the PKM can switch between different forward kinematics solutions without passing through a type 2 singularity. It was shown in [104] that this 'cuspidality' is sufficient for singularity-free assembly mode change of PKM. However, it was later shown in [97] that the existence of a cusp is not a necessary condition. Currently singularity-free assembly mode change is known for very few 6-DOF PKM only. The practical significance of singularity-free transition between different forward kinematics solution must be investigated for the specific PKM at hand. For novel PKM, which possess non-unique forward as well as inverse kinematics solutions, the cuspidality could be an advantage to be considered for the design.

15.4 Kinematic Singularities of the Mechanism

The input-output formulation (3) does not describe the instantaneous kinematics of the overall mechanism. Irrespective of its interaction, the PKM kinematics is defined by constraints imposed to its members. The γ kinematic loops impose geometric constraints on the joint variables ϑ of the form (42). The *configuration space (c-space)* of the PKM is the variety defined by the constraints as

$$V := \{\vartheta \,|\, \mathbf{g}(\vartheta) = \mathbf{0}\} \tag{126}$$

and $\vartheta \in V$ is called an *admissible configuration* (or simply *configuration*) of the PKM. The velocity loop constraints (43) define the admissible velocities (first-order motions) $\dot{\vartheta} \in \ker \mathbf{G}(\vartheta)$ at a given configuration ϑ. The constraints can be derived by combining the forward kinematics relations (9) of the $L = \gamma + 1$ limbs, which yields the constraint Jacobian

$$\mathbf{G} := \begin{pmatrix} \mathbf{J}_{t(1)} & -\mathbf{J}_{t(2)} & \mathbf{0} & \cdots & & \mathbf{0} \\ \mathbf{0} & \mathbf{J}_{t(2)} & -\mathbf{J}_{t(3)} & \cdots & & \mathbf{0} \\ \vdots & \vdots & \vdots & \ddots & & \vdots \\ \mathbf{0} & \mathbf{0} & \mathbf{0} & \cdots & \mathbf{J}_{t(\gamma)} & -\mathbf{J}_{t(\gamma+1)} \end{pmatrix}. \tag{127}$$

Generically, i.e. for 'most' configurations $\vartheta \in V$, the c-space is locally a smooth manifold. A configuration ϑ where V is not a manifold is called a *c-space singularity*. If ϑ is a c-space singularity, then the rank of $\mathbf{G}$ drops, and there is no neighborhood of ϑ in V where the rank is constant.

If at regular configurations the $m \times N$ constraint Jacobian $\mathbf{G}$ does not have full rank m, the PKM is *overconstrained*. Many lower-mobility PKM are overconstrained, e.g. the Tripteron in Fig. 4, the 3-DOF Delta in Fig. 5, the IRSBot-2 in Fig. 8, and the H4 in Fig. 3. The 10×12 Jacobian of the Tripteron, for instance, has rank 9 in regular configurations, which shows that its (instantaneous) DOF is $\delta = 3$. Configurations where $\mathbf{G}$ is not full rank are collectively called *constraint singularities*. Clearly, not all constraint singularities are c-space singularities.

The motion of a PKM corresponds to a curve in V, and the motion characteristic of a PKM is encoded in the c-space geometry. When passing through a c-space singularity, this characteristic changes drastically. C-space singularities are often points where different submanifolds of V intersect, and each of them corresponds to a motion mode of the PKM. An example is the 3-DOF 3U$\underline{\mathrm{R}}$U PKM presented in [189]. In its different motion modes, the platform can perform spatial rotations, translations, planar motions, or is immobile. The motion modes of a PKM can even be of different dimensions, so that the PKM can change its DOF δ. Such PKM are called kinematotropic [180].

Singularities affect the inherent stability of PKM, and investigating what can happen at and near to a singularity is a vital aspect of designing and controlling a PKM. This boils down to analyzing possible instantaneous and finite motions through a singularity. Instantaneous motions at any configuration are readily deduced from the rank of the instantaneous joint screw system, i.e. the rank of the constraint Jacobian (127). The Jacobian (43) can be partitioned as $\mathbf{G}_{\mathrm{pass}}\dot{\vartheta}_{\mathrm{p}} + \mathbf{G}_{\mathrm{a}}\dot{\vartheta}_{\mathrm{act}} = \mathbf{0}$ according to variables of passive and actuated joints. The instantaneous PKM kinematics is described by the model

$$\mathbf{A}\left(\boldsymbol{\vartheta}\right)\mathbf{V} = \mathbf{B}\left(\boldsymbol{\vartheta}\right)\dot{\boldsymbol{\vartheta}}_{\mathrm{act}} \tag{128}$$

$$\mathbf{C}\left(\boldsymbol{\vartheta}\right)\dot{\mathbf{u}} = \mathbf{D}\left(\boldsymbol{\vartheta}\right)\dot{\boldsymbol{\vartheta}} \tag{129}$$

$$\mathbf{0} = \mathbf{G}_{\mathrm{p}}\left(\boldsymbol{\vartheta}\right)\dot{\boldsymbol{\vartheta}}_{\mathrm{pass}} + \mathbf{G}_{\mathrm{a}}\left(\boldsymbol{\vartheta}\right)\dot{\boldsymbol{\vartheta}}_{\mathrm{act}} \tag{130}$$

where $\mathbf{u}$ are input coordinates (usually these are the actuator coordinates $\mathbf{u} = \boldsymbol{\vartheta}_{\mathrm{act}}$). Situations where the matrices $\mathbf{A}, \mathbf{B}, \mathbf{C}, \mathbf{D}$, and $\mathbf{G}_{\mathrm{p}}, \mathbf{G}_{\mathrm{a}}$ change rank indicate various types of singularities with different consequences for the stability and controllability of the PKM. All possible situations are listed in [187, 188], where the following six types are distinguished: i) Redundant Input (RI), ii) Redundant Output (RO), iii) Impossible Input (II), iv) Impossible Output (IO), v) Redundant Passive Motion (RPM), and vi) Increased Instantaneous Mobility (IIM). There are 21 possible combinations.

To investigate the finite local geometry of V, and thus the finite motions, higher-order local analysis methods were developed [42, 101, 116, 121]. The crucial point is that all higher-order relations can be determined algebraically in terms of joint screws [120]. Basic idea of these approaches is to determine the velocities that correspond to finite motions through a configuration $\boldsymbol{\vartheta}$, which is formalized by the concept of a kinematic tangent cone [121]. None of these local analysis method capture stationary singularities (e.g. cusps). In such situations one must resort to a local approximation of V. The defining polynomial equations are also determined algebraically in terms of the joint screws [127]. A local analysis reveals what happens near a specific configuration. An exhaustive understanding demands a global analysis of the c-space. This is facilitated by an algebraic formulation of the constraints (42) and application of computational algebraic geometry [68, 69]. The encountered complexity can be mitigated by appropriate preparation of the equations [171].

16   Dexterity and Elastic Stiffness

PKM exhibit drastically varying kinematic dexterity and elastic stiffness within the workspace, eventually degenerating in singular configurations. Compared to SM, the variation within the workspace is more pronounced for PKM [145]. Therefore, maximizing and homogenizing kinematic dexterity and stiffness was addressed in order to design optimal PKM for different applications.

Dexterity measures are introduced upon the forward or inverse kinematics Jacobian, as for SM (see chapter 10). If the PKM is non-redundant, $\mathbf{J}_{\mathrm{IK}}$ is square, and the inverse condition number $1/\kappa\left(\mathbf{J}_{\mathrm{IK}}\right) = 1/\kappa\left(\mathbf{J}_{\mathrm{FK}}\right)$ represents such a measure. If the PKM is kinematically redundant, the measure $1/\sqrt{\kappa(\mathbf{J}_{\mathrm{IK}}\mathbf{J}_{\mathrm{IK}}^{T})} = 1/\sqrt{\kappa(\mathbf{J}_{\mathrm{FK}}^{T}\mathbf{J}_{\mathrm{FK}})}$ is

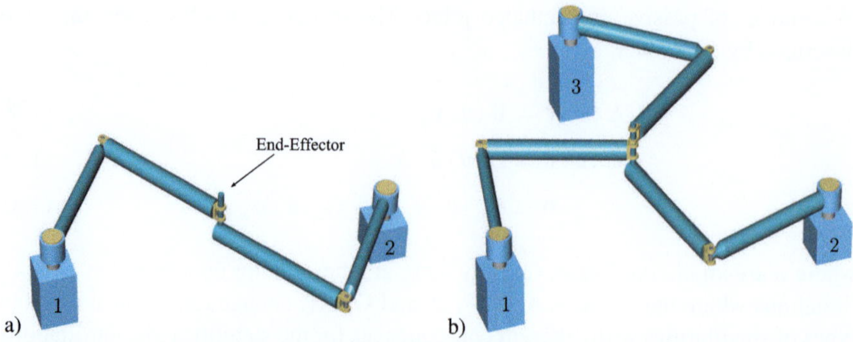

Fig. 21 (**a**) Non-redundant planar 2-DOF $\underline{R}R/\underline{R}RR$ PKM in a type 1 singularity, (**b**) Its redundantly actuated extension, the 2-DOF $\underline{R}R/2\underline{R}RR$

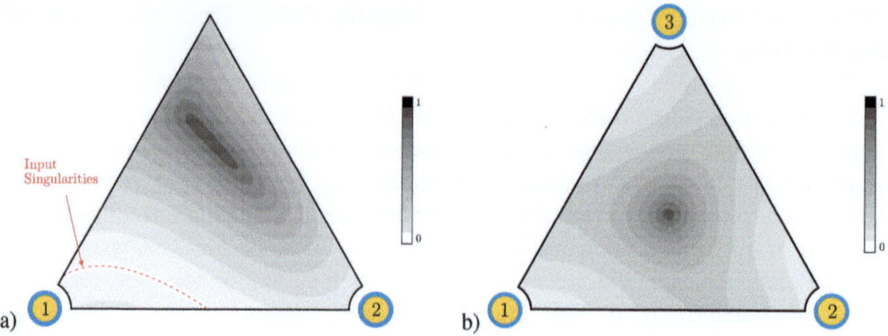

Fig. 22 Dexterity distribution of the non-redundant (**a**) and redundantly actuated planar 2 DOF PKM (**b**)

used. As for SM the problem of scaling rotations and translations remains [8]. The difference of the kinematics of PKM compared to SM, must be taken into account, which allows defining amended dexterity measures [109, 137, 150].

For RA-PKM a dexterity measure is $1/\sqrt{\kappa(\mathbf{J}_{\mathrm{IK}}^T\mathbf{J}_{\mathrm{IK}})} = 1/\sqrt{\kappa(\mathbf{J}_{\mathrm{FK}}\mathbf{J}_{\mathrm{FK}}^T)}$. Actuation redundancy is a means to eliminate singularities and to homogenize kinematic dexterity. Figure 21a shows a non-redundant planar 2-DOF PKM, and Fig. 21b a redundantly actuated version. The distribution of dexterity measure in the workspace is shown in Fig. 22. The non-redundant PKM exhibits an output-singularity where the dexterity measure becomes zero, while the RA-PKM shows a symmetric non-vanishing dexterity measure with increased homogeneity.

Denote with k_i the lumped stiffness of the actuated joint i (accounting for all elastic effects of motor and gear associated with the free motion). The task space (tangential) stiffness matrix of the PKM is then

$$\mathbf{K}_t = \mathbf{J}_{\mathrm{IK}}^T\mathbf{K}_\vartheta\mathbf{J}_{\mathrm{IK}} \tag{131}$$

with $\mathbf{K}_\vartheta := \text{diag}\,(k_1, \ldots, k_m)$. The stiffness distribution is clearly affected by the varying transmission property. Notice that this simple stiffness model does not take into account the elasticities of passive joints.

17 Synthesis

17.1 Specifics of PKM

The kinematic design of manipulators consists in the type synthesis (kinematic topology) of the mechanism followed by the dimensional synthesis (particular geometry). A PKM, being a closed loop mechanism, may possess different working modes (or assembly modes). That is, the mobility and motion pattern of the platform may be different in different parts of its workspace (Sect. 15). The selection of working modes is a crucial aspect in the kinematic design phase, which involves analysis of singularities (were different working modes meet). The allocation of drives is already part of the kinematic design since this determines the input-output relation and dexterity. The dynamics modeling finally allows assessing the dynamic capabilities of the PKM. As the topology, kinematics, and mass distribution are interrelated, these steps are repeated.

The forward and inverse kinematics of PKM is one of the hurdles in operation of PKM, and simplification of the input-output relations is one design goal. In the best case, the input-output relation is linear and the forward kinematics decoupled, as for the PKM presented in [26, 48, 52, 80, 81, 87, 88].

17.2 Type Synthesis

Type synthesis is often understood as the process of finding all possible PKM that generate platform motions with a specified DOF. Eventually more practically relevant is the synthesis of PKM generating platform motions with specified motion patterns [89]. Type synthesis thus amounts to enumerating all possible arrangements of joints and links, i.e. the kinematic topologies. In contrast to the synthesis of general closed loop mechanisms [105], the design of PKM takes into account the inherent modularity and the fact that most PKM comprise structurally identical limbs.

17.2.1 Methods Based on Instantaneous Screw Systems

A first attempt to synthesize closed loop mechanisms based on the instantaneous joint screw system was presented by Hunt [66]. The screw based synthesis has developed into a systematic theory [64, 89]. For each limb, a joint arrangement is determined so that the joint screws generate a screw system which would lead to the desired instantaneous platform motion. Basic premise of this theory is that the instantaneous mobility of the synthesized PKM is identical to the finite mobility.

17.2.2 Methods Based on Displacement Groups

This method operates on the level of finite motions rather than on instantaneous kinematics [7, 60, 61, 156]. The finite motion (mobility) of the PKM platform is the finite motion that can be collectively performed by the terminal links (virtual platforms) of all limbs. The platform motion (neglecting possible limits of the motion range) is assumed to form a displacement group $\mathscr{L}$, i.e. a subgroup of $SE(3)$ [154]. The synthesis method stems from the fact that the subgroup $\mathscr{L}$ can be obtained by intersection of L displacement subgroups $\mathscr{L}_l, l = 1, \ldots, L$ according to $\mathscr{L} = \mathscr{L}_1 \cap \cdots \cap \mathscr{L}_L$. The first step of the synthesis procedure is to select possible displacement subgroups $\mathscr{L}_l, l = 1, \ldots, L$ whose intersection is the desired displacement subgroup $\mathscr{L}$. Then, for each of these subgroups $\mathscr{L}_l$, a (simple) kinematic chain is synthesized so that the motion space of its terminal link is $\mathscr{L}_l$. This kinematic chain is then called the mechanical generator of this subgroup. The crucial steps of this method are the selection of candidate subgroups $\mathscr{L}_l$, and the design of mechanical generators, i.e. the synthesis of a limb with a prescribed motion pattern. Notice that the displacement group of the platform can be a subgroup of the displacement group of a limb, $\mathscr{L}_l \subset \mathscr{L}$ (as in case of the Tripteron in Figs. 4 and 14). A limb, i.e. the mechanical generator of $\mathscr{L}_l$, can possibly be a closed loop mechanism itself (as in case of the IRSBot-2 in Fig. 8 or the Delta robot in Fig. 2).

This synthesis procedure is facilitated by the fact that the intersections of two displacement subgroups are known [43]. As the subgroups (or their generators) can be described in a coordinate-free manner by means of geometric vectors, it represents an elegant synthesis approach. It was used in [59, 61, 78, 149, 156] to synthesize novel PKM. The synthesis method was developed for limbs generating displacement subgroups, noticing that the possible intersections of $SE(3)$ subgroups are known. The prescribed platform motion may generally not form a displacement subgroup but a submanifold of $SE(3)$, however. Furthermore, limbs can be used that do not generate a subgroup. The method will have to be developed further in order to allow for this more general approach.

17.2.3 Method Based on Linear Transformations

A related approach to the PKM synthesis is the method based on linear transformations developed in [50]. This theory builds upon the linear transformations constructed from relevant Jacobian matrices. The so-called evolutionary morphology is used to find all possible arrangements of joints and links that generate a desired motion. Depending on whether the transformation matrices involved are computed numerically or symbolically, the synthesis method generates all possible mechanisms with desired instantaneous or finite mobility, respectively.

17.2.4 Mobility

PKM are classified regarding the DOF and motion patterns (mobility) of their platform, which presents the principle synthesis criterion. A secondary kinematic design aspect could be the actual motions performed by the other members of the PKM, as they are relevant for limiting the housing volume (relative to the workspace) and for collision avoidance. Spatial 6 DOF PKM can be synthesized by means of any limb arrangement as long as the limb allows for spatial motions. The synthesis becomes more involved for lower mobility PKM. Most lower mobility PKM are (by design) 'exceptional' (according to the classification of Hervé [58]). The task space Jacobian in (15) of any of their limbs has full rank, but the overall constraint Jacobian in (127) is rank deficient, and the mobility of each of each of the γ loops is determined by the intersection of the two motion spaces corresponding to the two limbs forming the loop. Overconstrained PKM, whose task space Jacobians are not full rank (in regular configurations) are rare, and mostly of theoretical value. Furthermore, the special conditions imposed on the geometry of such 'paradoxically' overconstrained PKM [58] are difficult to satisfy in practice.

17.3 Dimensional Synthesis

The final synthesis step involves the specification of the actual geometric dimensions and assigning the dynamic properties due to the mass distribution, which yields a particular PKM with specific properties. This is called dimensional synthesis. For the selected topology and type of PKM, the optimal geometric dimensions are determined taking into account the workspace and kinematic dexterity. The workspace must include certain points or volumes that are relevant for the class of applications the PKM is intended for, while the workspace itself should be as large as possible. At the same time the transmission properties should be 'optimal' within the workspace, which are assessed using dexterity indexes. Here, optimal means that the dexterity should be as high as possible but also homogenous with the workspace, and there should be no singularities in the workspace. The geometric embodiment of the PKM implies inertia properties of the individual parts of the PKM and thus determines the overall dynamic properties of the PKM. Also these dynamic properties are to be optimized so to respect limitations of the drive system (e.g. maximal motor torques and joint loads, deflections of elastic components). A PKM intended for fast pick-and-place operations may be optimized so to reduce the weight of its moving elements (e.g. the Delta robot), whereas a PKM intended for lifting and handling heavy objects may be optimized for maximum load (e.g. Gough-Stewart platform used as flight simulator).

Clearly, the dimensional synthesis leads to a multi-objective optimization problem which possesses a Pareto-optimal rather then a unique solution. In practice, the synthesis and the selection of optimal designs is an iterative process involving kinematics and dynamics modeling and analysis, and the selection of the particular PKM design must further take into account the feasibility, reliability, and costs of its manufacturing.

18 Exercises

1. Draw a topological graph for the following PKM

   a. the planar PKM in Fig. 7,
   b. the H4 PKM in Fig. 3,
   c. the Agile Eye in Fig. 6.

2. Derive a closed form solution for the inverse geometric problem of the redundantly actuated planar 3-DOF PKM in Fig. 9. To this end, introduce a reference frame on the platform and a global reference, so that the position of the platform is described by $\mathbf{r}_\mathrm{p} = (x, y)^T$. The task is to determine the joint angles $\boldsymbol{\vartheta}_\mathrm{act} = (\vartheta_1, \vartheta_2, \vartheta_3, \vartheta_4)^T$ of the four actuated joints for given position $\mathbf{r}_\mathrm{p}$.

3. Derive a closed form solution for the inverse geometric problem of the 6-DOF Gough-platform using the parameters in Fig. 1.

   a. Introduce a reference frame on platform and ground,
   b. For given platform pose $\mathbf{C}_\mathrm{p}$ determine the attachment points of the six struts at the platform. The solution can be expressed in terms of the distance of these points to the space-fixed attachment points at the ground.

4. For the Gough-platform with parameters according to Fig. 23, determine explicitly the inverse kinematics Jacobian (6). Show that a PKM with $\varphi_\mathrm{p} = \varphi_\mathrm{B}$, is at a

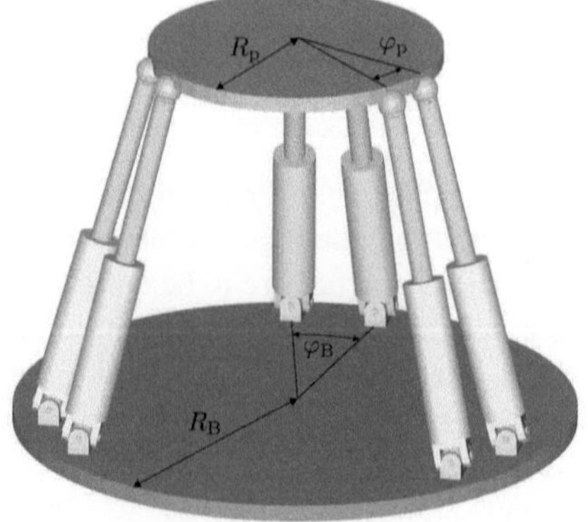

Fig. 23 Sketch of a Gough-platform with parameters describing the attachment points of the six struts

singularity when the circular platform is not horizontal aligned and positioned such that its center is directly above the center of the circular base, as shown in Fig. 23. Can you explain this without resorting to the equations?

5. The platform of the 3-DOF Tripteron in Fig. 4 can perform spatial motions. The PKM is actuated by three linear drives.

    a. Introduce actuator coordinates $\boldsymbol{\vartheta}_{act} \in \mathbb{R}^3$, and platform coordinates $\mathbf{r}_p \in \mathbb{R}^3$.

    b. Introduce a global inertial frame. For a limb (see also Fig. 14), determine the joint screw coordinates w.r.t. this frame.

    c. For limb l, derive the axis coordinates $\mathbf{W}_{act(l)}$ of the screws that are reciprocal to the screws of the three passive R-joint but not to that of the actuated prismatic joint.

    d. Derive the inverse kinematics Jacobian (4).

6. Figure 24 shows a planar 2-DOF 2$\underline{R}$RR PKM. Its 'platform' is just the bolt connecting the two limbs. Only the platform translation can be controlled by the

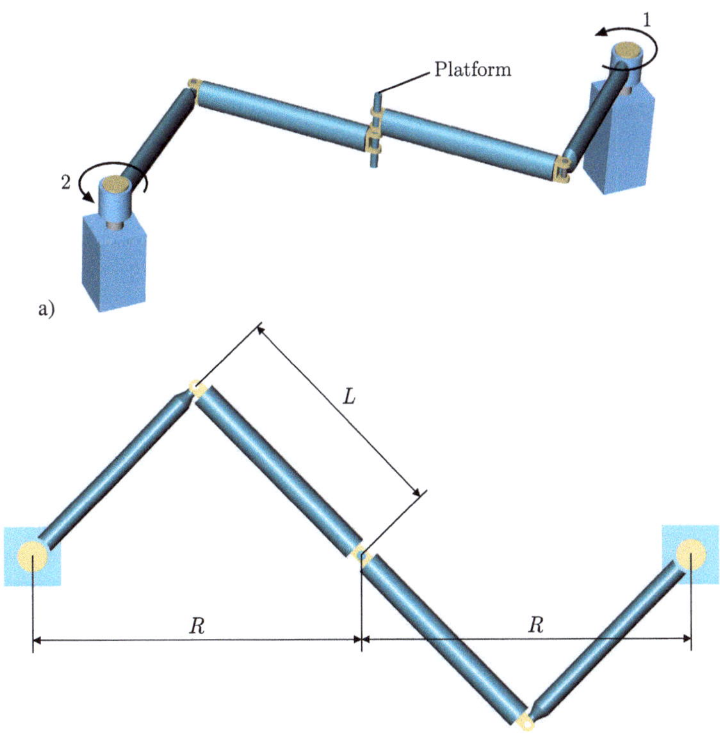

Fig. 24 (a) Planar 2$\underline{R}$RR PKM in reference configuration $\mathbf{q}_0$. (b) A links have the length L. The distance of the drives is $2R$, with $R = L/\sqrt{2}$

actuators, and two translation coordinates are used as output. For this exercise, you may want to use a computer algebra software.

a. Draw the topological graph for this PKM.
b. Define a global frame, and introduce coordinates $\mathbf{x} = \mathbf{r}_\mathrm{p} \in \mathbb{R}^2$ to describe the platform position.
c. What is the DOF of a separated limb including the platform? Is this PKM equimobile? For such a separated limb:
   i. Introduce the vector $\boldsymbol{\vartheta}_{(l)} \in \mathbb{R}^3$ of joint variables,
   ii. Derive the forward kinematics mapping $\mathbf{x} = f_{(l)}(\boldsymbol{\vartheta}_{(l)})$,
   iii. Derive the task space Jacobian in (15),
   iv. Derive the inverse kinematics mapping of the manipulator $\boldsymbol{\vartheta}_{\mathrm{act}} = f_{\mathrm{IK}}(\mathbf{x})$, which can be expressed in closed form. Here, $\boldsymbol{\vartheta}_{\mathrm{act}}$ comprises the angles of the two actuated joints.
d. Derive the velocity inverse kinematics (16) of a separated limb including the platform.
e. Use the result from Exercise 6d to assemble the matrix (20).
f. With the matrix (20), find the inverse kinematics Jacobian in (5).
g. What is the DOF of a separated limb without platform?
For such a separated limb:
   i. Introduce joint variables $\bar{\boldsymbol{\vartheta}}_{(l)}$,
   ii. Derive the inverse kinematics Jacobian in (45).
h. Derive the EOM (48) for the disconnected platform. Remember that for this PKM only the platform translation is considered.
i. For a separated limb, derive the EOM (47). Use any method that you like, e.g. Lagrange equations [9, 96, 128]. For simplicity, the links can be modeled as slender beams with homogenous mass distributions and circular cross sections.
j. Construct the EOM (51) in terms of EE-coordinates, either using the matrix form (50) or the explicit form (52), (53).
k. Find the joint coordinate vector $\boldsymbol{\vartheta}_0$ that corresponds to the reference configuration in Fig. 24, in which the platform position is $\mathbf{x}_0$.
   i. Check that this vector satisfies the geometric loop closure constraints.
   ii. Now assume the platform has to move along the horizontal axis. This is described by $\mathbf{x} = \mathbf{x}_0 + \Delta\mathbf{x}$. For a small increment (e.g. 1 mm) solve the geometric inverse kinematics problem of the manipulator using (22) to compute the corresponding $\Delta\boldsymbol{\vartheta}$. After one step check whether the geometric constraints (i.e. the forward kinematics of each limb) are satisfied. Remember that you already have the closed form solution from Exercise 6(c)iv.
   iii. Can you show that $\boldsymbol{\vartheta}_0$ is a singularity? Can you identify which type of singularity (as defined in Sect. 15) this is?

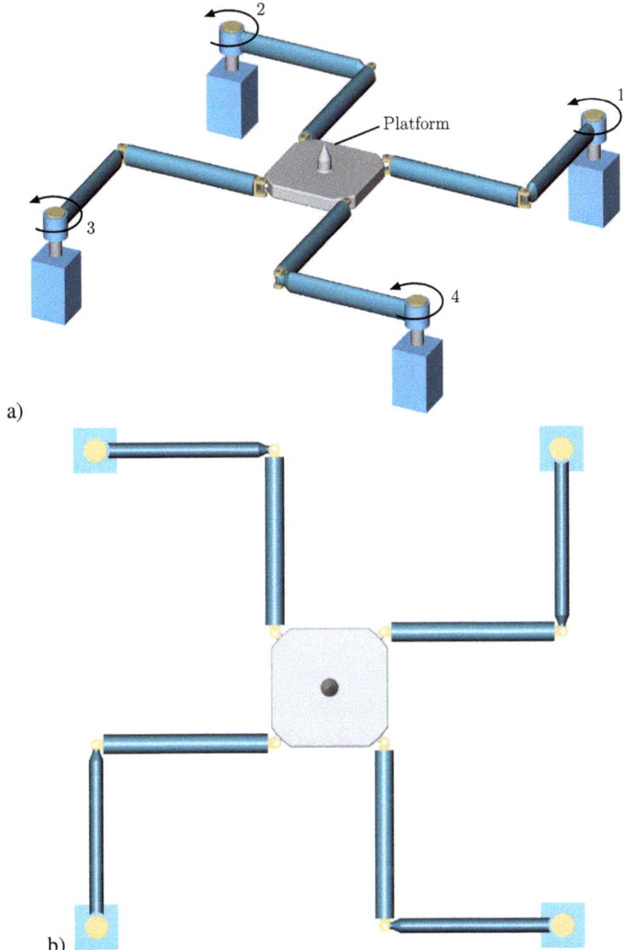

Fig. 25 (**a**) Redundantly actuated planar 4$\underline{R}$RR PKM in reference configuration $\mathbf{q}_0$. (**b**) Top view

7. For the 2-DOF planar PKM of Exercise 6,

   a. Derive the velocity forward kinematics (59).
   b. Derive the EOM (60) in actuator coordinates.

8. For the 2-DOF planar PKM of Exercise 6, use the EOM in task space coordinates to derive a solution for the inverse dynamics problems.
9. Figure 25 shows a planar 3-DOF PKM. Its platform can perform planar motions. Repeat a-j of Exercise 6 (with appropriate change of dimensions of $\mathbf{x}_p$ and ϑ). You may again want to use a computer algebra software.

10. The PKM from Exercise 9 is now equipped with an additional limb, which yields the redundantly actuated 3-DOF 4$\underline{R}$RR PKM in Fig. 25. Repeat all tasks of Exercise 9.
    Try not to derive the kinematics and the EOM separately for each of the four limbs, but reuse those of a 'prototypical' limb.
11. The redundantly actuated planar PKM in Fig. 9 comprises two FCs.

    a. Draw its topological graph, and introduce two FCs.
    b. If you were to repeat tasks b–j of Exercise 6, how could you reuse the results of Exercise 9?

12. Suggest a PKM model for a woodpecker. Explain the necessity for redundant actuation. Can you explain the 'pecking frequency' and ways to modulate it?
13. Give a few examples for technical and/or biological system that comprise modules resembling a PKM.
14. If you were to write a software for simulating the kinematics of PKM, which basic modules would this comprise? How would you exchange data, in particular how would you let the user define the kinematics?
    How would you extend that software if it was to simulate the PKM dynamics?
15. What are the practical challenges when implementing centralized model-based control schemes?

19 Conclusion and Further Reading

After almost three decades of research and development, PKM are now established as alternative robotic solutions for various industrial applications. Their modeling, design, and control can be built upon the approaches for serial manipulators, but special formulations are needed taking into account the closed loop kinematics, as summarized in this chapter. The chapter did not dive into the specific methods for modeling the kinematics and dynamics. The reader may apply any formulation he or she is familiar with. However, a modern and holistic approach to kinematics and dynamics modeling are the Lie group methods (or better simply say matrix methods). An introduction to these so-called geometric methods can be found in the textbooks [96, 128, 154] and the papers [118, 119]. A tailored approach using these techniques applied to the PKM modeling was reported in [123].

Cross-References

This chapter contains cross-references to content published in other chapters belonging to the four "Robotic Goes MOOC" books: KNO3, DES2, INT2, IMP1, IMP2. The full list of chapters abbreviations is available in the Preface.

References

1. K. Sébastien, G. Olivier, P. François, Control of a 3-DOF Over-Actuated Parallel Mechanism, *International Design Engineering Technical Conferences and Computers and Information in Engineering Conference (IDETC), vol. 5: 27th Biennial Mechanisms and Robotics Conference* (2002). https://doi.org/10.1115/DETC2002/MECH-34343
2. Iftomm terminology/english. Mech. Mach. Theory **38**(7), 607–682 (2003). https://doi.org/10.1016/S0094-114X(03)00011-9. Standardization of Terminology
3. B.N.R. Abadi, M. Farid, M. Mahzoon, Redundancy resolution and control of a novel spatial parallel mechanism with kinematic redundancy. Mech. Mach. Theory **133**, 112–126 (2019). https://doi.org/10.1016/j.mechmachtheory.2018.11.014
4. H. Abdellatif, B. Heimann, Computational efficient inverse dynamics of 6-dof fully parallel manipulators by using the lagrangian formalism. Mech. Mach. Theory **44**(1), 192–207 (2009). https://doi.org/10.1016/j.mechmachtheory.2008.02.003
5. H. Abdellatif, M. Grotjahn, B. Heimann, Independent identification of friction characteristics for parallel manipulators. ASME J. Dyn. Syst. Meas. Control **129**(3), 294–302 (2007)
6. R.I. Alizade, N.R. Tagiyev, J. Duffy, A forward and reverse displacement analysis of a 6-dof in-parallel manipulator. Mech. Mach. Theory **29**(1), 115–124 (1994). https://doi.org/10.1016/0094-114X(94)90024-8
7. J. Angeles, The qualitative synthesis of parallel manipulators. J. Mech. Des. **126**(4), 617–624 (2004). https://doi.org/10.1115/1.1667955
8. J. Angeles, Is there a characteristic length of a rigid-body displacement? Mech. Mach. Theory **41**(8), 884–896 (2006). https://doi.org/10.1016/j.mechmachtheory.2006.03.010. Special issue on CK 2005, International Workshop on Computational Kinematics
9. J. Angeles, *Fundamentals of Robotic Mechanical Systems*, 3rd edn. (Springer, Berlin, 2007)
10. J. Angeles, S. Lee, The formulation of dynamical equations of holonomic mechanical systems using a natural orthogonal complement. ASME J. Appl. Mech. **9**(5), 243–244 (1988). https://doi.org/10.1115/1.3173642
11. H. Asada, J. Slotine, *Robot Analysis and Control* (Wiley, New York, 1986)
12. C.G. Atkeson, C.H. An, J.M. Hollerbach, Estimation of inertial parameters of manipulator loads and links. Int. J. Robot. Res. **5**(3), 101–119 (1986)
13. F. Behi, Kinematic analysis for a six-degree-of-freedom 3-prps parallel mechanism. IEEE J. Robot. Autom. **4**(5), 561–565 (1988). https://doi.org/10.1109/56.20442
14. P. Ben-Horin, M. Shoham, Singularity analysis of a class of parallel robots based on Grassmann–Cayley algebra. Mech. Mach. Theory **41**(8), 958–970 (2006). https://doi.org/10.1016/j.mechmachtheory.2006.03.008. Special issue on CK 2005, International Workshop on Computational Kinematics
15. M. Bennehar, A. Chemori, F. Pierrot, V. Creuze, Extended model-based feedforward compensation in l1 adaptive control for mechanical manipulators: design and experiments. Front. Robot. AI **2**, 32 (2015). https://doi.org/10.3389/frobt.2015.00032
16. M. Bennehar, A. Chemori, F. Pierrot, S. Krut, *Control of Redundantly Actuated PKMs for Closed-Shape Trajectories Tracking with Real-Time Experiments* (De Gruyter Oldenbourg, Berlin, 2017), pp. 17–34
17. I. Bonev, The True Origins of Parallel Robots. Last Accessed 26 Sept 2020. http://www.parallemic.org/Reviews/Review007.html
18. I. Bonev, D. Zlatanov, C. Gosselin, Singularity analysis of 3-dof planar parallel mechanisms via screw theory. ASME J. Mech. Des. **125**(3), 573–581 (2003)
19. S. Briot, W. Khalil, *Dynamics of Parallel Robots* (Springer, Berlin, 2015)
20. S. Briot, M. Gautier, S. Krut, Dynamic parameter identification of actuation redundant parallel robots: application to the DualV, in *2013 IEEE/ASME International Conference on Advanced Intelligent Mechatronics* (2013), pp. 637–643
21. S. Briot, S. Krut, M. Gautier, Dynamic parameter identification of overactuated parallel robots. J. Dyn. Syst. Meas. Control **137**(11) (2015). 111002

22. H. Bruyninckx, J. De Schutter, *A Class of Fully Parallel Manipulators with Closed-Form Forward Position Kinematics* (Springer Netherlands, Dordrecht, 1996), pp. 411–420. https://doi.org/10.1007/978-94-009-1718-7_41

23. Y. Byung-Ju, R.A. Freeman, Simultaneous stiffness generation and internal load distribution in redundantly actuated mechanisms, in *Fifth International Conference on Advanced Robotics 'Robots in Unstructured Environments*, vol. 1 (1991), pp. 802–807

24. J. Calamia, Artifacts from the first 2000 years of computing. IEEE Spectrum **48**(5), 34–40 (2011)

25. K.L. Cappel, F.I.: Motion simulator. US Patent No. 3295224, US41637164A (filed: Dec 7, 1964; patented: Jan 3, 1967)

26. M. Carricato, V. Parenti-Castelli, Singularity-free fully-isotropic translational parallel mechanisms. Int. J. Robot. Res. **21**(2), 161–174 (2002). https://doi.org/10.1177/027836402760475360

27. S. Cha, T.A. Lasky, S.A. Velinsky, Singularity avoidance for the 3-rrr mechanism using kinematic redundancy, in *Proceedings 2007 IEEE International Conference on Robotics and Automation* (2007), pp. 1195–1200. https://doi.org/10.1109/ROBOT.2007.363147

28. S.H. Cha, T. Lasky, S. Velinsky, Determination of the kinematically redundant active prismatic joint variable ranges of a planar parallel mechanism for singularity-free trajectories. Mech. Mach. Theory **44**(5), 1032–1044 (2009). https://doi.org/10.1016/j.mechmachtheory.2008.05.010

29. S.H. Cha, T. Lasky, S. Velinsky, Determination of the kinematically redundant active prismatic joint variable ranges of a planar parallel mechanism for singularity-free trajectories. Mech. Mach. Theory **44**(5), 1032–1044 (2009). https://doi.org/10.1016/j.mechmachtheory.2008.05.010

30. D. Chablat, P. Wenger, Architecture optimization of a 3-dof parallel mechanism for machining applications, the orthoglide. IEEE Trans. Robot. Autom. **19**(3), 403–410 (2003)

31. D. Chakarov, Study of the antagonistic stiffness of parallel manipulators with actuation redundancy. Mech. Mach. Theory **39**(6), 583–601 (2004). https://doi.org/10.1016/j.mechmachtheory.2003.12.001

32. S. Charentus, Modélisation et commande d'un robot manipulateur redondant composé de plusieurs plateformes de Stewart. These de Doctorat, Universite Paul Sabatier, Toulouse (1990)

33. R. Clavel, Delta, a fast robot with parallel geometry, in *18th Int. Symp. Industrial Robots*, Lausanne (1988), pp. 91–100

34. O. Company, F. Pierrot, S. Krut, C. Baradat, Par2: a spatial mechanism for fast planar two-degree-of-freedom pick-and-place applications. Meccanica **46**, 239–248 (2011)

35. M.R. Cutkosky, P.K. Wright, Active control of a compliant wrist in manufacturing tasks. J. Eng. Ind. **108**(1), 36–43 (1986). https://doi.org/10.1115/1.3187038

36. P. Dietmaier, The Stewart-Gough platform of general geometry can have 40 real postures, in *Advances in Robot Kinematics: Analysis and Control*, ed. by J. Lenarčič, M.L. Husty (Springer Netherlands, Dordrecht, 1998), pp. 7–16

37. I. Ebrahimi, J. Carretero, R. Boudreau, 3-PRRR redundant planar parallel manipulator: inverse displacement, workspace and singularity analyses. Mech. Mach. Theory **42**, 1007–1016 (2007)

38. G. Ecorchard, R. Neugebauer, P. Maurine, Elasto-geometrical modeling and calibration of redundantly actuated pkms. Mech. Mach. Theory **45**(5), 795–810 (2010). https://doi.org/10.1016/j.mechmachtheory.2009.12.008

39. J.M. Escorcia-Hernández, A. Chemori, H. lito Aguilar-Sierra, J. Arturo Monroy-Anieva, A new solution for machining with ra-pkms: Modelling, control and experiments. Mech. Mach. Theory **150**, 103864 (2020)

40. R. Featherstone, *Rigid Body Dynamics Algorithms* (Springer, Berlin, 2008)

41. H. Furuya, K. Higashiyama, Dynamics of closed linked variable geometry truss manipulators. Acta Astronaut. **36**(5), 251–259 (1995). https://doi.org/10.1016/0094-5765(95)00104-8

42. J. Gallardo-Alvarado, J. Rico-Martinez, Jerk influence coefficients, via screw theory, of closed chains. Meccanica **36**, 213–228 (2001)
43. C. Galletti, Mobility analysis of single-loop kinematic chains: an algorithmic approach based on displacement groups. Mech. Mach. Theory **29**(8), 1187–1204 (1994)
44. V. Garg, S.B. Nokleby, J.A. Carretero, Wrench capability analysis of redundantly actuated spatial parallel manipulators. Mech. Mach. Theory **44**(5), 1070–1081 (2009). https://doi.org/10.1016/j.mechmachtheory.2008.05.011
45. C. Germain, S. Caro, S. Briot, P. Wenger, Singularity-free design of the translational parallel manipulator irsbot-2. Mech. Mach. Theory **64**, 262–285 (2013). https://doi.org/10.1016/j.mechmachtheory.2013.02.005
46. C. Gibson, K. Hunt, Geometry of screw systems–1: screws: genesis and geometry. Mech. Mach. Theory **25**(1), 1–10 (1990)
47. C. Gibson, K. Hunt, Geometry of screw systems–2: classification of screw systems. Mech. Mach. Theory **25**(1), 11–27 (1990)
48. G. Gogu, Fully-isotropic t3r1-type parallel manipulators, in *On Advances in Robot Kinematics*, ed. by J. Lenarčič, C. Galletti (Springer Netherlands, Dordrecht, 2004), pp. 265–272
49. G. Gogu, Structural synthesis of fully-isotropic translational parallel robots via theory of linear transformations. Eur. J. Mech. A/Solids **23**(6), 1021–1039 (2004). https://doi.org/10.1016/j.euromechsol.2004.08.006
50. G. Gogu, *Structural Synthesis of Parallel Robots: Part 1: Methodology*. Solid Mechanics and Its Applications (Springer Netherlands, Dordrecht, 2009)
51. C.M. Gosselin, J. Angeles, Singular analysis of closed-loop kinematic chains. IEEE Trans. Robot. Autom. **6**(3), 281–290 (1990)
52. C. Gosselin, X. Kong, Kinematics and singularity analysis of a novel type of 3-crr 3-dof translational parallel manipulator. Int. J. Robot. Res. **21**(9), 791–798 (2002)
53. C. Gosselin, T. Laliberte, A. Veillette, Singularity-free kinematically redundant planar parallel mechanisms with unlimited rotational capability. IEEE Trans. Robot. **31**(2), 457–467 (2015). https://doi.org/10.1109/TRO.2015.2409433
54. V. Gough, Contribution to discussion to papers on research in automobile stability and control and in tire performance, in *Proc. Institution of Mechanical Engineers* (1956), pp. 392–395
55. H. Hanafusa, M.A. Adli, Effect of internal forces on stiffness of closed mechanisms, in *Fifth International Conference on Advanced Robotics 'Robots in Unstructured Environments'*, vol. 1 (1991), pp. 845–850
56. K. Hanahara, Y. Tada, Dynamic behavior of truss-type parallel mechanism with actuated wire members, in *IEEE Int. Conf. Adv. Rob. (ICAR)*, June 30–Juli 3, 2003, Coimbra (2003), pp. 1793–1798
57. F. Hao, J.M. McCarthy, Conditions for line-based singularities in spatial platform manipulators. J. Robot. Syst. **15**(1), 43–55 (1998)
58. J. Herve, Intrinsic formulation of problems of geometry and kinematics of mechanisms. Mech. Mach. Theory **17**(3), 179–184 (1982)
59. J.M. Herve, Design of parallel manipulators via the displacement group, in *9th World Congress on the Theory of Machines and Mechanisms*, Milan (1995), pp. 2079–2082
60. J. Herve, The lie group of rigid body displacements, a fundamental tool for mechanisms design. Mech. Mach. Theory **34**(5), 719–730 (1999)
61. J.M. Herve, F. Sparacino, Structural synthesis of 'parallel' robots generating spatial translation, in *Fifth International Conference on Advanced Robotics 'Robots in Unstructured Environments'*, vol. 1 (1991), pp. 808–813
62. M. Honegger, A. Codourey, E. Burdet, Adaptive control of the hexaglide, a 6 dof parallel manipulator, in *Proceedings of International Conference on Robotics and Automation*, vol. 1 (1997), pp. 543–548
63. M. Honegger, R. Brega, G. Schweitzer, Application of a non-linear adaptive controller to a 6 dof parallel manipulator, in *IEEE International Conference on Robotics and Automation (ICRA)*, San Fransisco (2000), pp. 1930–1935

64. Z. Huang, Q. Li, H. Ding, *Differential Geometry Theory of Parallel Mechanisms* (Springer Science+Business Media, Berlin, 2013)
65. K. Hunt, *Kinematic Geometry of Mechanisms* (Clarendon Press, Oxford, 1978)
66. K.H. Hunt, Structural kinematics of in-parallel-actuated robot-arms. J. Mech. Transm. Autom. Des. **105**(4), 705–712 (1983). https://doi.org/10.1115/1.3258540
67. M. Husty, An algorithm for solving the direct kinematics of general Stewart-Gough platforms. Mech. Mach. Theory **31**(4), 365–379 (1996). https://doi.org/10.1016/0094-114X(95)00091-C
68. M.L. Husty, D.R. Walter, *Mechanism Constraints and Singularities—The Algebraic Formulation* (Springer International Publishing, Cham, 2019), pp. 101–180. https://doi.org/10.1007/978-3-030-05219-5_4
69. M. Husty, M. Pfurner, H.P. Schröcker, K. Brunnthaler, Algebraic methods in mechanism analysis and synthesis. Robotica **25**(6), 661–675 (2007)
70. C. Innocenti, V. Parenti-Castelli, Singularity-free evolution from one configuration to another in serial and fully-parallel manipulators. ASME J. Mech. Des. **120**, 73–99 (1998)
71. A. Jain, Graph theoretic foundations of multibody dynamics, part i: structural properties. Multibody Syst. Dyn. **26**, 307–333 (2011)
72. A. Jain, *Robot and Multibody Dynamics* (Springer US, New York, 2011)
73. A. Jain, Multibody graph transformations and analysis. Nonlinear Dyn. **67**, 2153–2170 (2012)
74. S.A. Joshi, L.W. Tsai, Jacobian analysis of limited-dof parallel manipulators. ASME J. Mech. Des. **124**, 254–258 (2020)
75. H.K. Jung, C.D. Crane, R.G. Roberts, Stiffness mapping of planar compliant parallel mechanisms in a serial arrangement, in *Advances in Robot Kinematics*, ed. by J. Lennarčič, B. Roth (Springer Netherlands, Dordrecht, 2006), pp. 85–94
76. D. Kanaan, P. Wenger, S. Caro, D. Chablat, Singularity analysis of lower mobility parallel manipulators using Grassmann–cayley algebra. IEEE Trans. Robot. **25**(5), 995–1004 (2009)
77. A. Karger, Architecture singular planar parallel manipulators. Mech. Mach. Theory **38**(11), 1149–1164 (2003)
78. M. Karouia, J.M. Hervé, A three-dof tripod for generating spherical rotation, in *Advances in Robot Kinematics*, ed. by J. Lenarčič, M.M. Stanišić (Springer Netherlands, Dordrecht, 2000), pp. 395–402
79. W. Khalil, E. Dombre, *Modeling Identification and Control of Robots* (Springer, Berlin, 2004)
80. H.S. Kim, L.W. Tsai, *Evaluation of a Cartesian Parallel Manipulator* (Springer Netherlands, Dordrecht, 2002), pp. 21–28. https://doi.org/10.1007/978-94-017-0657-5_3
81. H. Kim, L. Tsai, Design optimization of a cartesian parallel manipulator. ASME J. Mech. Des. **125**(1), 43–51 (2003)
82. J. Kim, F.C. Park, S.J. Ryu, J. Kim, J.C. Hwang, C. Park, C.C. Iurascu, Design and analysis of a redundantly actuated parallel mechanism for rapid machining. IEEE Trans. Robot. Autom. **17**(4), 423–434 (2001)
83. J. Kim, J.C. Hwang, J.S. Kim, C.C. Iurascu, F.C. Park, Y.M. Cho, Eclipse ii: a new parallel mechanism enabling continuous 360-degree spinning plus three-axis translational motions. IEEE Trans. Robot. Autom. **18**(3), 367–373 (2002)
84. S. Kock, W. Schumacher, A parallel x-y manipulator with actuation redundancy for high-speed and active-stiffness applications, in *IEEE International Conference on Robotics and Automation (ICRA)*, vol. 3 (1998), pp. 2295–2300
85. S. Kock, W. Schumacher, Control of a fast parallel robot with a redundant chain and gearboxes: experimental results, in *IEEE International Conference on Robotics and Automation (ICRA)*, vol. 2 (2000), pp. 1924–1929
86. S. Kock, W. Schumacher, A mixed elastic and rigid-body dynamic model of an actuation redundant parallel robot with high-reduction gears, in *IEEE International Conference on Robotics and Automation (ICRA)*, vol. 2 (2000), pp. 1918–1923
87. X. Kong, C. Gosselin, A class of 3-dof translational parallel manipulators with linear input-output equations, in *Proceedings of the Workshop on Fundamental Issues and Future Research Directions for Parallel Mechanisms and Manipulators*, October 3–4, 2002, Quebec, ed. by C. Gosselin, I. Ebert-Uphoff (2002), pp. 25–32

88. X. Kong, C.M. Gosselin, *Type Synthesis of Linear Translational Parallel Manipulators* (Springer Netherlands, Dordrecht, 2002), pp. 453–462. https://doi.org/10.1007/978-94-017-0657-5_48
89. X. Kong, C.M. Gosselin, *Type Synthesis of Parallel Mechanisms* (Springer, Berlin, 2007)
90. J. Kotlarski, T. Do Thanh, B. Heimann, T. Ortmaier, Optimization strategies for additional actuators of kinematically redundant parallel kinematic machines, in *2010 IEEE International Conference on Robotics and Automation* (2010), pp. 656–661. https://doi.org/10.1109/ROBOT.2010.5509982
91. W. Kuhlbusch, W. Moritz, J. Lückel, S. Toepper, Triplanar - a new parallel 6-dof robot for highly precise measurement and process tasks. IFAC Proc. Vol. **33**(26), 463–467 (2000). https://doi.org/10.1016/S1474-6670(17)39187-5. IFAC Conference on Mechatronic Systems, Darmstadt, Germany, 18–20 September 2000
92. R. Kurtz, V. Hayward, Multiple-goal kinematic optimization of a parallel spherical mechanism with actuator redundancy. IEEE Trans. Robot. Autom. **8**(5), 644–651 (1992)
93. Y.H. Lee, Y. Han, C.C. Iuraşcu, F.C. Park, Simulation-based actuator selection for redundantly actuated robot mechanisms. J. Robot. Syst. **19**(8), 379–390 (2002). https://doi.org/10.1002/rob.10047
94. S.H. Lee, J.H. Lee, B.J. Yi, S.H. Kim, Y.K. Kwak, Optimization and experimental verification for the antagonistic stiffness in redundantly actuated mechanisms: a five-bar example. Mechatronics **15**(2), 213–238 (2005). https://doi.org/10.1016/j.mechatronics.2004.07.008
95. V. Luigi, De.S. Joris, Force control, Springer handbook of robotics, 195–220 (2016)
96. K.M. Lynch, F.C. Park, *Modern Robotics* (Cambridge University Press, Cambridge, 2017)
97. E. Macho, O. Altuzarra, C. Pinto, A. Hernandez, *Transitions Between Multiple Solutions of the Direct Kinematic Problem* (Springer Netherlands, Dordrecht, 2008), pp. 301–310. https://doi.org/10.1007/978-1-4020-8600-7_32
98. F. Marquet, S. Krut, O. Company, F. Pierrot, Archi: a new redundant parallel mechanism-modeling, control and first results, in *Proceedings 2001 IEEE/RSJ International Conference on Intelligent Robots and Systems (IROS)*, vol. 1 (2001), pp. 183–188
99. J.R. Martinez, J. Duffy, Classification of screw systems–i. one- and two-systems. Mech. Mach. Theory **27**(4), 459–470 (1992)
100. J.R. Martinez, J. Duffy, Classification of screw systems–ii. three-systems. Mech. Mach. Theory **27**(4), 471–490 (1992)
101. J.R. Martinez, J. Gallardo, J. Duffy, Screw theory and higher order kinematic analysis of open serial and closed chains. Mech. Mach. Theory **34**(4), 559–586 (1999)
102. M.T. Masouleh, C. Gosselin, M. Husty, D.R. Walter, Forward kinematic problem of 5-rpur parallel mechanisms (3t2r) with identical limb structures. Mech. Mach. Theory **46**(7), 945–959 (2011). https://doi.org/10.1016/j.mechmachtheory.2011.02.005
103. V. Mata, N. Farhat, M. Diaz-Rodriguez, A. Valera, A. Page, Dynamic parameter identification for parallel manipulators, in *Parallel Manipulators, towards New Applications*, ed. by H. Wu. IntechOpen (2008). https://doi.org/10.5772/5424
104. P. McAree, R. Daniel, An explanation of never-special assembly changing motions for 3-3 parallel manipulators. Int. J. Robot. Res. **18**(6), 556–574 (1999)
105. J.M. McCarthy, G.S. Soh, *Geometric Design of Linkages*, 2nd edn. (Springer, Berlin, 2011)
106. J. Merlet, Singularity configurations of parallel manipulators and Grassman geometry. Int. J. Robot. Res. **10**(2), 123–134 (1991)
107. J.P. Merlet, Direct kinematics and assembly modes of parallel manipulators. Int. J. Robot. Res. **11**(2), 150–162 (1992). https://doi.org/10.1177/027836499201100205
108. J.P. Merlet, Direct kinematics of parallel manipulators. IEEE Trans. Robot. Autom. **9**(6), 842–846 (1993). https://doi.org/10.1109/70.265928
109. J.P. Merlet, Jacobian, manipulability, condition number, and accuracy of parallel robots. J. Mech. Des. **128**(1), 199–206 (2005). https://doi.org/10.1115/1.2121740
110. J. Merlet, *Parallel Robots* (Springer, Berlin, 2006)
111. K. Miura, H. Furuya, K. Suzuki, Variable geometry truss and its application to deployable truss and space crane arm. Acta Astronaut. **12**(7), 599–607 (1985). https://doi.org/10.1016/0094-5765(85)90131-6. Congress of the International Federation of Astronautics

112. A. Müller, Internal preload control of redundantly actuated parallel manipulators –its application to backlash avoiding control. IEEE Trans. Robot. **21**(4), 668–677 (2005)
113. A. Müller, Stiffness control of redundantly actuated parallel manipulators, in *Proceedings 2006 IEEE International Conference on Robotics and Automation, ICRA 2006* (2006), pp. 1153–1158
114. A. Müller, Consequences of geometric imperfections for the control of redundantly actuated parallel manipulators. IEEE Trans. Robot. **26**(1), 21–31 (2010)
115. A. Müller, On the terminology and geometric aspects of redundant parallel manipulators. Robotica **31**(1), 137–147 (2013). https://doi.org/10.1017/S0263574712000173
116. A. Müller, Higher-order analysis of kinematic singularities of lower pair linkages and serial manipulators. ASME J. Mech. Robot. **10**(1), 011008 (2018)
117. A. Müller, Kinematic topology and constraints of multi-loop linkages. Robotica **36**(11), 1641–1663 (2018). https://doi.org/10.1017/S0263574718000619
118. A. Müller, Screw and lie group theory in multibody dynamics –recursive algorithms and equations of motion of tree-topology systems. Multibody Syst. Dyn. **42**(2), 219–248 (2018). https://doi.org/10.1007/s11044-017-9583-6
119. A. Müller, Screw and lie group theory in multibody dynamics —motion representation and recursive kinematics of tree-topology systems. Multibody Syst. Dyn. **43**(1), 1–34 (2018). https://doi.org/10.1007/s11044-017-9582-7
120. A. Müller, An overview of formulae for the higher-order kinematics of lower-pair chains with applications in robotics and mechanism theory. Mech. Mach. Theory **142**, 103594 (2019). https://doi.org/10.1016/j.mechmachtheory.2019.103594
121. A. Müller, *Local Investigation of Mobility and Singularities of Linkages* (Springer International Publishing, Cham, 2019), pp. 181–229. https://doi.org/10.1007/978-3-030-05219-5_5
122. A. Müller, An O(n)-algorithm for the higher-order kinematics and inverse dynamics of serial manipulators using spatial representation of twists. IEEE Robot. Autom. Lett. **6**(2), 397–404 (2020)
123. A. Müller, Dynamics modeling of topologically simple parallel manipulators: a geometric approach. ASME Appl. Mech. Rev. **72**(3), 27 pp. (2020)
124. A. Müller, A constraint embedding approach for dynamics modeling of parallel kinematic manipulators with hybrid limbs. Robot. Auton. Syst. **155**, 20 pp. (2022)
125. A. Müller, Dynamics of parallel manipulators with hybrid complex limbs —modular modeling and parallel computing. Mech. Mach. Theory **167**, 104549 (36 pp.) (2022)
126. A. Müller, T. Hufnagel, Projection method for the elimination of contradicting control forces in redundantly actuated pkm, in *IEEE Int. Conf. Rob. Automat. (ICRA)*, Shanghai (May 9–13, 2011), pp. 3218–3223
127. A. Müller, P. López-Custodio, J. Dai, Identification of non-transversal bifurcations of linkages, in *44 Mechanisms and Robotics Conference (MR)/ASME International Design Engineering Technical Conferences (IDETC)*, August 16–19, 2020, St. Louis (2018)
128. R. Murray, Z. Li, S. Sastry, *A Mathematical Introduction to Robotic Manipulation* (CRC Press, Boca Raton, 1994)
129. F. Naccarato, P. Hughes, Inverse kinematics of variable-geometry truss manipulators. J. Robot. Syst. **8**(2), 249–266 (1991). https://doi.org/10.1002/rob.4620080207
130. S. Nokleby, R. Fisher, R. Podhorodeski, F. Firmani, Force capabilities of redundantly-actuated parallel manipulators. Mech. Mach. Theory **40**(5), 578–599 (2005). https://doi.org/10.1016/j.mechmachtheory.2004.10.005
131. J.F. O'Brien, J.T. Wen, Redundant actuation for improving kinematic manipulability, in *IEEE International Conference on Robotics and Automation (ICRA)*, vol. 2 (1999), pp. 1520–1525
132. A.L. Orekhov, N. Simaan, Directional stiffness modulation of parallel robots with kinematic redundancy and variable stiffness joints. ASME J. Mech. Robot. **11**, 051003 (2019)
133. R. Ortega, M.W. Spong, Adaptive motion control of rigid robots: a tutorial. Automatica **25**(6), 877–888 (1989). https://doi.org/10.1016/0005-1098(89)90054-X

134. B. Padmanabhan, V. Arun, C.F. Reinholtz, Closed-form inverse kinematic analysis of variable-geometry Truss manipulators. J. Mech. Des. **114**(3), 438–443 (1992). https://doi.org/10.1115/1.2926571

135. C. Parenti, C. Innocenti, Spatial open kinematic chains: singularities, regions and subregions, in *Proc. 7th CISM-IFTOMM Romansy* (1988), pp. 400–407

136. V. Parenti-Castelli, C. Innocenti, Position analysis of robot manipulators: regions and subregions, in *Int. Conf. on Advances in Robot Kinematics* (1988), pp. 150–158

137. F.C. Park, J.W. Kim, Manipulability of closed kinematic chains. J. Mech. Des. **120**(4), 542–548 (1998). https://doi.org/10.1115/1.2829312

138. F. Park, J. Bobrow, S. Ploen, A lie group formulation of robot dynamics. Int. J. Robot. Res. **14**(6), 609–618 (1995). https://doi.org/10.1177/027836499501400606

139. E. Pernette, S. Henein, I. Magnani, R. Clavel, Design of parallel robots in microrobotics. Robotica **15**(4), 417–420 (1997). https://doi.org/10.1017/S0263574797000519

140. F. Pierrot, O. Company, H4: a new family of 4-dof parallel robots, in *1999 IEEE/ASME International Conference on Advanced Intelligent Mechatronics* (Cat. No.99TH8399) (1999), pp. 508–513

141. F. Pierrot, T. Shibukawa, From hexa to hexam, in *Parallel Kinematic Machines*, ed. by C.R. Boër, L. Molinari-Tosatti, K.S. Smith (Springer London, London, 1999), pp. 357–364

142. F. Pierrot, V. Nabat, O. Company, S. Krut, P. Poignet, Optimal design of a 4-dof parallel manipulator: from academia to industry. IEEE Trans. Robot. **25**(2), 213–224 (2009)

143. F. Pierrot, S. Krut, A. Saenz, O. Company, V. Nabat, C. Baradat, Two-degree-of-freedom parallel manipulator (2011). Patent number: EP2252437 (B1) WO 2009/089916 (A1). Extension : 24/11/10 AR070196 (A1) AT521457 (T) CA2712260 (A1) CN101977737 (A) ES2375074 (T3) JP2011509837 (A) TW200932457 (A) US2011048159 (A1)

144. A. Plummer, A general co-ordinate transformation framework for multi-axis motion control with applications in the testing industry. Control Eng. Pract. **18**(6), 598–607 (2010). https://doi.org/10.1016/j.conengprac.2010.02.015

145. G. Pond, J.A. Carretero, Quantitative dexterous workspace comparison of parallel manipulators. Mech. Mach. Theory **42**(10), 1388–1400 (2007). https://doi.org/10.1016/j.mechmachtheory.2006.10.004

146. M. Raghavan, The Stewart platform of general geometry has 40 configurations. J. Mech. Des **115**(2), 277–282 (1993). https://doi.org/10.1115/1.2919188

147. C. Reinholz, D. Gokhale, Design and analysis of variable geometry truss robot, in *9th Applied Mechanisms Conf.*, Oklahoma (1987), pp. 1–5

148. C. Reymond, United States Granted Patent, US 4976582, A "Device For The Movement And Positioning Of An Element In Space", Date: 11 Dec 1990. https://lens.org/089-198-012-036-824

149. J.M. Rico, J. Cervantes-Sanchez, A. Tadeo-Chávez, G. Pérez-Soto, J. Rocha-Chavarria, A comprehensive theory of type synthesis of fully parallel platforms, in *30th Annual Mechanisms and Robotics Conference, International Design Engineering Technical Conferences and Computers and Information in Engineering Conference*, vol. 2, Parts A and B (2006), pp. 1067–1078. https://doi.org/10.1115/DETC2006-99070

150. A. Rosyid, B. El-Khasawneh, A. Alazzam, Review article: performance measures of parallel kinematics manipulators. Mech. Sci. **11**(1), 49–73 (2020). https://doi.org/10.5194/ms-11-49-2020

151. R. Sarma, S. Kramer, V. Ramamurti, The dynamic equations of motion and actuation scheme for the tetrahedron based variable geometry truss manipulator, in *22nd Biennial Mechanisms Conf.*, vol. DE-45, Scottsdale, Sep. 13–16 (1992), pp. 173–178

152. G. Sartori Natal, A. Chemori, F. Pierrot, Dual-space control of extremely fast parallel manipulators: payload changes and the 100g experiment. IEEE Trans. Control Syst. Technol. **23**(4), 1520–1535 (2015)

153. Y. Seguchi, M. Tanaka, T. Yamaguchi, Y. Sasabe, H. Tsuji, Dynamic analysis of a truss-type flexible robot arm. JSME Int. J. Ser. 3, Vib. Control Eng. Eng. Ind. **33**(2), 183–190 (1990). https://doi.org/10.1299/jsmec1988.33.183

154. J. Selig, *Geometric Fundamentals of Robotics* (Springer, Berlin, 2005)
155. B. Siciliano, O. Khatib, *Springer Handbook of Robotics*, 2nd edn. (Springer, Berlin, 2016)
156. F. Sparacino, J.M. Herve, Synthesis of parallel manipulators using lie-groups y-star and h-robot, in *Proceedings of 1993 IEEE/Tsukuba International Workshop on Advanced Robotics* (1993), pp. 75–80
157. A. Spinos, D. Carroll, T. Kientz, M. Yim, Variable topology truss: Design and analysis, in *2017 IEEE/RSJ International Conference on Intelligent Robots and Systems (IROS)* (2017), pp. 2717–2722
158. D. Stewart, A platform with 6 degrees of freedom. Proc. Inst. Mech. Eng. **180**(15), 371–386 (1965)
159. M. Subramaniam, S.N. Kramer, The inverse kinematic solution of the tetrahedron based variable-geometry truss manipulator. J. Mech. Des. **114**(3), 433–437 (1992). https://doi.org/10.1115/1.2926570
160. M. Tanaka, Large-scale framed structure as parallel mechanism with hyper-redundancy. Adv. Robot. **8**(6), 573–587 (1993). https://doi.org/10.1163/156855394X00266
161. J. Tong, J. Somerset, Control, performance and applications of antagonist actuated manipulator joints, in *1985 American Control Conference* (1985), pp. 63–64
162. L.W. Tsai, *Kinematics of A Three-Dof Platform with Three Extensible Limbs* (Springer Netherlands, Dordrecht, 1996), pp. 401–410. https://doi.org/10.1007/978-94-009-1718-7_40
163. L.W. Tsai, The Jacobian analysis of a parallel manipulator using reciprocal screws, in *Advances in Robot Kinematics: Analysis and Control*, ed. by J. Lenarčič, M. Husty (Springer, Dordrecht, 1998), pp. 327–336
164. L.W. Tsai, *Robot Analysis: The Mechanics of Serial and Parallel Manipulators* (Wiley, New York, 1999)
165. L.W. Tsai, *Robot Analysis: The Mechanics of Serial and Parallel Manipulators* (Wiley, New York, 1999)
166. L. Tsai, F. Tahmasebi, Synthesis and analysis of a new class of six-degree-of-freedom parallel minimanipulators. J. Robot. Syst. **10**(5), 561–580 (1993)
167. M. Valasek, V. Bauma, Z. Sika, T. Vampola, Redundantly actuated parallel structures-principle, examples, advantages, in *3rd Parallel Kinematics Seminar Chemnitz* (2002), pp. 993–1009
168. M. Valasek, K. Belda, M. Florian, Control and calibration of redundantly actuated parallel robots, in *3rd Parallel Kinematics Seminar Chemnitz*, Chemnitz (2002), pp. 411–427
169. V. Van Der Wijk, S. Krut, F. Pierrot, J. Herder, Generic method for deriving the general shaking force balance conditions of parallel manipulators with application to a redundant planar 4-RRR parallel manipulator, in *IFToMM 2011 World Congress: The 13th World Congress in Mechanism and Machine Science*, Guanajuato (2011), pp. A12–523
170. V. Van Der Wijk, S. Krut, F. Pierrot, J. Herder, Design and experimental evaluation of a dynamically balanced redundant planar 4-rrr parallel manipulator. Int. J. Robot. Res. **32**(6), 744–759 (2013)
171. D.R. Walter, M.L. Husty, On implicitization of kinematic constraint equations, in *Machine Design & Research (CCMMS 2010)*, vol. 26, Shanghai (2010), pp. 218–226
172. C. Wampler, Inverse kinematic functions for redundant manipulators, in *Proceedings. 1987 IEEE International Conference on Robotics and Automation*, 31 March–3 April 1987, Raleigh, vol. 4 (1987), pp. 610–617. https://doi.org/10.1109/ROBOT.1987.1087950
173. C.W. Wampler, Forward displacement analysis of general six-in-parallel SPS (Stewart) platform manipulators using soma coordinates. Mech. Mach. Theory **31**(3), 331–337 (1996). https://doi.org/10.1016/0094-114X(95)00068-A
174. L. Wang, J. Wu, J. Wang, Z. You, An experimental study of a redundantly actuated parallel manipulator for a 5-dof hybrid machine tool. IEEE/ASME Trans. Mechatron. **14**(1), 72–81 (2009)
175. X. Wang, L. Baron, G. Cloutier, Topology of serial and parallel manipulators and topological diagrams. Mech. Mach. Theory **43**, 754–770 (2008)

176. P. Wenger, Classification of 3r positioning manipulators. ASME J. Mech. Des. **120**, 327–332 (1998)
177. P. Wenger, D. Chablat, Workspace and assembly-modes in fully parallel manipulators: a descriptive study, in *Advances in Robot Kinematics*. ed. by J. Lenarčič, M. Husty (Kluwer Academic Publisher, Dordrecht, 1998)
178. J. Wittenburg, Dynamics of multibody systems –a brief review. Acta Astron. **20**, 89–92 (1989). https://doi.org/10.1016/0094-5765(89)90057-X
179. J. Wittenburg, *Dynamics of Multibody Systems*, 2nd edn. (Springer, Berlin, 2008)
180. K. Wohlhart, Kinematotropic linkages, in *Recent Advances in Robot Kinematics*, ed. by J. Lenarčič, V. Parent-Castelli (Kluwer, 1996), pp. 359–368
181. K.H. Wurst, Linapod — machine tools as parallel link systems based on a modular design, in *Parallel Kinematic Machines*, ed. by C.R. Boër, L. Molinari-Tosatti, K.S. Smith (Springer London, London, 1999), pp. 377–394
182. L.J. Xu, S.W. Fan, H. Li, Analytical model method for dynamics of n-celled tetrahedron-tetrahedron variable geometry truss manipulators. Mech. Mach. Theory **36**(11), 1271–1279 (2001). https://doi.org/10.1016/S0094-114X(01)00050-7
183. B.J. Yi, R.A. Freeman, D. Tesar, Open-loop stiffness control of overconstrained mechanisms/robotic linkage systems, in *Proceedings, 1989 International Conference on Robotics and Automation*, vol. 3 (1989), pp. 1340–1345
184. B.J. Yi, W. Cho, R.A. Freeman, Open-loop stability of overconstrained parallel robotic systems, in *Proceedings, IEEE International Conference on Robotics and Automation*, vol. 2 (1990), pp. 1350–1355
185. B.J. Yi, R.A. Freeman, D. Tesar, Force and stiffness transmission in redundantly actuated mechanisms: the case for a spherical shoulder mechanism. Robot. Spatial Mech. Mech. Syst. **45**, 163–172 (1994)
186. K.E. Zanganeh, R. Sinatra, J. Angeles, Kinematics and dynamics of a six-degree-of-freedom parallel manipulator with revolute legs. Robotica **15**(4), 385–394 (1997). https://doi.org/10.1017/S0263574797000477
187. D. Zlatanov, R. Fenton, B. Benhabib, A unifying framework for classification and interpretation of mechanism singularities. ASME J. Mech. Des. **117**(4), 566–572 (1995)
188. D. Zlatanov, R. Fenton, B. Benhabib, Identification and classification of the singular configurations of mechanisms. Mech. Mach. Theory **36**(6), 743–760 (1998)
189. D. Zlatanov, I. Bonev, C. Gosselin, Constraint singularities as c-space singularities, in *Advances in Robot Kinematics: Theory and Application*, Caldes de Malavella, ed. by J. Lenarčič, F. Thomas (Kluwer Academic Publishers, Dordrecht, 2002), pp. 183–192

Elastic Robots

Jörn Malzahn, Freia I. Muster, and Torsten Bertram

Abstract The chapter focuses on dynamics modelling and touches also design aspects of elastic robots. This chapter assumes the reader to be familiar with the basics of kinematics, dynamics, and control of ideally rigid robots. It aims to equip the reader with knowledge and tools to analyse and quantify the various desired and undesired effects of elasticity on the robot operation. Furthermore, the chapter provides techniques for dealing with undesired elasticity and criteria to intentionally synthesise elastic structural members.

In its eight sections, the chapter introduces into general terminology, mathematics and dynamics of elastic multi-body systems. It covers elastic joints and transmissions as well as robots with elastic links. Advanced topics guiding the reader to current and emerging fields of research include considerations and criteria for intentional elastic robot design as well as variable impedance actuation.

1 Introduction

The research on elastic robots dates back to the early seventies [2]. It has originally been driven by the desire to find solutions for the compensation of undesired side effects of elasticity. The idea of intentionally introducing elastic components came up in the nineties [29]. Since then elastic robotics has become a very active field of research. The field ranges from moderately compliant robots with force/torque sensing capabilities for safe physical human-robot interaction [43], to very soft worm like continuum robots [30]. Research towards variable impedance actuators with online-controllable physical elasticity properties [40] aims at energy efficient

J. Malzahn (✉)
Istituto Italiano di Technologia, Genoa, Italy
e-mail: joern.malzahn@tu-dortmund.de; jorn.malzahn@iit.it

F. I. Muster · T. Bertram
Institute of Control Theory and Systems Engineering, TU Dortmund, Dortmund, Germany
e-mail: freia.muster@tu-dortmund.de; torsten.bertram@tu-dortmund.de

© Springer Nature Switzerland AG 2025
B. Siciliano (ed.), *Robotics Goes MOOC*,
https://doi.org/10.1007/978-3-319-75823-7_4

and impact resilient robots capable of both, soft and gentle physical interactions as well as powerful explosive movements.

1.1 Is This Robot Elastic?

The material properties and geometry intrinsically define the elastic behaviour of solid objects such as the links and joint shafts of robots. Along with these intrinsic parameters, there are extrinsic requirements in terms of speed and accuracy defined by the intended robot application.

A massive industrial robot might at first glance be considered rigid. This impression might hold, if this robot is used to place comparably lightweight work-pieces on a conveyor belt. The same robot might, however, display clearly noticeable elastic deformations when manipulating heavier work-pieces. An application designer aiming to realise a high precision assembly task has higher accuracy requirements and might consider the same robot "too elastic" for the new task.

The video [23] represented by Fig. 1 demonstrates the actually present elasticity in a supposedly rigid industrial robot. Moderate forces manually applied to the end effector visibly displace the robot from its original position. The elastic behaviour predominantly occurs due to the finite rigidity of gears and transmission belts. However, the elasticity can also be measured in the robot links. The experiments in the video demonstrate that the de-facto present link deflection can be measured and exploited for active impedance control of the supposedly rigid robot.

Fig. 1 Elastic deformations on a supposedly rigid industrial robot arm. A video demonstrating gravity compensated low gain active impedance control of this robot based on link strain measurements is available online (youtube link). Courtesy of Institute of Control Theory and Systems Engineering, TU Dortmund University

We learn from this example that the proper answer to the question *"Is this robot elastic?"* must be: *"Sure, it always is!"*.

1.2   A (Not So) Formal Definition of Elastic Robots

In literature, robots are commonly considered elastic, if they are made by so-called soft materials characterized by low values for the Young's modulus (see next Sect. 2). A formal definition of elastic robots based on a single material property is misleading. This should be clear after the example in the previous subsection. To give another tangible example: an aluminium bar can be quite stiff with little bending under a certain load. The same amount of aluminium shaped into a foil will bend quite easily under the same force. In general, the material strength and a body's geometry equally contribute to the stiffness properties of an object, as the Young's modulus. This will be detailed in Sect. 2.4.

A definition of the common notion of an elastic robot is nevertheless helpful. The incorporation of structural elasticity into the equations of motion of a robot is associated with substantial efforts and potentially time consuming and error prone. Also the computational complexity might increase such that simulation times are significantly prolonged and also embedded control becomes demanding. From the practical perspective it is therefore important to know, if a robot has to be considered elastic or not. While a formal definition of the term "'Elastic Robot'" is difficult, as per the examples given above, the practitioner might be satisfied by the following not so formal definition:

A robot is considered an elastic robot, if at least one of the following perspectives are true:

Accuracy Perspective: The purely rigid-body kinematics and dynamics cannot model the robot's actual behaviour with the accuracy requested by the application. As a consequence, also the control accuracy and response time remains below the requirements due to deformations and oscillations. Considering elasticity of one or multiple of the structural members improves the model accuracy and thereby the control performance. This is a perspective of a classical industrial robot automation scenario, where the robot setup is optimized for positioning accuracy and minimum cycle times.

Force Measurement Perspective: The elastic deformations are measurable and the load-dependent deformation characteristics can be calibrated, such that the forces and torques acting on the elastic members can be estimated within the accuracy requirements of a given force/torque control application. This is a perspective of modern collaborative robots intended for physical interaction with humans and objects in less structured environments.

Resonance Perspective: Structural resonances of the robot can be excited during nominal robot operation. This is again a perspective of a classical industrial robot automation scenario under minimization of cycle times, but also a more modern perspective of efficient passive dynamic walking robots.

Energy Saving Perspective: A relevant amount of the work done by the robot during cyclic task execution is temporarily stored in the elastic deformation of structural members of the robot. This is a modern perspective on efficient robots exploiting the intrinsic robot dynamics for periodic and explosive movement tasks.

Impact Absorption Perspective: A relevant amount of impact energy is instantaneously transformed into elastic deformation of structural members. This is a perspective mostly from walking robots with intentionally introduced series elastic actuators.

Force Equilibration Perspective The structural members nestle up against any object in contact with the robot, which leads to an intrinsic equilibration of forces across a continuous contact area. This is a modern soft-robotics perspective.

1.3 Chapter Outline

The chapter starts from basic concepts of elasticity in solid bodies in the next Sect. 2. It then introduces the dynamics of single mass-spring-damper systems along with chains of multiple mass-spring-damper systems in Sect. 3. Such systems are the fundamental building blocks often used to understand more complex elastic multi-body systems. The chapter further details the modelling of individual elasticity in the robot joints and transmission systems in Sect. 4 as well as elasticity the robot links in Sect. 5. The general equations of motion for multi-degree-of-freedom robots with elastic elements and common model simplifications are discussed in Sect. 6.

The last two sections represent and outlook towards more advanced topics and further reading on considerations for the intentional elastic element design in Sect. 7 and Variable Impedance Actuation principles in Sect. 8.

2 Elasticity of Objects

The term *elasticity* describes the reversible deformation of objects under the influence of static and dynamic forces acting on them. Irreversible deformations of objects due to static and dynamic forces acting on them are called *plastic* deformations. The magnitude, frequency, and decay of elastic deformations are characterised by the object material, geometry, and mechanical support.

2.1 Stress and Strain

It is desirable to characterise the elastic behaviour of engineering materials independent of the eventual component geometry or size. Therefore, planar load and deformations of a volume element of a body are normalised by the cross section area or element length. Forces f passing through the cross-section, as indicated in Fig. 2a, stress the component material inversely proportional to the cross-section area A. An infinitesimal force $\mathrm{d}f$ acting through the cross-section area $\mathrm{d}A$ of a volume element causes the stress σ:

$$\mathrm{d}f = \sigma \, \mathrm{d}A . \tag{1}$$

For homogeneous bodies, normal stress $\sigma_\perp$ is created by the force component $f_\perp$ normal to the cross-section and a force component $f_\parallel$ in parallel to the cross-section causes shear stress $\sigma_\parallel$, such that:

$$\sigma_\perp = \frac{f_\perp}{A} , \quad \text{and} \quad \sigma_\parallel = \frac{f_\parallel}{A} . \tag{2}$$

The stressed component undergoes deformations that change length l and shear sections of the material. Thus, when examining a physical body under stress, translational and rotational deformations can be observed. Strain ε illustrated in Fig. 2b is the change in length Δl normalised by the undeformed length l_0:

$$\varepsilon = \frac{\Delta l}{l_0} . \tag{3}$$

The shear ι illustrated in Fig. 2c denotes the small angular distortion:

$$\iota = \arctan\left(\frac{\Delta x}{y}\right) \approx \frac{\Delta x}{y} . \tag{4}$$

Therein, Δx and y denote the relative displacement and the associated height, respectively. Non-planar stress, strain and shear in three dimensional space can be described using tensor algebra as detailed e.g. in [16].

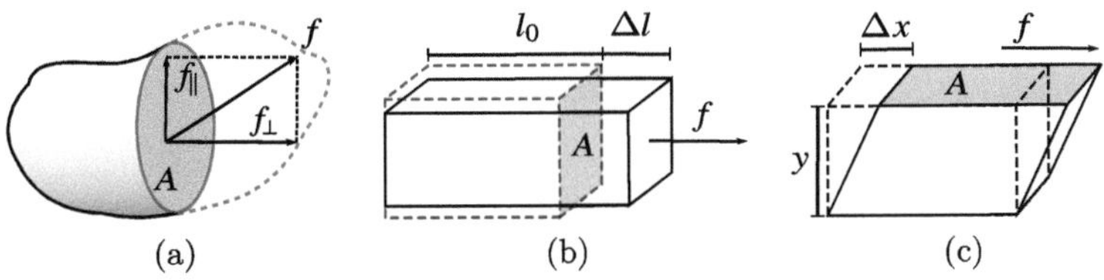

Fig. 2 Normal force $f_\perp$ (**a, b**) and shear force $f_\parallel$ (**a, c**) creating stress in the cross section area A

2.2 Elastic Moduli

The elastic *moduli* represent the rigidity of a material independent of an eventual component geometry. The Young's modulus, also known as E-modulus E, characterises the strain along an axis in reaction to tensile forces acting along that axis. Normalising the tensile forces by the cross-section area of the considered volume element, this yields the stress-strain expression:

$$\sigma_\perp = E\,\varepsilon\,, \tag{5}$$

also referred to as *Hooke's Law*. Analogously, the shear modulus G describes the volume-preserving material deformation in response to opposing forces in parallel to the material cross-section. It is defined as the quotient of shear stress and shear strain:

$$\sigma_\| = G\,\iota\,. \tag{6}$$

The larger the moduli E and G, the smaller is the material deformation under identical stress.

While the stress σ in a material in general develops as a nonlinear function of the strain ε, the direct proportionality is a good approximation for low strain levels in most materials. Figure 3 qualitatively displays the stress-strain relations of various materials. The linear stress-strain relationship holds in particular for ductile metals up to the yield strain ε_y.

Typically, the cross-section area of the volume element shrinks while the component deforms with increasing stress. If the material is strained parallel to one axis $\varepsilon_\|$, it contracts along perpendicular axes and shows the negative strain $-\varepsilon_\perp$. Both quantities are related by the Poisson number ν_ε:

$$\varepsilon_\perp = -\nu_\varepsilon\,\varepsilon_\|\,, \tag{7}$$

which is another material characteristic parameter. The terms *engineering stress* and *engineering strain* refers to contraction being neglected for simplicity, while the actual *true* stress or *true* strain is computed considering the variation in cross-section.

2.3 Material Strength and Fatigue

Another important type of material parameters describing robot component elasticity are static and dynamic stress and strain limits defining the "strength" of a material to withstand loads. Within certain bounds, the limits can be influenced by thermal treatment of the component. The understanding of strength limits is paramount for

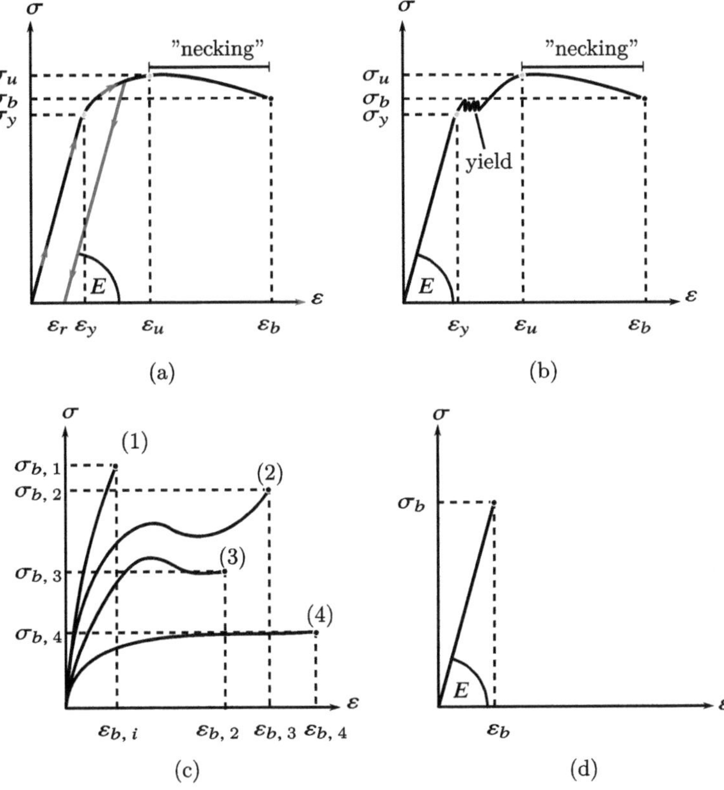

Fig. 3 Stress-strain curves for different materials (redrawn from [31]). (**a**) Ductile metal with smooth yield transition. (**b**) Ductile metal with pronounced yield plateau. (**c**) Polymers. (**d**) Brittle materials

elastic robot component material selection and design. A brief introduction into the terminology and physical principles is given in this subsection. For an in-depth study of the topic, the reader is referred to [32].

2.3.1 Static Strength Limits

When reaching the linear elastic yield stress limit σ_y, the strain begins to no longer develop in direct proportion to the stress. Further increasing of the yield strength leads to irreversible nonlinear plastic deformation. This is illustrated in Fig. 3a, where after stress release, a residual strain ε_r remains. The highest stress that can build up in the material is the ultimate stress limit σ_u. Beyond that point, any effort to increase the stress releases into a material deformation. "Necking" can occur which is a rapid elongation of the material along with a pronounced cross-sectional contraction. Eventually, breakage is to be expected at the breakage strain ε_b. Figure 3 qualitatively illustrates typical stress-strain curves for different materials. Figure 3a and b indicate the yield, ultimate and failure stresses as well as strains for ductile metals with and without pronounced yield transition plateau. The more complex curves for different polymers are exemplified in Fig. 3c. Brittle materials display a very linear stress-strain curve, whereas their yield stress and ultimate stress coincide with the breakage point.

2.3.2 Dynamic Strength Limits

The fatigue strength limit is a stress amplitude σ_f that, if periodically applied for a specified number of load cycles, leads to a highly probable material breakage.

The impact strength is measured by the resilience modulus U_y, which is the strain energy per unit volume absorbed by the material before creating a permanent deformation. It computes as the integral of the stress σ over the strain ε up to the yield point:

$$U_y = \int_0^{\varepsilon_y} \sigma \, d\varepsilon \,. \tag{8}$$

For linear stress-strain relationship and pure tensile stress $\sigma_\parallel$, this turns into:

$$U_y = \frac{1}{2} E \, \varepsilon_y^2 \,, \quad \text{or} \quad U_y = \frac{1}{2} \frac{1}{E} \sigma_y^2 \,. \tag{9}$$

Analogously, the toughness modulus U_f can be defined by the strain energy per unit volume absorbed by the material before a fracture occurs, through replacing the integral bound ε_y by ε_f:

$$U_f = \int_0^{\varepsilon_f} \sigma \, d\varepsilon \,. \tag{10}$$

The relations equivalently hold for shearing deformations with $\sigma = \sigma_\parallel$, where the shear modulus G substitutes the Young's modulus E.

2.4 Component Geometry

The previous subsections have addressed material specific properties of elastic robot components independent of the component's geometry. The cross-sectional geometry of an elastic component defines its area moment of inertia, which expresses the property of a shape to resist a deformation. This will be further detailed within the scope of elastic joint shafts in Sect. 4 and within the scope of elastic robot links in Sect. 5. It explains for instance, why an I-beam (a beam with an I-shaped cross-section) can be lighter than a solid bar of the same material while displaying equal stiffness in the loaded direction.

In extension to material strength limits, the geometry defines the amount of energy that can be absorbed within the volume of the elastic component. This is quantified by the energy capacity W_y of the equivalent spring element representing the component:

$$W_y = \int_0^{x_{max}} f \, dx \, . \tag{11}$$

As an example, consider a homogeneous uniform spring with constant cross-section area $A(x) = A$ under pure normal loading $f = f_\perp$. The stress in the cross-section A is pure normal stress $\sigma = \sigma_\perp$. The spring is loaded in the linear elastic regime, such that Hooke's law applies.

Under these conditions, the integral for the energy capacity W_y can be expanded by the cross-section area A and the undeformed spring length l_0:

$$W_y = \int_0^{x_{max}} \frac{f}{A} \, A l_0 \, \frac{dx}{l_0} \, . \tag{12}$$

The ratio $\frac{f}{A}$ is the constant stress σ and the product $A l_0$ is the undeformed spring's volume V. The normalised infinitesimal displacement $\frac{dx}{l_0}$ can be replaced by the infinitesimal strain $d\varepsilon$ and, after applying Hooke's law (5), the energy capacity W_y:

$$W_y = \frac{V}{E} \int_0^{\sigma_y} \sigma \, d\sigma = \frac{V}{2E} \sigma^2 = V U_y \, . \tag{13}$$

The energy capacity of a uniform homogeneous spring under pure normal load is product of the purely material specific resilience modulus U_y and the geometry dependent spring volume V.

Again, the energy capacity can analogously be computed for pure shearing deformations with $\sigma = \sigma_\parallel$ and the shear modulus G substituting the Young's modulus E.

For springs with non-uniform cross-sections of arbitrary spring geometries, the integral (11) must be solved by substitution of the force f according to the stress distribution (1) and subsequent integration over the spring volume. To simplify comparison of different spring geometries for the same material, the volumetric efficiency factor η_A is introduced in the computation of the spring's energy capacity [13]:

$$W_y = \eta_A \frac{V}{2E} \sigma^2 \, . \tag{14}$$

The volumetric efficiency factor η_A determines how well a spring shape design uses the space it occupies to store energy and references it to a solid uniform spring of

Table 1 Stiffness, energy absorption and volumetric efficiency for spring elements of different geometry (source: [13])

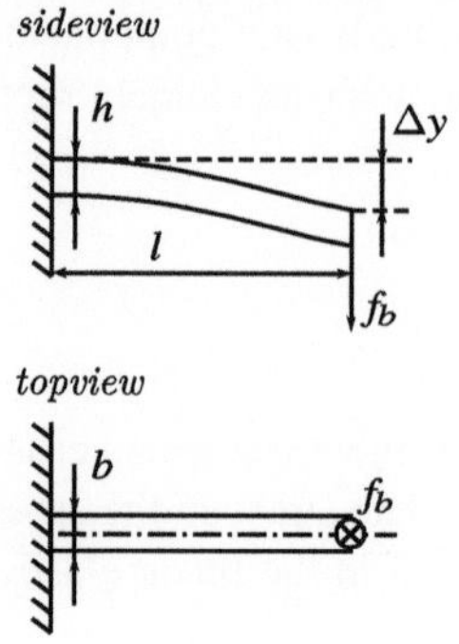

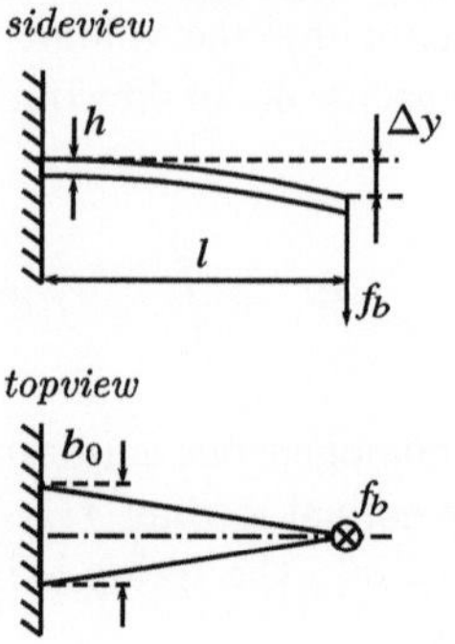

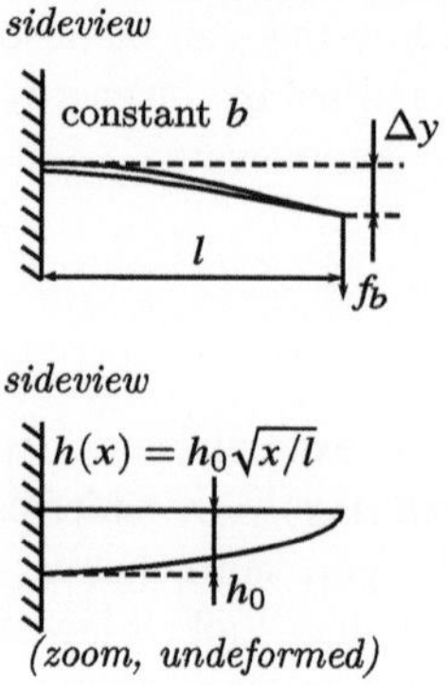

Shape:	Rectangular	Triangular	Parabolic
k	$\frac{bh^3}{4}\frac{E}{l^3}$	$\frac{b_0 h^3}{6}\frac{E}{l^3}$	$\frac{b h_0^3}{8}\frac{E}{l^3}$
W_y	$\frac{bh}{18}\frac{l}{E}\sigma_y^2$	$\frac{b_0 h}{12}\frac{l}{E}\sigma_y^2$	$\frac{b h_0}{9}\frac{l}{E}\sigma_y^2$
η_A	$\frac{1}{9}$	$\frac{1}{3}$	$\frac{1}{3}$

the same material and volume. This renders it a purely geometric characterisation of a spring. Examples for different spring geometries are provided in Table 1.

3 Dynamics of Elastic Elements

3.1 *The Single Mass-Spring-Damper System*

Structural members of elastic robots are often modelled as linear mass-spring-damper systems such as depicted in Fig. 4. This is because the members have a wide linear deflection range, deformations are small or linearised about an operating point. Very frequently the linear mass-spring-damper system is a good initial approximation to iteratively refine models and designs. Especially in robot control, linear approximations represent a trade-off between model complexity and model accuracy. This section concentrates on linear motions. However, the analogous equations exist with interchanged variables for rotary mass-spring-damper systems.

Fig. 4 Schematic of a linear mass-spring-damper system with stiffness k, viscous damping d and mass m

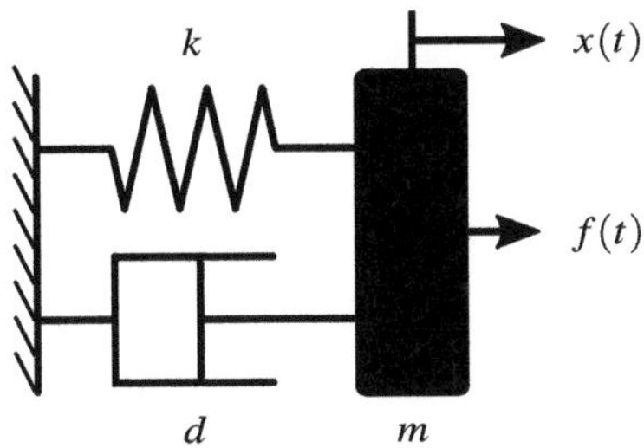

3.1.1 Equation of Motion

The differential equation of motion for a single mass-spring-damper system derives to:

$$f(t) = m\,\ddot{x}(t) + d\,\dot{x}(t) + k\,x(t)\,. \tag{15}$$

In literature, the equation is often presented in the parametric form:

$$K_P\,f(t) = \ddot{x}(t) + 2\zeta\,\omega_0\,\dot{x}(t) + \omega_0^2\,x(t)\,. \tag{16}$$

This form results from (15) after division by the mass m and substituting:

$$\omega_0 = \sqrt{\frac{k}{m}}\,, \quad \zeta = \frac{d}{2k}\omega_0 = \frac{d}{2\sqrt{k\,m}}\,, \quad \text{and} \quad K_P = \frac{1}{m}\,. \tag{17}$$

The parameter ω_0 is the so-called natural frequency of the undamped free system motion $x(t)$ in response to initial conditions. The apparent damping d varies with the square of the natural frequency as visible in Fig. 5a. The variation is determined by the dimensionless damping ratio ζ. The damping ratio is analysed in more detail in Sect. 3.1.5.

3.1.2 Laplace-Domain Transfer Function

It is convenient to treat the oscillatory dynamics of elastic robotic components in the Laplace domain, where (16) corresponds to:

$$x(s) = \frac{K_P}{s^2 + 2\,\zeta\,\omega_0\,s + \omega_0^2}\,f(s) + \ldots$$

$$+ \frac{s + 2\,\zeta\,\omega_0}{s^2 + 2\zeta\,\omega_0\,s + \omega_0^2}\,x_0 + \frac{1}{s^2 + 2\,\zeta\,\omega_0\,s + \omega_0^2}\,\dot{x}_0\,. \tag{18}$$

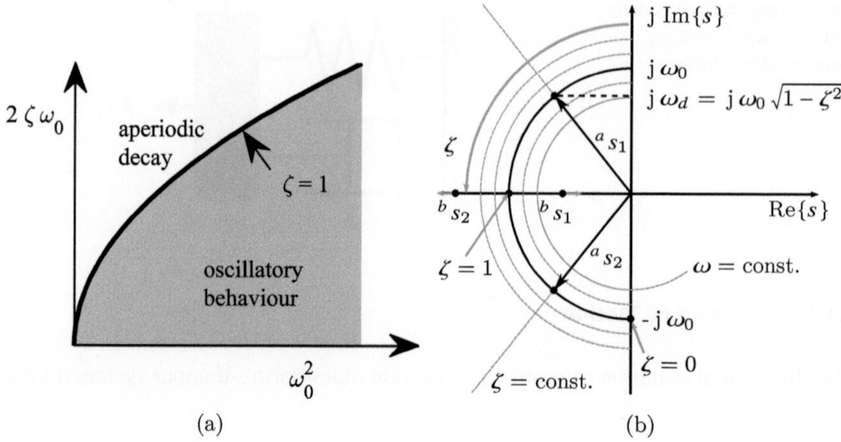

Fig. 5 (**a**) Damping-frequency parameter plane illustrating that the apparent damping is frequency dependent. The dependency is characterised by the damping ratio ζ. (**b**) Complex s-plane with relating locations of roots $s_{1,2}$ to damping ratios ζ and characteristic frequencies ω_0 (redrawn based on [26])

Therein, $x_0 = x(t = 0)$ and $\dot{x}_0 = \dot{x}(t = 0)$ denote the initial displacement and velocity. For vanishing initial conditions, we obtain the Laplace domain transfer function of the system:

$$G(s) = \frac{x(s)}{f(s)} = \frac{K_P}{s^2 + 2\zeta\,\omega_0\,s + \omega_0^2}\,.\tag{19}$$

The poles of (19) or, respectively, the complex roots of the characteristic equation to (15), have the form:

$$\begin{matrix} s_1 \\ s_2 \end{matrix} = -\zeta\,\omega_0 \pm \omega_d\,,\qquad \text{with}\qquad \omega_d = \omega_0\,\sqrt{\zeta^2 - 1}\,.\tag{20}$$

In analogy to the motion frequency ω_0 of the undamped free system, the parameter ω_d denotes the damped free system motion $x(t)$ in response to initial conditions.

Figure 5b shows the location of the roots (20) in the complex s-plane with respect to $0 \le \zeta$. Lines of equal damping ratio ζ are rays pointing away from the origin of the complex s-plane. The lines for equal values of the natural frequencies are concentric circles around the origin.

3.1.3 Time-Domain Responses

Either by directly solving the differential equation (15), or by using correspondence tables to transform (18) back into the time-domain, the general time response of the mass spring damper system to initial conditions x_0 and $\dot{x}_0$ as well as the excitation force $f(t)$ computes to:

$$x(t) = x_0 \, \frac{\omega_0}{\omega_d} \, \mathrm{e}^{-\zeta \omega_0 \, t} \cos(\omega_d \, t - \psi) + \ldots$$

$$+ \frac{\dot{x}_0}{\omega_0} \sin(\omega_d \, t) + \frac{K_P}{\omega_d} \int_0^t f(t') \, \mathrm{e}^{-\zeta \omega_0 (t - t')} \sin\left(\omega_d \left(t - t'\right)\right) \mathrm{d}t' . \tag{21}$$

The phase angle ψ is defined as $\psi = \arctan\left(\zeta \, \frac{\omega_0}{\omega_d}\right)$. The integral in the last summand is known as the convolution integral.

3.1.4 Responses to Common Excitations

Responses to external excitations f are most conveniently analysed in the Laplace domain using (18). The time domain solution requires the cumbersome (piecewise) solution of the convolution integral in (21). In the Laplace domain, the convolution corresponds to a simple multiplication of the transfer function with the Laplace representation of the excitation signal. The forward and inverse transitions between the Laplace and time domain can be done using correspondence tables.

The excitation signals commonly used to analyse elastic systems are the Dirac impulse and the step function. Following either mathematical approach, the closed-form response of a damped mass spring damper system to a Dirac impulse under vanishing initial conditions can be derived to:

$$g(t) = \frac{\omega_0}{\sqrt{1 - \zeta^2}} \, \mathrm{e}^{-\zeta \omega_0 t} \sin(\omega_d \, t), \quad t \geq 0 . \tag{22}$$

Similarly, the closed-form response to a unit-step function is:

$$h(t) = \left(1 - \mathrm{e}^{-\zeta \omega_0 t} \left(\frac{\zeta \omega_0}{\omega_d} \sin(\omega_d \, t) + \cos(\omega_d \, t)\right)\right) . \tag{23}$$

Both equations are valuable for understanding the potentials and challenges associated with elastic robot components.

3.1.5 Influence of Damping

The damping ratio ζ apparently has a significant impact on the dynamics of elastic robots. The absence of damping can simplify the mathematical analysis through simpler equations. However, an amount of intrinsic damping, and thereby rapid decay of structural oscillations, might be desired for a particular application. We classify five different cases of system behaviour based to the damping ratio ζ.

Growing Without Bounds: $\zeta < 0$

For negative damping ratios the characteristic values s_1 and s_2 have positive real parts and the system response to any excitations exponentially grows without bounds. This type of unstable system behaviour does not exist in passive mass-spring-damper systems and is almost never desired. A common example for this case is a control loop accidentally applied to a mass-spring-damper system.

The Undamped Case: $\zeta = 0$

In the absence of damping, the characteristic values s_1 and s_2 in (20) are purely imaginary (see Fig. 5b). Their absolute value equal the natural frequency ω_0. System responses of the undamped system are summarised and illustrated in Table 2.

The Underdamped Case: $0 < \zeta < 1$

In the underdamped case, the roots (20) have real and imaginary parts. The roots $^a s_{1,2}$ in Fig. 5b exemplify these roots. The oscillation frequency ω_d of

Table 2 Time-domain responses of the **undamped** single mass-spring-damper system

No Damping	$\zeta = 0$
Initial Excitation	$x(t) = C \cos(\omega_0 t - \varphi),\ C = \frac{1}{m}\sqrt{x_0^2 + \left(\frac{\dot{x}_0}{\omega_0}\right)^2},\ \varphi = \arctan\left(\frac{\dot{x}_0}{x_0\,\omega_0}\right)$ $\hat{t} = \frac{i\,\pi + \varphi}{\omega_0}$, with $i \in \mathbb{N}$ and $|x(\hat{t})| = C$
Impulse Response	$g(t) = \frac{\omega_0}{k}\sin(\omega_0 t),\ \hat{t} = \frac{(2i+1)\pi}{2\,\omega_0},\ |\hat{g}| = |g(\hat{t})| = \frac{\omega_0}{k}$
Step Response	$h(t) = \frac{1}{k}(1 - \cos(\omega_0 t)),\ \hat{t} = \frac{(2i+1)\pi}{\omega_0},\ \hat{h} = h(\hat{t}) = \frac{2}{k}$
$x(t)$:	$g(t):$ $h(t):$

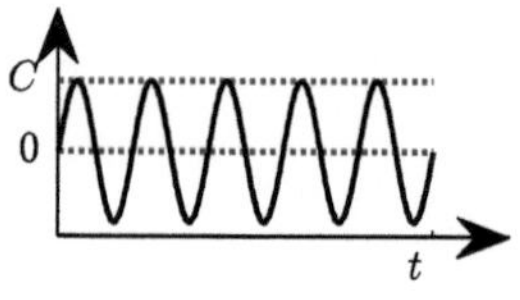

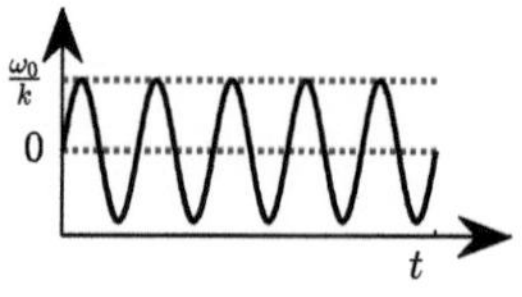

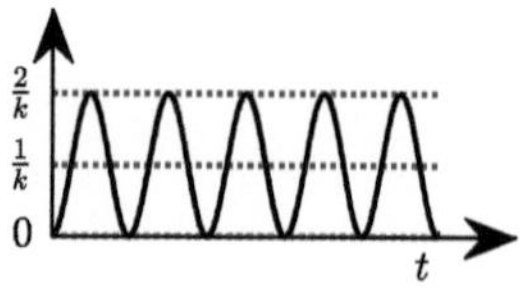

Table 3 Time-domain responses of the **underdamped** single mass-spring-damper system

Underdamping	$0 < \zeta < 1$
	$x(t) = C\, e^{-\zeta\,\omega_0\,t}\cos(\omega_d\,t - \varphi),$
Initial Excitation	$C = \sqrt{x_0^2 + \left(\dfrac{\zeta\omega_0 x_0 + \dot{x}_0}{\omega_d}\right)^2},\ \varphi = \arctan\dfrac{\zeta\omega_0 x_0 + \dot{x}_0}{\omega_d\, x_0}$
	$\hat{t} = \dfrac{1}{\omega_0\sqrt{1-\zeta^2}}\log\left(-\dfrac{\sqrt{1-2\zeta^2}}{\zeta-\sqrt{1-\zeta^2}}\right) + \tfrac{1}{2}\left(\log(1-a) - \log(1+a)\right)$
	with $a = \dfrac{\dot{x}_0 + \omega_0\, x_0\, \zeta}{\omega_0\, x_0\sqrt{1-\zeta^2}}$
	$g(t) = \dfrac{\omega_d}{k}\, e^{-\zeta\omega_0 t}\sin(\omega_d\, t),$
Impulse Response	$\hat{t} = -\dfrac{a}{\sqrt{1-z^2}},\ g(\hat{t}) = \hat{g} = \dfrac{\omega_d}{k}\, e^{\frac{\omega_0}{\sqrt{1-\zeta^2}}a}\sin\left(\dfrac{\omega_d}{\sqrt{1-\zeta^2}}a\right)$
	and $a = \log\left(\dfrac{\sqrt{\omega_0^2\,\zeta^2 + \zeta^2 - 1}}{\omega_0\,\zeta + \sqrt{1-\zeta^2}}\right)$
	$h(t) = \tfrac{1}{k}\left(1 - e^{-\zeta\omega_0 t}\left(\dfrac{\zeta\omega_0}{\omega_d}\sin(\omega_d\, t) + \cos(\omega_d\, t)\right)\right),$
Step Response	$\hat{t} = \dfrac{\pi}{\omega_d},\ h(\hat{t}) = \hat{h} = \tfrac{1}{k}\left(1 + e^{-\pi\frac{\zeta}{\sqrt{(1-\zeta^2)}}}\right)$

$x(t):$ $g(t):$ $h(t):$

the corresponding motion is the length of the projection of these roots onto the imaginary axis. The system oscillates with exponentially decaying amplitude. The decay is characterised by the logarithmic decrement $\zeta\,\omega_0$. Closed-form system responses of the underdamped system are summarised and illustrated in Table 3.

The Critically Damped Case: $\zeta = 1$

For the critically damped case, the roots (20) are identical and real-valued. This is exemplified in Fig. 5b.

This solution represents the fastest transition from the initial condition to the equilibrium without oscillations. Critically damped system responses are summarised and illustrated in Table 4.

The Overdamped Case: $\zeta > 1$

For the overdamped case, the roots (20) are real-valued and distinct as indicated by $^b s_{1,2}$ in Fig. 5b. The response to any excitation is an aperiodic transition from the

Table 4 Time-domain responses of the **critically damped** single mass-spring-damper system

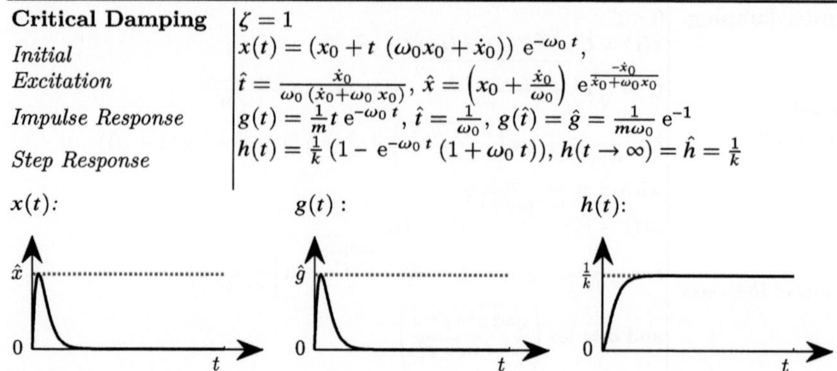

Critical Damping	$\zeta = 1$
Initial Excitation	$x(t) = (x_0 + t\,(\omega_0 x_0 + \dot{x}_0))\,e^{-\omega_0 t},$ $\hat{t} = \frac{\dot{x}_0}{\omega_0\,(\dot{x}_0 + \omega_0 x_0)},\ \hat{x} = \left(x_0 + \frac{\dot{x}_0}{\omega_0}\right)e^{\frac{-\dot{x}_0}{\dot{x}_0 + \omega_0 x_0}}$
Impulse Response	$g(t) = \frac{1}{m} t\,e^{-\omega_0 t},\ \hat{t} = \frac{1}{\omega_0},\ g(\hat{t}) = \hat{g} = \frac{1}{m\omega_0}\,e^{-1}$
Step Response	$h(t) = \frac{1}{k}\left(1 - e^{-\omega_0 t}\,(1 + \omega_0 t)\right),\ h(t \to \infty) = \hat{h} = \frac{1}{k}$

$x(t):$ $g(t):$ $h(t):$

Table 5 Time-domain responses of the **overdamped** single mass-spring-damper system

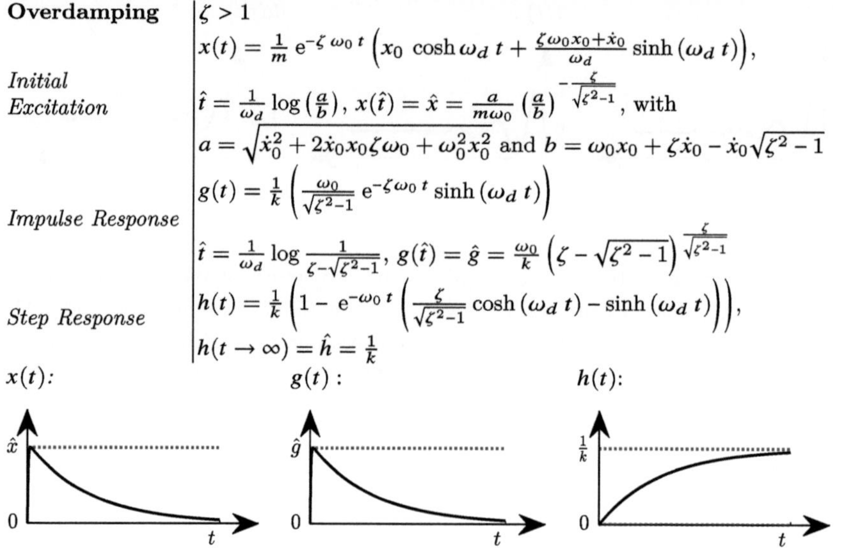

Overdamping	$\zeta > 1$
Initial Excitation	$x(t) = \frac{1}{m}\,e^{-\zeta\,\omega_0 t}\left(x_0 \cosh \omega_d t + \frac{\zeta \omega_0 x_0 + \dot{x}_0}{\omega_d}\sinh(\omega_d t)\right),$ $\hat{t} = \frac{1}{\omega_d}\log\left(\frac{a}{b}\right),\ x(\hat{t}) = \hat{x} = \frac{a}{m\omega_0}\left(\frac{a}{b}\right)^{-\frac{\zeta}{\sqrt{\zeta^2-1}}},$ with $a = \sqrt{\dot{x}_0^2 + 2\dot{x}_0 x_0 \zeta \omega_0 + \omega_0^2 x_0^2}$ and $b = \omega_0 x_0 + \zeta \dot{x}_0 - \dot{x}_0\sqrt{\zeta^2 - 1}$
Impulse Response	$g(t) = \frac{1}{k}\left(\frac{\omega_0}{\sqrt{\zeta^2-1}}\,e^{-\zeta \omega_0 t}\sinh(\omega_d t)\right)$ $\hat{t} = \frac{1}{\omega_d}\log\frac{1}{\zeta - \sqrt{\zeta^2-1}},\ g(\hat{t}) = \hat{g} = \frac{\omega_0}{k}\left(\zeta - \sqrt{\zeta^2-1}\right)^{\frac{\zeta}{\sqrt{\zeta^2-1}}}$
Step Response	$h(t) = \frac{1}{k}\left(1 - e^{-\omega_0 t}\left(\frac{\zeta}{\sqrt{\zeta^2-1}}\cosh(\omega_d t) - \sinh(\omega_d t)\right)\right),$ $h(t \to \infty) = \hat{h} = \frac{1}{k}$

$x(t):$ $g(t):$ $h(t):$

initial condition to the equilibrium. Table 5 summarises and illustrates overdamped system responses.

3.2 Multi-Degree-of-Freedom Systems

Elastic robots are often comprised of multiple elastic members. Together they form chains of mass-spring-damper systems. More complex elastic members may require a more detailed consideration with multiple masses, springs and dampers in series, parallel or branched connections to obtain accurate models. This subsection demonstrates, how solutions for series chains of mass-spring-damper systems can be obtained by an approach called modal analysis. The idea is to transform a system of coupled differential equations into a system of independent modal equations using the modal matrix as a coordinate transformation matrix. The approach is briefly introduced here as a recipe. A more detailed derivation can be found in [26].

1. *Derive Equations of Motion*

   Using e.g. the Newton-Euler or Euler-Lagrange formalism the equations of motion can be derived in the matrix form:

$$\mathbf{f}(t) = \mathbf{M}\,\ddot{\mathbf{x}}(t) + \mathbf{D}\,\dot{\mathbf{x}}(t) + \mathbf{K}\,\mathbf{x}(t)\,. \tag{24}$$

   The matrices $\mathbf{M}$, $\mathbf{D}$ and $\mathbf{K}$ are the mass, damping and stiffness matrices. All three matrices are symmetric $n \times n$ matrices. The symmetry is a fundamental property of oscillating systems. Moreover, the mass matrix $\mathbf{M}$ is positive definite. The stiffness matrix $\mathbf{K}$ is positive definite if the mass cannot undergo rigid body motions. It is positive semi-definite, if at least one mass can perform rigid body motions.

   The vectors $\ddot{\mathbf{x}}$, $\dot{\mathbf{x}}$ and $\mathbf{x}$ are the vectors of acceleration, velocity and position of each of the n masses as exemplified in Fig. 6. With this coordinate choice, the mass matrix $\mathbf{M}$ is diagonal. If instead the spring deflection is chosen as coordinates, the stiffness matrix becomes diagonal, while the mass matrix will be non-diagonal. The initial conditions of all masses are collected by $\mathbf{x}_0$ and $\dot{\mathbf{x}}_0$. The vector $\mathbf{f}$ holds the non-conservative forces exciting the mass elements.

2. *Determine Natural Modes*

   The squares ω_i^2 of the n natural frequencies ω_i are determined as the roots of the characteristic polynomial:

$$\det\left(\mathbf{K} - \omega^2\mathbf{M}\right) = 0 \tag{25}$$

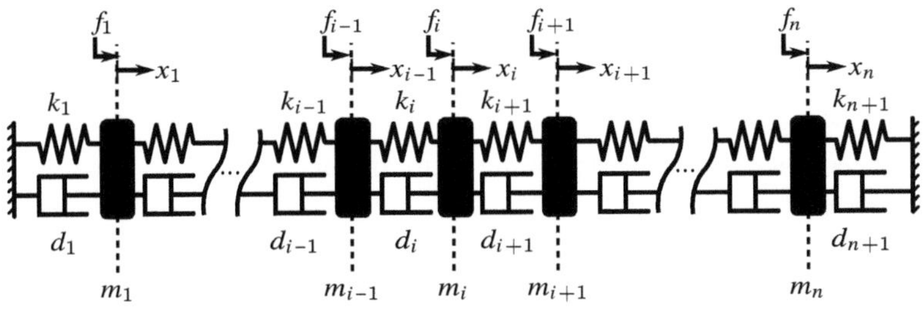

Fig. 6 Multi-degree-of-freedom mass-spring-damper system composed of n masses

to the eigenvalue problem $\left(\mathbf{K} - \omega^2\mathbf{M}\right)\,\boldsymbol{\phi} = \mathbf{0}$. For each natural frequency ω_i, non-trivial modal vectors $\boldsymbol{\phi}_i$, that satisfy the eigenvalue problem can be found. The i-th natural frequency together with the corresponding modal vector fully describe the i-th natural mode.

Single- and two-degree-of-freedom systems typically admit closed form solutions to the eigenvalue problem. For higher degrees-of-freedom, there exist numeric methods based on a Cholesky decomposition, which efficiently determine the modal vectors $\boldsymbol{\phi}$ exploiting the symmetry of the matrices $\mathbf{K}$ and $\mathbf{M}$. Sometimes, only an estimate of the lowest natural frequency ω_1 is of interest. Such an estimate and approximate solutions for higher order natural frequencies can be derived based on the so-called Rayleigh quotient detailed in [26].

3. *Normalise Modal Vectors*

The eigenvalue problem is homogeneous, so that any scalar multiple of any modal vector $\boldsymbol{\phi}_i$ will again satisfy the eigenvalue problem. As an important property, the modal vectors for each natural frequency are mutually orthogonal with respect to the mass matrix $\mathbf{M}$ as well as the stiffness matrix $\mathbf{K}$. A convenient normalisation of the modal vectors is:

$$\boldsymbol{\Phi}^\top \mathbf{M} \boldsymbol{\Phi} = \mathbf{I}\,. \tag{26}$$

This also leads to the result:

$$\boldsymbol{\Phi}^\top \mathbf{K} \boldsymbol{\Phi} = \boldsymbol{\Omega}\,, \qquad \text{with} \qquad \boldsymbol{\Omega} = diag\left\{\omega_i^2\right\}\,. \tag{27}$$

The basis of modal vectors $\boldsymbol{\phi}$ diagonalises the mass and stiffness matrix.

4. *Transform Coordinates*

The displacement vector $\mathbf{x}(t)$ in (24) is the superposition of constant modes $\boldsymbol{\phi}_i$ weighted with their temporal amplitudes $v_i(t)$:

$$\mathbf{x}(t) = \boldsymbol{\phi}_1\,v_1(t) + \boldsymbol{\phi}_2\,v_2(t) + \boldsymbol{\phi}_3\,v_3(t) + \cdots + \boldsymbol{\phi}_n\,v_n(t) = \sum_{i=1}^{n}\boldsymbol{\phi}_i\,v_i(t) = \boldsymbol{\Phi}\,\boldsymbol{v}\,. \tag{28}$$

The temporal amplitudes $v_i(t)$ are also called modal coordinates. The modal vectors form the new coordinate basis $\boldsymbol{\Phi}$ to express the system motion in modal coordinates.

The coordinate basis transforms the vector of non-conservative forces $\mathbf{f}$ into the vector $\boldsymbol{\eta}$ of modal forces:

$$\boldsymbol{\eta} = \boldsymbol{\Phi}^\top \mathbf{f}\,. \tag{29}$$

5. *Diagonalise and Modally Decouple*

Applying the coordinate transformation (28) to (24) and left multiplying with $\mathbf{\Phi}^\top$ yields:

$$\mathbf{\Phi}^\top \mathbf{f}(t) = \mathbf{\Phi}^\top \mathbf{M}\, \mathbf{\Phi}\, \ddot{\mathbf{v}}(t) + \mathbf{\Phi}^\top \mathbf{D}\, \mathbf{\Phi}\, \dot{\mathbf{v}}(t) + \mathbf{\Phi}^\top \mathbf{K}\, \mathbf{\Phi}\, \mathbf{v}(t)\,. \tag{30}$$

a. *No Damping:* If the system exhibits no internal damping, using (26) as well as (27) simplifies (30) to:

$$\boldsymbol{\eta}(t) = \ddot{\mathbf{v}}(t) + \mathbf{\Omega}\, \mathbf{v}(t)\,. \tag{31}$$

This is a system of n decoupled modal equations with a diagonal frequency matrix $\mathbf{\Omega}$. Each modal equation has a structure identical to the undamped simple mass-spring-damper system (16) with solutions for the free motion and common excitation signals summarised in Table 2.

It follows that:

$$v_i(t) = \boldsymbol{\phi}_i^\top \mathbf{M}\, \mathbf{x}(t) \quad \rightarrow \quad \mathbf{v} = \mathbf{\Phi}^\top \mathbf{M}\mathbf{x}(t)\,, \tag{32}$$

$$\omega_i^2\, v_i(t) = \boldsymbol{\phi}_i^\top \mathbf{K}\, \mathbf{x}(t) \quad \rightarrow \quad \mathbf{\Omega}\, \mathbf{v}(t) = \mathbf{\Phi}^\top \mathbf{K}\, \mathbf{x}(t)\,, \tag{33}$$

$$v_i(0) = \boldsymbol{\phi}_i^\top \mathbf{M}\mathbf{x}(0)\,, \quad \text{and} \quad \dot{v}_i(0) = \boldsymbol{\phi}_i^\top \mathbf{M}\dot{\mathbf{x}}(0)\,. \tag{34}$$

b. *Proportional Damping:* While the basis of modal vectors $\mathbf{\Phi}$ diagonalises the mass and stiffness matrices, in general, it cannot diagonalise the damping matrix $\mathbf{D}$. However, if the damping matrix has a special form, i.e. it can be expressed as a linear combination of the mass and the stiffness matrix:

$$\mathbf{D} = p_1\mathbf{M} + p_2\mathbf{K}\,. \tag{35}$$

This leads to the system of decoupled damped modal equations:

$$\ddot{\mathbf{v}}(t) + \mathbf{\Psi}\, \dot{\mathbf{v}}(t) + \mathbf{\Omega}\mathbf{v}(t) = \boldsymbol{\eta}(t)\,, \tag{36}$$

with the diagonal modal damping matrix $\mathbf{\Psi}$:

$$\mathbf{\Psi} = diag\,\{2\,\zeta_i\omega_i\}\,, \quad \text{where} \quad 2\zeta_i\omega_i = p_1 + p_2\,\omega_i^2\,. \tag{37}$$

With proportional damping, each of the independent modal equations has the identical structure to the simple mass-spring-damper system (16).

c. *General Viscous Damping:* In the general case of viscous damping, the modal matrix $\mathbf{\Phi}$ does not diagonalise the damping matrix $\mathbf{D}$. The analytic solution of the equations of motion then requires rewriting and solving (24) in state space form. Eventually, the resulting analytic expressions in state space form must be evaluated numerically. A detailed description of this path is described in [26].

6. *Solve System of Differential Equations*
   After modal decoupling of the equations of motion, each modal equation can be solved separately in time or frequency domain applying the techniques discussed in Sect. 3.1. Eventually, the system motion is obtained by inserting the solutions of the modal equations in (28).
   The i-th modal response to initial conditions is:

$$v_i(t) = v_{0,i} \frac{\omega_i}{\omega_{d,i}} \, e^{-\zeta_i \omega_i \, t} \cos\left(\omega_{d,i} \, t - \psi_i\right) + \frac{\dot{v}_{0,i}}{\omega_i} \sin\left(\omega_{d,i} \, t\right). \tag{38}$$

Applying (28) considering also the equivalences (34) yields:

$$\mathbf{x}(t) = \sum_{i=1}^{n} \boldsymbol{\phi}_i \left(\boldsymbol{\phi}_i^{\top} \mathbf{M} \mathbf{x}(0) \, \frac{\omega_i}{\omega_{d,i}} \, e^{-\zeta_i \omega_i \, t} \cos\left(\omega_{d,i} \, t - \psi_i\right) + \dots \right.$$

$$\left. + \boldsymbol{\phi}_i^{\top} \mathbf{M} \dot{\mathbf{x}}(0) \frac{1}{\omega_i} \sin\left(\omega_{d,i} \, t\right) \right). \tag{39}$$

Similarly, the modal response to external excitations computes to:

$$v_i(t) = \frac{K_P}{\omega_{d,i}} \int_0^t \eta(t') \, e^{-\zeta_i \omega_i \, (t-t')} \sin\left(\omega_{d,i} \, (t - t')\right) \mathrm{d}t'. \tag{40}$$

Applying (28) along with (29) gives the displacement response $\mathbf{x}(t)$ to the external excitations $\mathbf{f}$:

$$\mathbf{x}(t) = \sum_{i=1}^{n} \boldsymbol{\phi}_i \left(\frac{K_P}{\omega_{d,i}} \int_0^t \boldsymbol{\Phi}^{\top} \mathbf{f}(t') \, e^{-\zeta_i \omega_i \, (t-t')} \sin\left(\omega_{d,i} \, (t - t')\right) \mathrm{d}t' \right).$$

$$\tag{41}$$

The total system response in the presence of initial conditions and external excitations is the superposition of (39) and (41).

4   Linear-Elastic Joints

Transmission or reduction elements such as belts, long shafts, cables, harmonic drives, or cycloidal gears introduce joint elasticity. This elasticity must be considered, if the employed robot does not match the accuracy specifications under load dependent deformations or displays prolonged settling times past the application's demand. Joint elasticity is often also intentionally introduced. In such cases, the deflection of the elastic member is measured and together with the knowledge of the member's stiffness permits to infer the mechanical torque delivered to the joint. This section concentrates on rotational robot joints where the elasticity is introduced in the joint shaft. The elastic effects in most other transmission elements can be treated analogously. The principles can be equivalently applied to prismatic elastic joints.

4.1 Joint Shaft Torsion

Elastic joint shafts can often be modelled as uniform rods with cylindrical cross section of inner radius r. The symmetry ensures the sheared cross sections to remain flat and parallel. According to the assumptions, the torsion appears in the linear elastic regime of the shaft material. It emerges from the torques exerted about the x-axis of shaft through the driving actuator as well as the robot link structure, load and interaction forces.

Figure 7 depicts the cylindrical volume element of a shaft. The element has the length $\mathrm{d}x$ and the torsion moment τ_x causes the tangential cross section shear ι:

$$\iota = r \frac{\mathrm{d}\theta_x}{\mathrm{d}x} \; . \tag{42}$$

Using (6), the tangential stress σ in the cross section computes to:

$$\sigma = G\, r \, \frac{\partial \theta_x}{\partial x} \; . \tag{43}$$

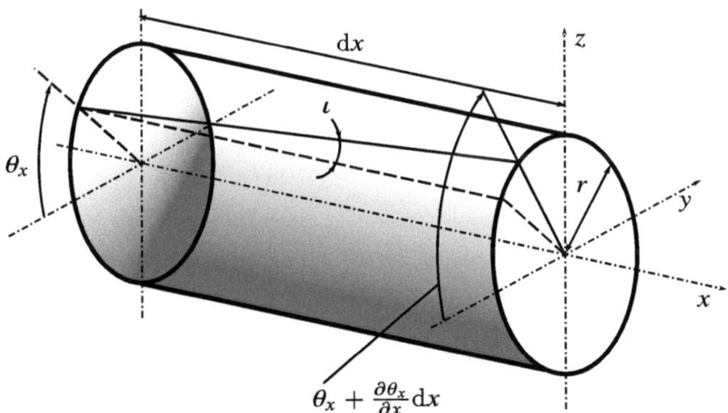

Fig. 7 Shearing of a cylindrical shaft element

Equivalent to (2), the torsional moment τ_x is the torsional stress σ integrated over the cross section area. For the circular cross section with inner radius r_i and outer radius r_o this leads to:

$$\tau_x = 2\pi \int_{r_i}^{r_o} \sigma r^2 \, dr = T \frac{\partial \theta_x}{\partial x} . \tag{44}$$

For a solid shaft we have $r_i = 0$ and $r_o = r$, while for a hollow shaft we would have $r_i \neq 0$. The torsion constant T depends on the shear modulus G of the material and the cross section dimensions:

$$T = \frac{\pi}{2} \left(r_o^4 - r_i^4 \right) G . \tag{45}$$

The torsion rate $\dot{\theta}_x$ is the time derivative of the torsion angle θ_x:

$$\dot{\theta}_x = \frac{\partial \theta_x}{\partial t} \tag{46}$$

and, together with the time derivative of the torsion moment (44), yields the partial differential equation:

$$\frac{\partial \tau_x}{\partial t} = T \frac{\partial \dot{\theta}_x}{\partial x} . \tag{47}$$

The spatial change in the torsion moment τ_x is related to the torsional acceleration $\frac{\partial \dot{\theta}_x}{\partial t}$ of the cross section element via the inertia per unit length $\mathcal{I}$:

$$\frac{\partial \tau_x}{\partial x} = \mathcal{I} \frac{\partial \dot{\theta}_x}{\partial t} . \tag{48}$$

The inertia per unit length for a uniform cylindrical shaft of density ρ computes to:

$$\mathcal{I} = 2\pi \rho \int_{r_i}^{r_o} r^3 \, dr = \frac{\pi}{2} \rho \left(r_o^4 - r_i^4 \right) . \tag{49}$$

Combining (47) and (48) leads to the torsional wave equation:

$$T \frac{\partial^2 \tau_x}{\partial x^2} = \mathcal{I} \frac{\partial^2 \tau_x}{\partial t^2} , \quad \text{and} \quad T \frac{\partial^2 \dot{\theta}_x}{\partial x^2} = \mathcal{I} \frac{\partial^2 \dot{\theta}_x}{\partial t^2} . \tag{50}$$

This second order partial differential equation describes the temporal and spatial propagation of a torsion moment wave $\tau_x(x, t)$ along a shaft. The identical relation describes the propagation of a torsion rate wave $\dot{\theta}_x(x, t)$.

4.2 *Torsion Wave Propagation*

The equations (50) represent two equivalent second order partial differential equations describing the propagation of torsion waves within the elastic joint shaft. The waves propagate with the wave speed:

$$c_\theta = \sqrt{\frac{T}{\mathcal{I}}}.$$

(51)

For rotation symmetric cross sections, the geometric dependencies in this expression cancel and the wave speed reduces to $c_\theta = \sqrt{\frac{G}{\rho}}$.

4.3 *Equivalent Spring Element*

The equations of motion (50) capture the fact that elastic properties are actually never absolutely concentrated in one point, but rather continuously distributed along the entire joint shaft. In practice however, the mass of the elastic joint shaft section is typically negligibly small compared to the effective inertia of the remaining transmission as well as the link and load inertia. Another argument is the high torsion wave propagation speed of commonly used engineering materials for joint shafts in relation to the actual joint shaft length.

As a consequence, the variation of the torsion moment τ_x along the shaft axis in (48) vanishes and (47) is sufficient to derive an equivalent lumped spring element for the elastic shaft.

To exemplify the derivation of the equivalent spring constant, consider a uniform cylindrical shaft as shown by Fig. 8. The shaft is of length l and clamped at $x = 0$, such that $\theta_x(0) = 0$. The shaft is subject to a torque τ, which equals the torsional moment at $x = l$ and effects a torsional deflection $\theta_x(l) = \Delta\theta_x$. The integration of (44) along the shaft length leads to:

$$\theta_x(l) = \Delta\theta_x = \tau \int_0^l \frac{1}{T}\, dx' = \tau\, \frac{l}{T}.$$

(52)

Fig. 8 Torsion of a clamped cylindrical shaft

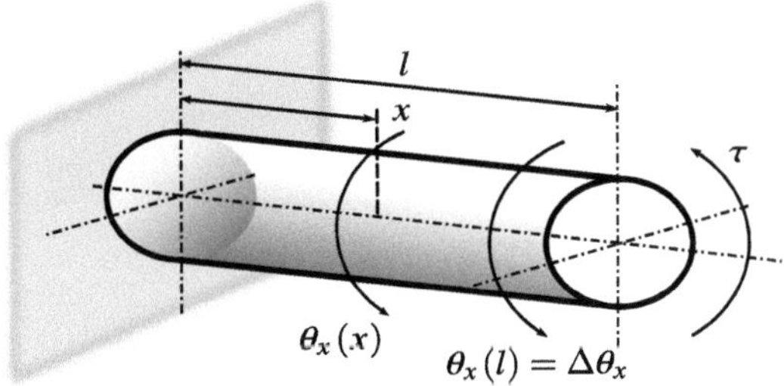

The equivalent torsional spring stiffness of the uniform shaft is therefore:

$$k = \frac{\tau}{\Delta\theta_x} = \frac{T}{l} = \frac{\pi}{2}\left(r_o^4 - r_i^4\right)\frac{G}{l}.$$

(53)

4.4 Non-uniform Joint Shafts

Often, the cross section radius of the elastic joint shaft varies along the longitudinal shaft axis. If the radius is piecewise constant along the shaft axis as sketched in Fig. 9a, a good approximation for the equivalent spring stiffness can be obtained by treating the shaft as a series connection of multiple springs. The equivalent stiffness for a number of n uniform shaft sections computes to:

$$k = \left(\frac{1}{k_1} + \frac{1}{k_2} + \frac{1}{k_3} + \cdots + \frac{1}{k_n}\right)^{-1} = \left(\sum_{i=1}^{n}\frac{1}{k_i}\right)^{-1}.$$

(54)

In a similar way, the stiffness of parallel springs, as exemplified in Fig. 9b, can be grouped to one equivalent stiffness as:

$$k = k_1 + k_2 + k_3 + \cdots + k_n = \sum_{i=1}^{n} k_i.$$

(55)

This is often done to share the same load among different elastic members.

From relation (54), it can be seen that the connection of a very stiff shaft section (large k, vanishing $1/k$) with a rather soft shaft (small k) section essentially results in an overall equivalent shaft stiffness close to the one of the softer section. If the elastic joint shaft cannot be reasonably subdivided into uniform sections, the integration of (43) over the varying cross section area to obtain the torsion moment τ_x becomes more involved and must in general be carried out numerically. It is however a common practice for elastic joint transmission design, to combine an intentionally soft section of the joint shaft with a much stiffer shaft section. This simplifies the elastic joint analysis as given by (54), since the analysis can then focus only on the

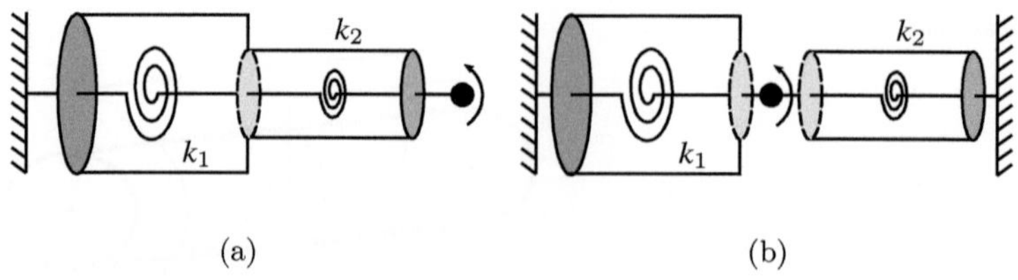

Fig. 9 Piecewise uniform torsional shafts in series (a) and in parallel (b)

soft section. Moreover, the approach enables a more precise mechanical support of the joint shaft and an easier placement of the deflection sensor for accurate torque measurements.

5   Linear-Elastic Links

In conventional industrial manufacturing, robot link elasticity becomes relevant for precise robot manipulation, when very short cycle times shall be achieved or heavy payloads have to be handled. Other applications are long-reach inspection robots or very lightweight arm designs, e.g. arms for nuclear waste tank inspection and maintenance or remote manipulators used in space.

This section derives the equations of motion for elastic links, discusses the challenges related to the underlying boundary value problem and gives an overview of modelling techniques for the dynamics of multi-elastic-link systems.

5.1   Link Bending Equation

In literature, elastic links are typically modelled as Euler-Bernoulli beams. The Euler-Bernoulli beam theory uses to the following assumptions:

1. Beam deflections are small in comparison to the beam length.
2. The neutral fibre always maintains constant length.
3. The shear of a differential beam element is negligible compared to transversal bending deformation.
4. Hooke's law applies. Deflections remain in the linear elastic and reversible material regime.
5. Beam cross-sections remain perpendicular to the neutral fibre, so that any rotation of a differential beam element is negligible with respect to its translation.

Figure 10 illustrates a planar differential beam element of length $\mathrm{d}x$.

5.1.1   Bending Quantities

The deflection of the beam element is measured as the y-coordinate of a coordinate frame at the left end of the element with the x-axis coinciding with the undeformed neutral fibre, as shown in Fig. 10a. Figure 10b shows the same beam element under bending. For the small bending angle θ_b, the spatial derivative of the bending curve simplifies to:

$$\theta_b(x,\ t) \approx \tan\left(\theta_b(x,\ t)\right) = \frac{\partial y\,(x,\ t)}{\partial x}\,. \tag{56}$$

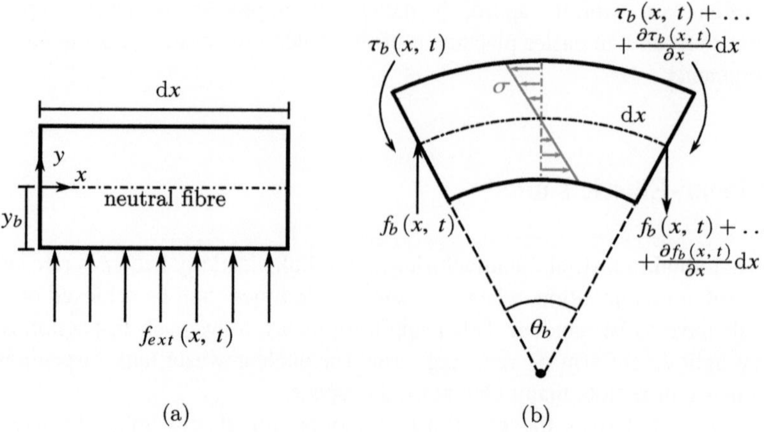

Fig. 10 Undeformed beam element (**a**) and bent differential beam element (**b**)

The constant length assumption for the neutral fibre implies that all fibres in the beam element are concentric arc segments. The link surface fibre at a distance y_b to the neutral fibre experiences the strain $\varepsilon(x, t)$:

$$\varepsilon(x, t) = y_b \frac{\partial \theta_b(x, t)}{\partial x} = y_b \frac{\partial^2 y(x, t)}{\partial x^2}. \tag{57}$$

With negligible shear, the infinitesimal stress σ is pure normal stress growing linearly with the distance y to the normal fibre:

$$\sigma(y) = c\, y. \tag{58}$$

The constant c is derived at the infinitesimal cross-section element of size dA, where the infinitesimal normal force df creates the bending moment dτ_b:

$$d\tau_b = y d f = y \sigma(y) dA. \tag{59}$$

Inserting (58) and integrating yields:

$$\tau_b = c\, \mathcal{I}_b \quad \text{with} \quad \mathcal{I}_b = \int y^2 dA. \tag{60}$$

The constant c is therefore the quotient of the bending moment τ_b with the area moment of inertia $\mathcal{I}_b$. Re-substitution in (58) yields:

$$\tau_b = \sigma(y) \frac{\mathcal{I}_b}{y}. \tag{61}$$

Hooke's law (5) relates the bending torque to the elastic link surface strain ε using (57):

$$\tau_b(x, t) = \frac{E\mathcal{I}_b(x)}{y_b}\varepsilon(x, t) = E\mathcal{I}_b(x)\frac{\partial^2 y(x, t)}{\partial x^2}. \tag{62}$$

The product $E\mathcal{I}_b(x)$ incorporates all material and geometry parameters that define the rigidity of a planar elastic link.

5.1.2 Force and Moment Balance

The force balance on the beam element with mass per unit length ρ_b in bending direction equates to:

$$f_b(x, t) - \left(f_b(x, t) + \frac{\partial f_b(x, t)}{\partial x}dx\right) + f_{ext}(x, t)dx = \rho_b(x)\,dx\frac{\partial^2 y(x, t)}{\partial t^2}, \tag{63}$$

whereas the torque balance about the coordinate origin of the beam element is:

$$\tau_b(x, t) - \left(\tau_b(x, t) + \frac{\partial \tau_b(x, t)}{\partial x}dx\right) + \frac{1}{2}f_{ext}dxdx + \dots$$
$$- \left(f_b(x, t) + \frac{\partial f_b(x, t)}{\partial x}dx\right)dx = 0. \tag{64}$$

For the infinitesimal beam element, second order terms of the infinitesimal element length dx can be neglected, such that both balances reduce to:

$$-\frac{\partial f_b(x, t)}{\partial x} + f_{ext}(x, t) = \rho_b(x)\frac{\partial^2 y(x, t)}{\partial t^2}, \tag{65}$$

$$f_b(x, t) = -\frac{\partial \tau_b(x, t)}{\partial x}. \tag{66}$$

Finally, inserting (66) into (65) and substituting (62) results in:

$$\rho_b(x)\frac{\partial^2 y(x, t)}{\partial t^2} - \frac{\partial^2}{\partial x^2}\left(E\mathcal{I}_b(x)\frac{\partial^2 y(x, t)}{\partial x^2}\right) = f_{ext}(x, t), \tag{67}$$

which is known as the Euler-Bernoulli beam equation. Just like the torsional wave equation (50) is identical for the torsion moment τ_x and torsion rate $\dot{\theta}_x$, the bending wave relation (67) can be equivalently written in terms of the bending angle θ_b, the bending force f_b or the bending torque τ_b.

5.2 Exact Solutions for Elastic Link Bending

Similar to the mass-spring-damper systems (28), the deflection in any point along the beam at any time is assumed to be expressible as the product of a shape function $\phi(x)$ and an amplitude function $v(t)$:

$$y(x, t) = \phi(x)\, v(t) \,. \tag{68}$$

The key difference to discrete mass-spring-damper systems (28) is that the shape function ϕ for the continuous elastic link is now a function of the location x along the link. In Sect. 3, this relation is implicitly present in the location of the mass elements in the system.

5.2.1 Homogeneous and Uniform Links

Assuming a homogeneous and uniform link renders $E\mathcal{I}_b$ along with ρ_b constant and inserting (68) into (67) with $f_{ext} = 0$ allows to write:

$$\frac{E\mathcal{I}_b}{\rho_b}\, \frac{\mathrm{d}^4\phi(x)/\,\mathrm{d}x^4}{\phi(x)} = \frac{\mathrm{d}^2 v(t)/\,\mathrm{d}t^2}{v(t)} = \omega^2 \,. \tag{69}$$

In this form, the spatial and temporal dependencies are separated on different sides of the equals sign. Each side represents an independent eigenvalue problem. Their solutions are equal and constant for all times t as well as at any point along the entire beam $0 < x < l$. The shared constant is the squared frequency ω^2.

The eigenvalue problem with respect to the time variable t:

$$\frac{\mathrm{d}^2 v(t)}{\mathrm{d}t^2} - v(t)\omega^2 = 0 \,, \tag{70}$$

subject to two initial conditions of the form $v(0)$ and $\frac{\mathrm{d}}{\mathrm{d}t}v(0)$ is called the initial value problem. Solutions to this equation are essentially discussed in Sect. 3.1.

The eigenvalue problem related to the spatial variable x:

$$\frac{\mathrm{d}^4\phi(x)}{\mathrm{d}x^4} - k_\omega^4 \phi(x) = 0 \,, \tag{71}$$

subject to four boundary conditions of the form $\frac{\mathrm{d}^j}{\mathrm{d}x^j}\phi(0)$ or $\frac{\mathrm{d}^j}{\mathrm{d}x^j}\phi(l)$ is known as the boundary value problem. The parameter k_ω is the so-called wave number:

$$k_\omega = \sqrt[4]{\omega^2 \frac{\rho_b}{E\mathcal{I}_b}} \,. \tag{72}$$

The general procedure to solve the Euler-Bernoulli equation is to first solve the boundary value problem and then look at the remaining initial value problem.

5.2.2 Bending Wave Propagation

The wave number k_ω is the spatial equivalent to the temporal frequency ω. It is a measure of the number of periods per unit length. A particular bending oscillation phase travels along the uniform homogeneous link with the bending wave speed c_b:

$$c_b = \frac{\omega}{k_\omega} = \sqrt[4]{\frac{E\mathcal{I}_b}{\rho_b}}\sqrt{\omega}\,. \tag{73}$$

Unlike torsion waves in elastic joint shafts, bending waves in elastic robot links display dispersive effects, i.e. bending excitations of different frequencies ω propagate at different speeds c_b.

5.2.3 Boundary Conditions

In order to exactly solve the Euler-Bernoulli equation for elastic robot link modelling, first the boundary value problem is considered. The integration of the forth order spatial derivative in (71) demands four integration constants to be determined with the knowledge of four boundary conditions. Often, two kinematic boundary conditions are formulated for each link end. Typical kinematic boundary conditions are collected in Table 6.

Alternatively, dynamic boundary conditions in the form of masses, lumped springs and dampers connected to elastic link ends can be considered. However, in multi-elastic-link systems, the effective inertia and spring constants "seen" by individual links can vary substantially with the joint configuration, while kinematic constraints in most cases can be considered to remain constant. The dynamic couplings with other links in the kinematic chain along with dynamic load masses and contacts of arbitrary compliance are more easily incorporated in the bending problem (67) through the external load force f_{ext}.

5.2.4 Link Shape Functions

It can be verified, that a shape function $\phi(x)$ of the form $\phi(x) = \hat{a}\,e^{\lambda x}$ satisfies the differential equation (71). Using this exponential form leads to the characteristic equation $\lambda^4 - k_\omega^4 = 0$ with the roots $\lambda_{1/2} = \pm k_\omega$, $\lambda_{3/4} = \pm jk_\omega$. These roots lead to the general solution of the form:

$$\phi(x) = {}^+\hat{a}\,e^{-jk_\omega x} + {}^-\hat{a}\,e^{jk_\omega x} + {}^+\hat{a}_N\,e^{-k_\omega x} + {}^-\hat{a}_N\,e^{k_\omega x}\,, \tag{74}$$

Table 6 Static boundary conditions for elastic links

Kinematic Boundary		y	θ_b	f_b	τ_b
free end		$\neq 0$	$\neq 0$	0	0
clamped end		0	0	$\neq 0$	$\neq 0$
pinned end		0	$\neq 0$	$\neq 0$	0
linear guide		$\neq 0$	0	0	$\neq 0$

with the integration constants ${}^+\hat{a}, {}^-\hat{a}, {}^+\hat{a}_N, {}^-\hat{a}_N$. Using:

$$
{}^+\hat{a} = \frac{\hat{a}_4 + \mathrm{j}\hat{a}_3}{2}, \quad
{}^-\hat{a} = \frac{\hat{a}_4 - \mathrm{j}\hat{a}_3}{2}, \quad
{}^+\hat{a}_N = \frac{\hat{a}_2 - \hat{a}_1}{2}, \quad
{}^-\hat{a}_N = \frac{\hat{a}_2 + \mathrm{j}\hat{a}_1}{2},
\tag{75}
$$

this general solution can be translated into the entirely real-valued form:

$$
\phi(x) = \hat{a}_1 \sinh\left(k_\omega x\right) + \hat{a}_2 \cosh\left(k_\omega x\right) + \hat{a}_3 \sin\left(k_\omega x\right) + \hat{a}_4 \cos\left(k_\omega x\right),
\tag{76}
$$

with the new real-valued integration constants $\hat{a}_1$, $\hat{a}_2$, $\hat{a}_3$, $\hat{a}_4$. In either case, the integration constants are determined using the known boundary conditions.

5.2.5 Example: Boundary Value Problem for Clamped-Free Cantilever

In literature, elastic robot links are frequently modelled as planar clamped-free cantilever beams. This motivates the procedure of solving the boundary value problem to be exemplified for a uniform homogeneous cantilever beam of length l in the following. Using Table 6 with (66) and (62), the boundary conditions are:

$$y(0) = 0 \quad \text{and} \quad \theta_b(0) = 0 = \left[\frac{\mathrm{d}}{\mathrm{d}x}y(x)\right]_{|x=0} \tag{77}$$

at the clamped link end, as well as:

$$\tau_b(l) = 0 = E\mathcal{I}_b\left[\frac{\mathrm{d}^2}{\mathrm{d}x^2}y(x)\right]_{|x=l} \quad \text{and} \quad f_b(l) = 0 = -E\mathcal{I}_b\left[\frac{\mathrm{d}^3}{\mathrm{d}x^3}y(x)\right]_{|x=l} \tag{78}$$

at the free link end. Differentiating (76) and applying these boundary conditions leads to the linear system of equations in the integration constants $\begin{bmatrix} \hat{a}_1 & \hat{a}_2 & \hat{a}_3 & \hat{a}_4 \end{bmatrix}^\top$:

$$\begin{bmatrix} 0 & 1 & 0 & 1 \\ k_\omega & 0 & k_\omega & 0 \\ k_\omega^2 \sinh(k_\omega l) & k_\omega^2 \cosh(k_\omega l) & -k_\omega^2 \sin(k_\omega l) & -k_\omega^2 \cos(k_\omega l) \\ k_\omega^3 \cosh(k_\omega l) & k_\omega^3 \sinh(k_\omega l) & -k_\omega^3 \cos(k_\omega l) & k_\omega^3 \sin(k_\omega l) \end{bmatrix} \begin{bmatrix} \hat{a}_1 \\ \hat{a}_2 \\ \hat{a}_3 \\ \hat{a}_4 \end{bmatrix} v(t) = 0. \tag{79}$$

The system (79) has a non-trivial solution for all times t, if the determinant of the matrix vanishes, which is equivalent to:

$$\cos(k_\omega l) = -1/\cosh(k_\omega l). \tag{80}$$

The roots $k_{\omega,i}l$ of this equation can only be determined numerically. A visualisation of the roots is obtained by plotting each side of the equation, as in Fig. 11. The points where the two graphs intersect determine the infinitely many solutions $k_{\omega,i}l$ that lead to non-vanishing amplitudes $\hat{a}_1$, $\hat{a}_2$, $\hat{a}_3$ and $\hat{a}_4$. Moreover, for a given link length l, the solutions of (80) define an infinite set of wave numbers $k_{\omega,i}$ with $i = 1, 2, 3, \ldots, \infty$. Through (72), each wave number is associated with a discrete frequency:

$$\omega_i = \sqrt{\frac{E\mathcal{I}_b}{\rho_b}} k_{\omega,i}^2 \tag{81}$$

and a corresponding mode shape function $\phi_i(x)$. The shape function $\phi_i(x)$ describes the link deformation at the location x along the link when oscillating with the frequency ω_i. It is characterized by the vector of integration parameters: $\begin{bmatrix} \hat{a}_1 & \hat{a}_2 & \hat{a}_3 & \hat{a}_4 \end{bmatrix}_i^\top$.

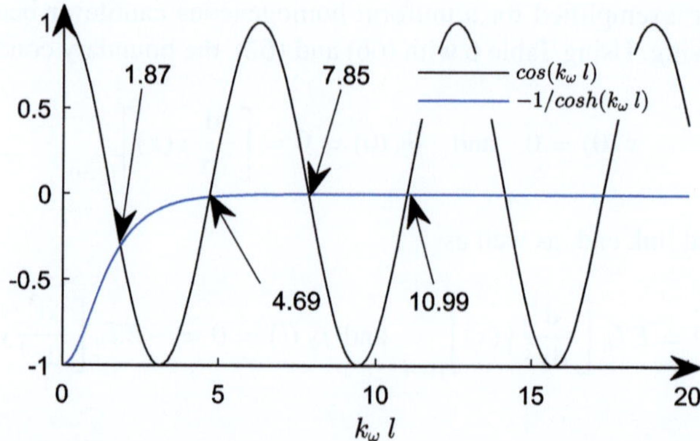

Fig. 11 Intersections of the two graphs determine the solutions to the boundary value problem of the clamped-free cantilever

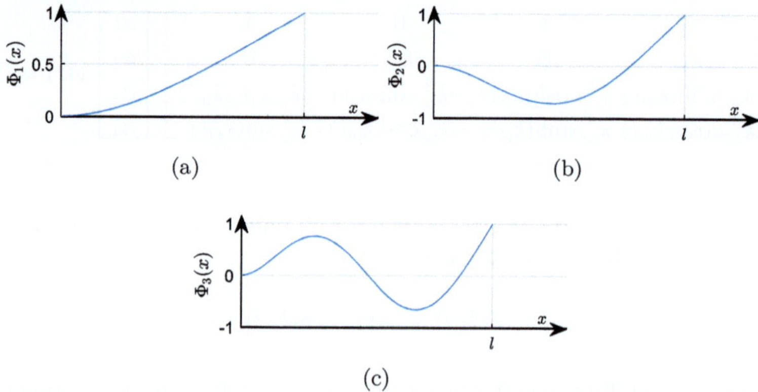

Fig. 12 First three mode shapes of the clamped-free cantilever. For better visualisation the mode shapes have been normalised, so that $\phi_i(l) = 1$. (**a**) 1st mode, (**b**) 2nd mode, (**c**) 3rd mode

Since the system of equations (79) is homogeneous, the mode shapes $\phi_i(x)$ are only defined up to a scale $\hat{a}_i$, such that:

$$\phi_i(x) = \hat{a}_i \Big[\sin\left(k_{\omega,\,i}\, x\right) - \sinh\left(k_{\omega,\,i}\, x\right) + \ldots$$

$$- \frac{\sin\left(k_{\omega,\,i}\, l\right) + \sinh\left(k_{\omega,\,i}\, l\right)}{\cos\left(k_{\omega,\,i}\, l\right) + \cosh\left(k_{\omega,\,i}\, l\right)} \left(\cos\left(k_{\omega,\,i}\, x\right) - \cosh\left(k_{\omega,\,i}\, x\right)\right) \Big]. \tag{82}$$

The first three mode shapes are visualised in Fig. 12. The triple formed by the wave number $k_{\omega,\,i}$, the natural frequency ω_i and the mode shape $\phi_i(x)$ are the parameters that fully describe the i-th bending mode of the link.

Fig. 13 Video: Mode shapes
of a cantilever beam (mode-
ShapeCantilever.mp4youtube
link). Courtesy of Institut für
Robotik, JKU Linz

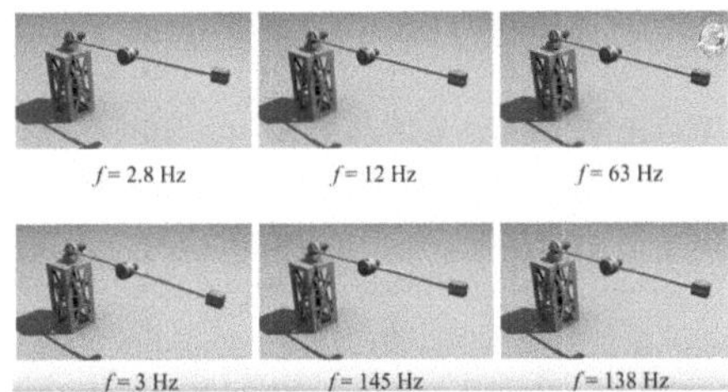

While this example concentrates on the clamped-free cantilever, the solution
principle is generic for arbitrary boundary value problems. More examples can be
found in [26] (Fig. 13).

5.2.6 Modal Orthogonality and Expansion Theorem

An important property of the shape functions $\phi_i(x)$ describing the beam bending is
their mutual orthogonality, as proven e.g. in [26]. The orthogonality condition is:

$$\int_0^l \phi_{i_1}(x) \frac{\mathrm{d}^2}{\mathrm{d}x^2} \left[E\mathcal{I}_b(x) \frac{\mathrm{d}^2}{\mathrm{d}x^2} \phi_{i_2}(x) \right] \mathrm{d}x = 0 \,,$$

$$i_1,\ i_2 = 1,\ 2,\ 3,\ \ldots,\ \infty \,;\ \omega_{i_1}^2 \neq \omega_{i_2}^2 \,. \tag{83}$$

One way to compute normalisation coefficients c_i for each mode shape is:

$$c_i = \int_0^l \rho_b(x) \phi_i(x) \phi(x) \,\mathrm{d}x \,,\ i = 1,\ 2,\ 3,\ \ldots,\ \infty \,, \tag{84}$$

so that also:

$$\omega_i^2 c_i = \int_0^l \phi_i(x) \frac{\mathrm{d}^2}{\mathrm{d}x^2} \left[E\mathcal{I}_b(x) \frac{\mathrm{d}^2 \phi(x)}{\mathrm{d}x^2} \right] \mathrm{d}x \,,\ i = 1,\ 2,\ 3,\ \ldots,\ \infty \,, \tag{85}$$

and eventually:

$$\phi(x) = \sum_{i=1}^{\infty} c_i \,\phi_i(x) \,. \tag{86}$$

Equations (84) to (86) represent the so-called modal expansion theorem for bending
beams. Following [26], it can be stated as:

*Any function $\phi(x)$ satisfying the boundary conditions of the link and ensuring
$\frac{\mathrm{d}^2}{\mathrm{d}x^2}\left[E\mathcal{I}_b(x) \frac{\mathrm{d}^2 \phi(x)}{\mathrm{d}x^2} \right]$ to be continuous, can be expanded in the absolutely and
uniformly convergent series of the eigenfunctions (86).*

5.2.7 Modal Differential Equations

To simplify the solution of the initial value problem, the orthogonality of the mode shapes is exploited. For the i-th mode, (67) is multiplied with the mode shape $\phi_i(x)$ and integrated over the link length:

$$\int_0^l \phi_i(x)\rho_b(x)\frac{\partial^2}{\partial t^2}y(x,\,t)\,\mathrm{d}x - \int_0^l \phi_i(x)\frac{\partial^2}{\partial x^2}\left[EI_b(x)\frac{\partial^2}{\partial x^2}y(x,\,t)\right]\mathrm{d}x = \ldots$$

$$= \int_0^l \phi_i(x)f_{ext}(x,\,t)\,\mathrm{d}x\,. \tag{87}$$

The bending deflection $y(x,\,t)$ expands to (86). The mutual orthogonality of mode shapes (83) effects that only integrals with products of equal mode index i have to be computed and (87) reduces to the modal ordinary differential equation:

$$m_i\frac{\mathrm{d}^2 v_i(t)}{\mathrm{d}t^2} - k_i v_i(t) = f_i(t)\,, \tag{88}$$

with the modal mass:

$$m_i = \int_0^l \phi_i(x)\rho_b\phi_i(x)\,\mathrm{d}x\,, \tag{89}$$

the modal stiffness:

$$k_i = \int_0^l \phi_i(x)EI_b\frac{\mathrm{d}^4}{\mathrm{d}x^4}\phi_i(x)\,\mathrm{d}x\,, \tag{90}$$

and the modal force:

$$f_i = \int_0^l \phi_i(x)f_{ext}(x,\,t)\,\mathrm{d}x\,. \tag{91}$$

To summarise, by exploiting the modal expansion (86) and orthogonality relation (83), the partial differential equation (67) can be converted into a system of mutually independent ordinary modal differential equations of the form (88). The system contains one modal differential equation per mode. Each can be solved with the techniques presented in Sect. 3.

5.3 Approximate Solutions for Elastic Link Bending

The previous subsection introduces the modal analysis as an approach to obtain exact solutions to elastic link bending problems. The solution is exemplified for a simple uniform cantilever. It is one of the rare examples that admit a closed-form solution like (82). The transcendent characteristic equation (80) must already be solved numerically. For real robots with more complex link geometries and non-uniform mass distributions, the existence of such closed-form solutions is a rare exception. In practice, the solution to link bending problems necessitates approximate solutions. Another challenge is to practically deal with the infinite number of modes. The implementation of control algorithms demands a finite set of simplified, real-time solvable model equations.

5.3.1 Equivalent Spring Elements

Analogous to an elastic joint shaft, the dynamics of an elastic link can be approximated by an equivalent massless spring, if the mass of the elastic link itself is small compared to the mass of the joints and the load of the robot. This means, the bending wave propagation speed (73) in the frequency range of interest is high compared to the elastic link length. The spatial variation of the bending variables along the link then becomes negligible and an equivalent spring element may be expected to yield good approximations.

The derivation of equivalent springs for bending beams is analogous to the procedure for torsional shafts. In practice these conditions are sufficiently met in static conditions when in contact with the environment. However, in free, dynamic scenarios, they are seldom met if robot link elasticity becomes relevant in an application. Equivalent springs are rather used as initial approximations to design elastic robot components, such as the elastic spokes of joint torque sensors. Against this background, Fig. 14 depicts frequently used equivalent spring approximations without derivation. The relations presented for combinations of series and parallel connections of equivalent torsional springs in Sect. 4.4 also hold here.

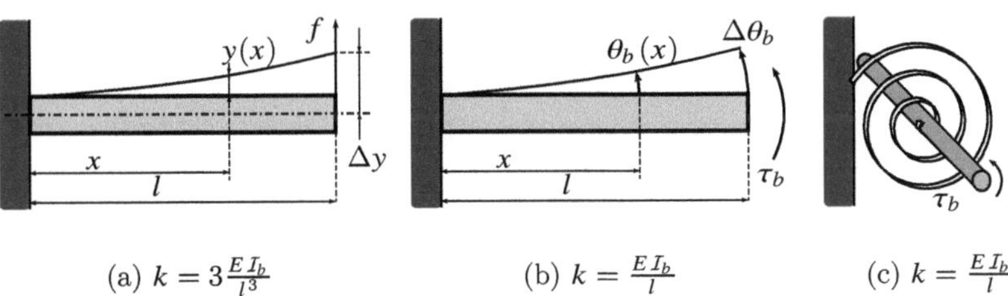

(a) $k = 3\frac{E I_b}{l^3}$ (b) $k = \frac{E I_b}{l}$ (c) $k = \frac{E I_b}{l}$

Fig. 14 Equivalent spring stiffness k for a clamped-free beam under lumped bending force (**a**) and bending torque (**b**) as well as a rotary leaf spring (**c**). The parameters are the Young's Modulus E, the constant area moment of inertia I_b and the spring length l

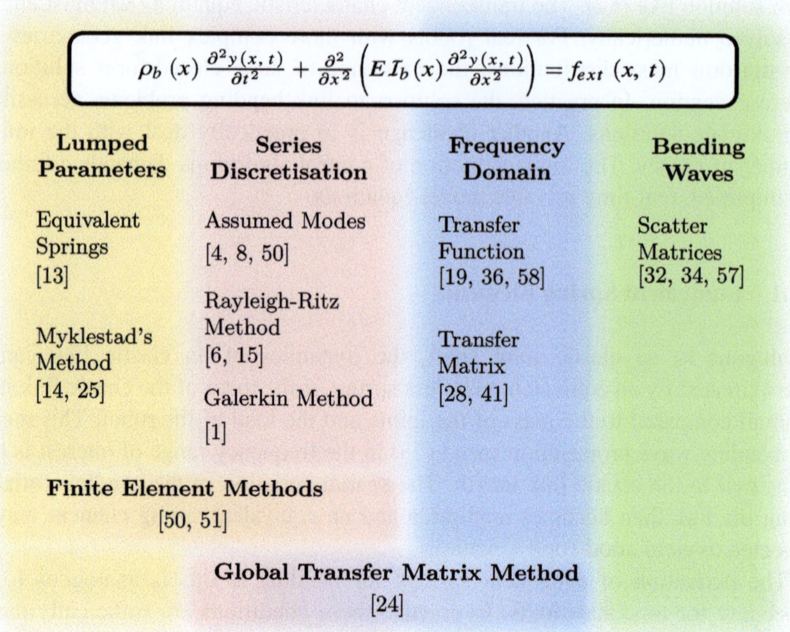

$$\rho_b\,(x)\,\frac{\partial^2 y(x,\,t)}{\partial t^2} + \frac{\partial^2}{\partial x^2}\left(E\,I_b\,(x)\,\frac{\partial^2 y(x,\,t)}{\partial x^2}\right) = f_{ext}\,(x,\,t)$$

Lumped Parameters	Series Discretisation	Frequency Domain	Bending Waves
Equivalent Springs [13]	Assumed Modes [4, 8, 50]	Transfer Function [19, 36, 58]	Scatter Matrices [32, 34, 57]
Myklestad's Method [14, 25]	Rayleigh-Ritz Method [6, 15]	Transfer Matrix [28, 41]	
	Galerkin Method [1]		

Finite Element Methods
[50, 51]

Global Transfer Matrix Method
[24]

Fig. 15 Overview of approximate techniques for elastic link modelling

5.3.2 Link Dynamics Approximations

Literature provides many approaches for approximate modelling solutions for elastic links. They can be grouped into five categories: series discretisation methods, lumped parameter methods, finite element techniques, frequency domain techniques and wave based approaches. While the approaches are too many to be detailed here, Fig. 15 provides examples for each of these classes. The material provided in this chapter equips the reader with the necessary background and tools to understand the content of any indicated reference. Further references and a detailed reference of the individual techniques is described in [26] and [8].

6 Multi-Degree-of-Freedom Robots

An elastic joint model assumes the predominant elasticity to occur in the mechanical interface between the motor and the rigid robot links. An elastic link model assumes the mechanical interface between the motor and the links to be rigid and the links to be elastic, instead. Robots with a combination of joint and link elasticity are conceivable of course. This section presents the equations of motion for isolated joint and link elasticity cases in order to highlight their common properties as well as their unique differences.

6.1 General Equations of Motion

Both, the Newton-Euler formalism and the Euler-Lagrange formalism can be used to derive the equations of motion for elastic joint robots (Euler-Lagrange [34], Newton-Euler [5]) as well as elastic link robots (Euler-Lagrange [3, 6], Newton-Euler [1]). The formalisms can be applied to robots with either revolute or prismatic joints, or both. For simplicity of notation and with only acceptable loss of generality, this section focuses on robots with only revolute joints driven by geared electrical drives.

Independent of the chosen formalism, the resulting equations of motion for multi-degree-of-freedom elastic robots have the same principal mathematical structure, which is similar to (24):

$$
\begin{bmatrix} \boldsymbol{\tau}_m \\ \boldsymbol{\tau}_\delta \end{bmatrix} = \overbrace{\begin{bmatrix} \mathbf{M}_m(\boldsymbol{\theta}_m, \boldsymbol{\delta}) & \mathbf{M}_{m\delta}(\boldsymbol{\theta}_m, \boldsymbol{\delta}) \\ \mathbf{M}_{m\delta}^\top(\boldsymbol{\theta}_m, \boldsymbol{\delta}) & \mathbf{M}_\delta(\boldsymbol{\theta}_m, \boldsymbol{\delta}) \end{bmatrix}}^{\mathbf{M}(\boldsymbol{\theta}_m, \boldsymbol{\delta})} \begin{bmatrix} \ddot{\boldsymbol{\theta}}_m \\ \ddot{\boldsymbol{\delta}} \end{bmatrix} + \cdots
$$

$$
+ \overbrace{\begin{bmatrix} \mathbf{C}_m(\boldsymbol{\theta}_m, \dot{\boldsymbol{\theta}}_m, \boldsymbol{\delta}, \dot{\boldsymbol{\delta}}) & \mathbf{C}_{m\delta}(\boldsymbol{\theta}_m, \dot{\boldsymbol{\theta}}_m, \boldsymbol{\delta}, \dot{\boldsymbol{\delta}}) \\ \mathbf{C}_{\delta m}(\boldsymbol{\theta}_m, \dot{\boldsymbol{\theta}}_m, \boldsymbol{\delta}, \dot{\boldsymbol{\delta}}) & \mathbf{C}_\delta(\boldsymbol{\theta}_m, \dot{\boldsymbol{\theta}}_m, \boldsymbol{\delta}, \dot{\boldsymbol{\delta}}) \end{bmatrix}}^{\mathbf{C}(\boldsymbol{\theta}_m, \dot{\boldsymbol{\theta}}_m, \boldsymbol{\delta}, \dot{\boldsymbol{\delta}})} \begin{bmatrix} \dot{\boldsymbol{\theta}}_m \\ \dot{\boldsymbol{\delta}} \end{bmatrix} + \underbrace{\begin{bmatrix} \mathbf{D}_m & \mathbf{0} \\ \mathbf{0} & \mathbf{D}_\delta \end{bmatrix}}_{\mathbf{D}} \begin{bmatrix} \dot{\boldsymbol{\theta}}_m \\ \dot{\boldsymbol{\delta}} \end{bmatrix} + \cdots
$$

$$
+ \underbrace{\begin{bmatrix} \mathbf{K}_m(\boldsymbol{\theta}_m, \boldsymbol{\delta}) & \mathbf{K}_{m\delta}(\boldsymbol{\theta}_m, \boldsymbol{\delta}) \\ \mathbf{K}_{m\delta}(\boldsymbol{\theta}_m, \boldsymbol{\delta}) & \mathbf{K}_\delta(\boldsymbol{\theta}_m, \boldsymbol{\delta}) \end{bmatrix}}_{\mathbf{K}(\boldsymbol{\theta}_m, \boldsymbol{\delta})} \begin{bmatrix} \boldsymbol{\theta}_m \\ \boldsymbol{\delta} \end{bmatrix} + \underbrace{\begin{bmatrix} \mathbf{g}_m(\boldsymbol{\theta}_m, \boldsymbol{\delta}) \\ \mathbf{g}_\delta(\boldsymbol{\theta}_m, \boldsymbol{\delta}) \end{bmatrix}}_{\mathbf{g}(\boldsymbol{\theta}_m, \boldsymbol{\delta})} . \tag{92}
$$

The $n_\theta \times 1$ vector $\boldsymbol{\theta}_m$ collects the motor angles for each of the n_θ robot joints. The $n_\delta \times 1$ vector $\boldsymbol{\delta}$ holds the coordinates added by the elastic degrees of freedom. In total, an elastic robot has $n_\theta + n_\delta$ degrees of freedom. All coordinates and their derivatives are functions of time. The time argument is omitted here for brevity.

The mass and stiffness matrices $\mathbf{M}$ and $\mathbf{K}$ are square matrices of dimension $(n_\theta + n_\delta) \times (n_\theta + n_\delta)$. In general, their elements are non-linear functions of the coordinates $\boldsymbol{\theta}_m$ and $\boldsymbol{\delta}$.

In addition to the $(n_\theta + n_\delta) \times (n_\theta + n_\delta)$ diagonal dissipative damping matrix $\mathbf{D}$ already present in the mass-spring-damper systems (24), the states of multi-degree-of-freedom robots are mutually coupled through non dissipative velocity dependent gyroscopic terms collected in the $(n_\theta + n_\delta) \times (n_\theta + n_\delta)$ matrix $\mathbf{C}$. Eventually, the load due to gravity is a non-linear function of the coordinate vector and enters the general equations of motion through the $(n_\theta + n_\delta) \times 1$ vector $\mathbf{g}$.

It is conceivable that all degrees of freedom are actuated by the $n_\theta \times 1$ vector $\boldsymbol{\tau}_m$ and the $n_\delta \times 1$ vector $\boldsymbol{\tau}_\delta$. Though, robot designs often do not incorporate actuators for direct control of the n_δ elastic degrees of freedom. The motivation is to reduce system complexity and cost. Instead, the controllers are augmented for such underactuated systems exploiting the dynamic couplings through inertia, gyroscopic and stiffness matrices to control the elastic degrees of freedom.

6.2 Elastic Joint Robots

For elastic joint robots the most common choice of coordinates to incorporate elastic degrees of freedom is the link angle $\delta_i := \theta_{l,\,i}$ as illustrated in Fig. 16. Most elastic joint robots do not feature a separate actuator to control the link side angle or the deflection directly, so that $\boldsymbol{\tau}_\delta = \mathbf{0}$. An exception are variable damping actuators discussed later in Sect. 8.2 of this chapter. The joint elasticity is modelled as an equivalent spring of stiffness k_i per joint, as outlined in Sect. 4.

6.2.1 Simplifying Assumptions

The specific equations of motion for elastic joint robots that have become standard in literature are subject to the assumptions introduced in [34]:

1. The joint rotor and gear inertia is symmetric about the joint's axis of rotation.
2. The kinetic energy of the motor shaft is mainly due to its own rotation, such that the joint rotor and gear motion is purely rotational about the joint's axis of rotation.

These assumptions allow several simplifications of (92). It appears reasonable to design elastic joint robots so that they satisfy the first assumption, as any eccentricity in the drive train will increase joint wear. Moreover, the joint axes can be considered as uniform cylinders. As a consequence, the vector of gravitational load $\mathbf{g}(\boldsymbol{\theta}_l)$ does no longer include any term due to the joint shaft's mass and is independent of the motor angle θ_m.

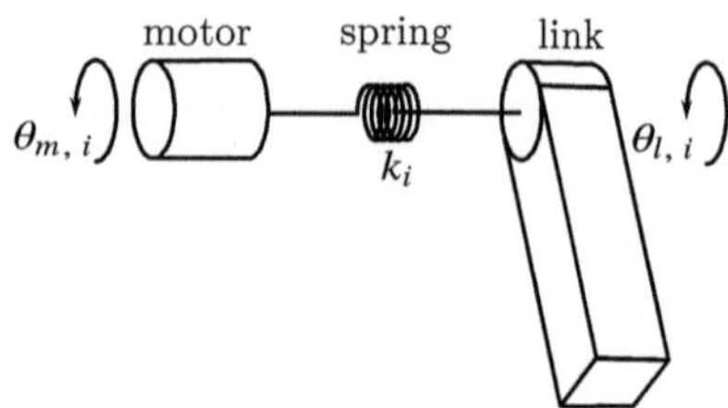

Fig. 16 A single elastic joint with rigid link (redrawn from [35])

The second assumption is justified by the typically high transmission ratios used in elastic joint robots. While the mass of the joint shaft is low compared to the overall robot mass, the high transmission ratio scales the apparent rotor inertia as well as the rotational speed about the joint axis up. The contribution of the scaled inertia and rotor speed to the overall kinetic energy become large compared to the contributions of the corresponding non-scaled shaft inertia rotating about and translating along remaining principal axes of motion of the robot. Consequentially, the rotor motion can be considered to be purely rotational about the joint axis. Therefore, the only rotor contribution to the robot inertia matrix $\mathbf{M}$ is the constant and diagonal matrix $\mathbf{M}_m$. The off-diagonal block matrices $\mathbf{M}_{m\delta}(\boldsymbol{\theta}_m, \boldsymbol{\theta}_l)$ and $\mathbf{M}_{m\delta}^{\top}(\boldsymbol{\theta}_m, \boldsymbol{\theta}_l)$ vanish. The same applies to any gyroscopic terms $\mathbf{C}_m$, $\mathbf{C}_{m\delta}$ and $\mathbf{C}_{\delta m}$ related to the rotor mass and inertia in $\mathbf{C}$.

Beyond these two assumptions, the damping is often considered to emerge from gearing friction only. Friction in the mechanical interface (bearings) between joint and link is comparably small. The engineering materials that form the elastic joint component are typically metals, that exhibit very low structural damping. In fact, many control problems of elastic joint robots actually emerge from the very low intrinsic joint damping [36]. Hence, the elements of the damping matrix $\mathbf{D}$ reduce to the motor damping block matrix $\mathbf{D}_m$.

Most frequently, the individual joint stiffnesses k_i are considered as constant, which renders the stiffness Matrix $\mathbf{K}$ coordinate independent.

6.2.2 Reduced Equations of Motion

Finally, the equations of motion for an elastic joint robot have the form:

$$
\begin{bmatrix} \boldsymbol{\tau}_m \\ \mathbf{0} \end{bmatrix} = \begin{bmatrix} \mathbf{M}_m & \mathbf{0} \\ \mathbf{0} & \mathbf{M}_\delta(\boldsymbol{\theta}_l) \end{bmatrix} \begin{bmatrix} \ddot{\boldsymbol{\theta}}_m \\ \ddot{\boldsymbol{\theta}}_l \end{bmatrix} + \begin{bmatrix} \mathbf{0} & \mathbf{0} \\ \mathbf{0} & \mathbf{C}_\delta(\boldsymbol{\theta}_l, \dot{\boldsymbol{\theta}}_l) \end{bmatrix} \begin{bmatrix} \dot{\boldsymbol{\theta}}_m \\ \dot{\boldsymbol{\theta}}_l \end{bmatrix} + \cdots
$$

$$
+ \begin{bmatrix} \mathbf{D}_m & \mathbf{0} \\ \mathbf{0} & \mathbf{0} \end{bmatrix} \begin{bmatrix} \dot{\boldsymbol{\theta}}_m \\ \dot{\boldsymbol{\theta}}_l \end{bmatrix} + \begin{bmatrix} -\mathbf{K} & \mathbf{K} \\ \mathbf{K} & -\mathbf{K} \end{bmatrix} \begin{bmatrix} \boldsymbol{\theta}_m \\ \boldsymbol{\theta}_l \end{bmatrix} + \begin{bmatrix} \mathbf{0} \\ \mathbf{g}_\delta(\boldsymbol{\theta}_l) \end{bmatrix} . \tag{93}
$$

In literature, the equations of motion (93) can often be found in the more compact form:

$$
\boldsymbol{\tau}_m = \mathbf{M}_m \, \ddot{\boldsymbol{\theta}}_m + \mathbf{D}_m \, \dot{\boldsymbol{\theta}}_m - \mathbf{K} \, (\boldsymbol{\theta}_m - \boldsymbol{\theta}_l) \, , \tag{94}
$$

$$
\mathbf{0} = \mathbf{M}_\delta(\boldsymbol{\theta}_l) \, \ddot{\boldsymbol{\theta}}_l + \mathbf{C}_\delta(\boldsymbol{\theta}_l, \dot{\boldsymbol{\theta}}_l) \, \dot{\boldsymbol{\theta}}_l + \mathbf{K} \, (\boldsymbol{\theta}_m - \boldsymbol{\theta}_l) + \mathbf{g}_\delta(\boldsymbol{\theta}_l) \, . \tag{95}
$$

As the stiffness increases and $\boldsymbol{\theta}_l \rightarrow \boldsymbol{\theta}_m$, $\dot{\boldsymbol{\theta}}_l \rightarrow \dot{\boldsymbol{\theta}}_m$ as well as $\ddot{\boldsymbol{\theta}}_l \rightarrow \ddot{\boldsymbol{\theta}}_m$, this form allows to see how the model converges towards the familiar rigid robot dynamics model:

$$
\boldsymbol{\tau}_m - \left(\mathbf{M}_\delta(\boldsymbol{\theta}_m) + \mathbf{M}_m \right) \ddot{\boldsymbol{\theta}}_m + \left(\mathbf{C}_\delta(\boldsymbol{\theta}_m, \dot{\boldsymbol{\theta}}_m) + \mathbf{D}_m \right) \dot{\boldsymbol{\theta}}_m + \mathbf{g}_\delta(\boldsymbol{\theta}_m) \, . \tag{96}
$$

6.3 Elastic Link Robots

The dynamics of multi-elastic-link robots tend to be more complicated compared to those of elastic joint robots. This is already apparent by the fact that, in practice, the equivalent spring approximation (see Sect. 5.3.1) does not yield good results and the Euler-Bernoulli equation (67) has to be solved for each link in some approximate way (Figs. 17 and 18).

The coordinates describing the multi-elastic-link robot degrees of freedom are the joint angles $\boldsymbol{\theta}_m$ along with the modal coordinates $\boldsymbol{\delta} := \boldsymbol{v}$. The joint angles $\boldsymbol{\theta}_m$ are identical to the joint coordinates used to write the dynamics of the equivalent rigid robot (96). If the robot is comprised of n_θ joint-link modules and the i-th link elastic dynamics is approximated by $n_{\omega,\,i}$ modal coordinates, the elastic link system has in total $n_\theta + n_\delta$ degrees of freedom, with:

$$n_\delta = \sum_{i=1}^{n_\theta} n_{\omega,\,i} \, . \tag{97}$$

Note that other choices of coordinates for the elastic degrees of freedom other than the modal coordinates are possible. Particular choices can be more efficient in combination with corresponding modelling techniques such as those listed in Fig. 15.

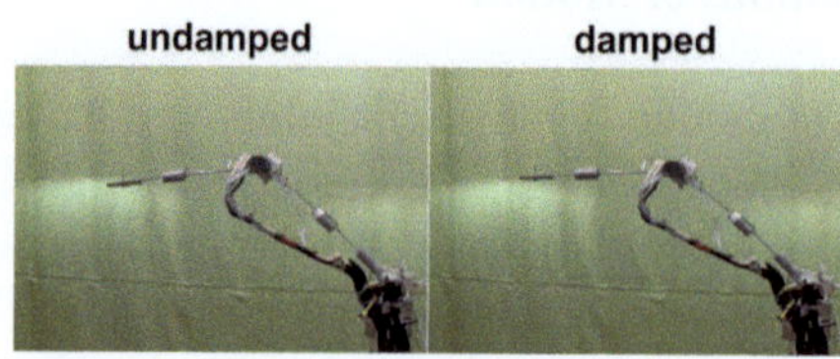

Fig. 17 Video: An example of an elastic link robot and interaction control (youtube link). Courtesy of Institute of Control Theory and Systems Engineering, TU Dortmund University

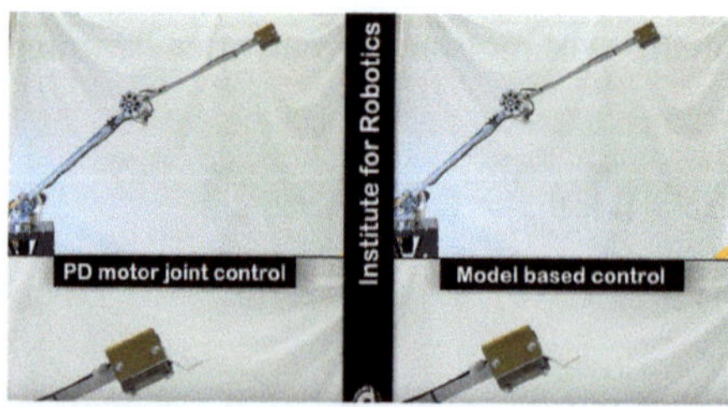

Fig. 18 Video: Fast tip positioning of an elastic link robot (youtube link). Courtesy of Institut für Robotik, JKU Linz

6.3.1 Simplifying Assumptions

For now, the link bending deformations are assumed to be limited to the plane of rigid motion. The more general case of bending deformations in multiple planes is briefly discussed from a control theoretic point of view in Sect. 6.3.4.

The most important assumption is the assumption of small link deflections. Not only does this assumption enable the derivation of the dynamics of each individual link as presented in Sect. 5, but it also helps to reduce the complexity of the multi-link equations to make them more graspable. In particular, it allows the gravitational load vector $\mathbf{g}_v$ associated with the modal coordinates to be expressed as a function of the joint angles $\boldsymbol{\theta}_m$ only. The joint related term $\mathbf{g}_m$ however remains with linear and trigonometric functions of $\boldsymbol{\theta}_m$ and v_i [22].

Then, following [22], the inertia matrix $\mathbf{M}$ is assumed to be a function of the joint angles $\boldsymbol{\theta}_m$ only. Moreover, the inertia sub-matrix $\mathbf{M}_v$, associated with only the modal coordinates, is constant. This emerges from a suitable orthonormalisation and transformation into modal coordinates based on the formulation of the boundary conditions. In this regard, the boundary conditions on the joint actuator side are usually chosen to be clamped boundary conditions (see Table 6). This choice can be enforced by high gain joint actuation controllers.

The matrix of gyroscopic terms $\mathbf{C}$ derives from the inertial terms and adds second order terms of the form $\dot{\theta}_{m,i}\,\dot{\theta}_{m,j}$ and $\dot{\theta}_{m,i}\,\dot{v}_j$. The elements of $\mathbf{C}$ thus depend on the joint angles as well as joint velocities.

The stiffness matrix $\mathbf{K}$ is a zero matrix except for the bottom right block $\mathbf{K}_v$, denoting diagonal modal stiffness terms for each link as derived in Sect. 5.2.7.

Just like elastic joint robots, the joints of elastic link robots experience a damping $\mathbf{D}_m$. In cases where link elasticity occurs as an unexpected problem, the inner structural damping in the links is typically small. Just like with elastic joint problems, the actual control problems emerge from the structural damping being quasi absent. Hence, the term $\mathbf{D}_v$ is often neglected for controller design. However, researchers have proposed lightweight elastic link designs based on composite materials. The friction between the composite laminate layers introduces structural damping that helps to passively solve oscillation problems for elastic link robots (see e.g. [11]). Another approach consists of adding a passively damping layer of material on the link surface [4].

6.3.2 Reduced Equations of Motion

With the assumptions and simplifications discussed above, the general equations of motion (92) reduce for elastic link robots to:

$$\begin{bmatrix} \boldsymbol{\tau}_m \\ \boldsymbol{\tau}_v \end{bmatrix} = \begin{bmatrix} \mathbf{M}_m & \mathbf{M}_{mv}(\boldsymbol{\theta}_m) \\ \mathbf{M}_{mv}^{\top}(\boldsymbol{\theta}_m) & \mathbf{M}_v \end{bmatrix} \begin{bmatrix} \ddot{\boldsymbol{\theta}}_m \\ \ddot{v} \end{bmatrix} + \cdots$$

$$+ \begin{bmatrix} \mathbf{C}_m(\boldsymbol{\theta}_m, \dot{\boldsymbol{\theta}}_m) & \mathbf{C}_{mv}(\boldsymbol{\theta}_m, \dot{\boldsymbol{\theta}}_m) \\ \mathbf{C}_{vm}(\boldsymbol{\theta}_m, \dot{\boldsymbol{\theta}}_m) & \mathbf{C}_v(\boldsymbol{\theta}_m, \dot{\boldsymbol{\theta}}_m) \end{bmatrix} \begin{bmatrix} \dot{\boldsymbol{\theta}}_m \\ \dot{v} \end{bmatrix} + \dots$$

$$+ \begin{bmatrix} \mathbf{D}_m & \mathbf{0} \\ \mathbf{0} & \mathbf{D}_v \end{bmatrix} \begin{bmatrix} \dot{\boldsymbol{\theta}}_m \\ \dot{v} \end{bmatrix} + \begin{bmatrix} \mathbf{0} & \mathbf{0} \\ \mathbf{0} & \mathbf{K}_v \end{bmatrix} \begin{bmatrix} \boldsymbol{\theta}_m \\ v \end{bmatrix} + \begin{bmatrix} \mathbf{g}_m(\boldsymbol{\theta}_m, v) \\ \mathbf{g}_v(\boldsymbol{\theta}_m) \end{bmatrix} . \tag{98}$$

6.3.3 Actuation of Elastic Link Robots

Many multi-elastic-link robots are solely actuated by the joint torques $\boldsymbol{\tau}_m$. This includes the cases where link elasticity emerges as an unexpected design issue as well as the elasticity aware elastic link robot design avoiding additional actuators to minimise system complexity and cost. In such scenarios, the elastic-link robot is underactuated and the vector $\boldsymbol{\tau}_v$ vanishes. The key difference with respect to elastic joint robots is that the off diagonal blocks of the inertia matrix can in general not be neglected as easily. Next, depending on the required number $n_{\omega, i}$ of oscillation modes per link, the degree of underactuation is higher. This imposes a more challenging control problem compared to the one of elastic joints.

To circumvent these control problems, additional actuators can be introduced into the link structure. An example for such actuators are piezoelectric patches distributed along the links (see e.g [7, 33]). In this case, $\boldsymbol{\tau}_v$ is non-zero. From an elastic link robot design perspective, the piezo patches are interesting, because the can serve as sensors as well as actuators. However, they must be carefully placed. If the patch location is close to the zero crossing of a particular mode shape, it becomes difficult to measure or influence the respective mode with that patch.

6.3.4 Controllability and Modal Accessibility

The question directly connected to the actuation of multi-elastic-link robots is the question of controllability. Especially, if the assumption of link bending only occurring in the plane of rigid motion does not hold and link mounted patch actuators do not exist, controllability of the oscillation modes is not generally guaranteed.

In the literature on elastic link robots, *controllability* is closely related to the controllability term from control theory. It refers to the overall feasibility to control the robot motion and to dampen undesired structural oscillation modes with the available actuators. More precisely, the term *modal accessibility* quantifies the potential of the present actuators to dampen individual oscillation modes.

For elastic joint robots, the torsional deflection axis always coincides with the joint actuation axis, so that intuitively a loss of controllability by reason of the elastic joint states is unlikely.

For elastic links however, it is well conceivable that oscillations may occur in certain bending planes that cannot directly be influenced by the joint actuators driving the robot. Figure 19 exemplifies such a case for the three DOF lightweight

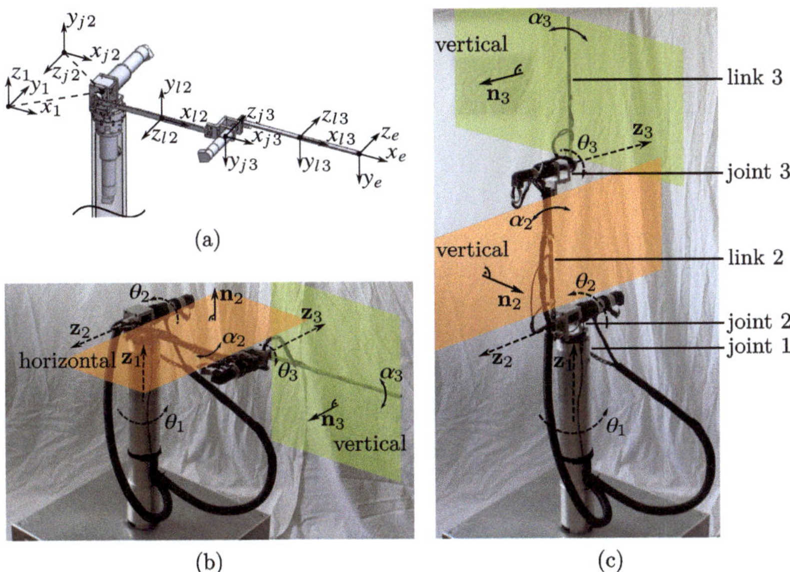

Fig. 19 Schematic sketch of TUDORA (**a**) along with two configurations with expected maximum (**b**) and minimum (**c**) controllability

robot test-bed TUDORA (**TU D**ortmund **O**mni-elastic **R**obot **A**dapted), designed to study the oscillation controllability of elastic link robots. The geometry and area moment of inertia are intentionally chosen, such that oscillations of link 2 and 3 primarily occur in a single plane respectively. The individual planes for the two elastic links are purposely oriented to remain perpendicular to each other in any joint configuration. The light orange area marks the oscillation plane of the second link, the light green area is the oscillation plane of the third link. Intuitively, in Fig. 19b it should be possible to control oscillations in both planes through the joint actuators. In contrast, in Fig. 19c controlling the oscillations in the light orange plane is expected to be at least difficult.

Different metrics to determine the oscillation controllability of elastic robots are found. The authors of [37, 38] derive a binary controllability indicator from the equations of motion of elastic link robots. A quantitative controllability index, i.e. "how well" the robot's oscillations can be controlled, is presented in [18, 19] and the concept of "modal accessibility" is introduced to rate the accessibility of individual oscillation modes.

The geometric controllability metric proposed in [17] evaluates the relative orientation of each joint axis z_i with respect to the normal vectors n_i of present link oscillation planes. This approach avoids computation of the entire system dynamics and is applicable to adapt oscillation damping schemes.

In summary, elastic robots should be carefully designed, such that oscillations only occur in planes which can be governed by the present actuators. Ideally, the design renders the joint actuators sufficient to fully control the elastic robot. Otherwise additional link-mounted actuators must be introduced into the system.

7 Design Considerations for Elastic Elements

This section illustrates the various robot design aspects of elastic robot components from different perspectives covering the benefits along with the challenges. The presentation uses simple mathematical models based on the tools we have become acquainted with in the previous section.

7.1 Elastic Elements as Force Sensors

The intentional integration of an elastic element as a sensor in a robotic actuator serves for the purpose of calculating the load force f_l based on spring deflection measurements. The principal setup is shown in Fig. 20. The spring with stiffness k and damping d_k is located between motor mass m_m and the load mass m_l. The motor generates the force f_m against ground. Movements of the motor mass are damped by d_b, which models linear proportional viscous friction effects. The load force f_l effects the spring displacement $\Delta x = x_l - x_m$:

$$f_l(t) = d_k \Delta \dot{x}(t) + k \, \Delta x(t) \,. \tag{99}$$

A careful mechanical spring design can minimise the spring damping d_k so that it can be neglected. In such a case only the spring displacement must be measured. The spring design is a challenging task, since it must meet the requirements with respect to the desired stiffness, strength, resolution, and linearity, while suppressing parasitic effects such as ripples, cross-talk to off-axial torque loads, noise, thermal and temporal drift.

The torque sensing principle (99) can be implemented in different ways as the deflection can be sensed differently. Researchers have explored accelerometers, contactless inductive transducers and photo-interrupters, but the most common approaches are strain gauge based and encoder based deflection sensing.

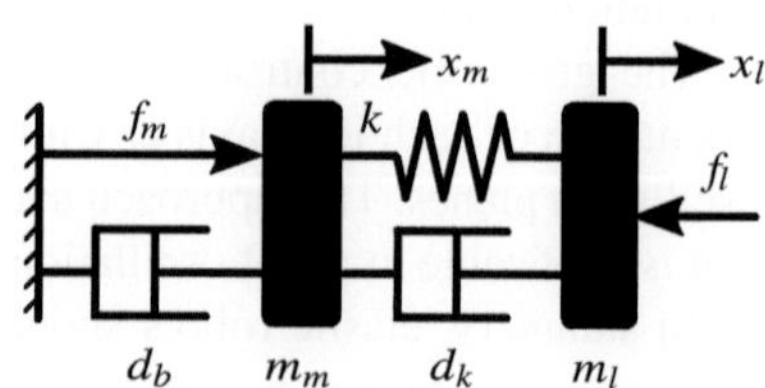

Fig. 20 Spring integrated between the motor and the load mass as a force sensor

For the perspective of maximising torque sensing accuracy, it seems that elastic elements should be designed as soft as possible. With the same deflection resolution, the feasible force/torque resolution will increase. However, the softer the compliant element, the larger the deflection under the same force. The limitation is the material strength, where softer structural members under the same load display higher stress densities. Next, the application might impose constraints on the maximum permissible member deflections.

7.2 *Elastic Elements for Impact Mitigation*

Novel robot applications in flexible industrial manufacturing or even household environments are dominated by reliable manipulation of objects with uncertainly known poses and safe collaborative physical interactions with humans. Inaccurate relative positioning between the robot and a work-piece as well as robot motions unforeseen by the human collaboration partner can lead to accidental collisions. In either case, elasticity can help mitigate the forces building up during such contacts. This concept extends to the point that elastic components even allow a robot to fulfil tasks, that would be very difficult to program and successfully accomplish with purely position controlled rigid robots. A human can blindly screw the cap onto a bottle by compliantly manipulating the cap on top of the bottle. Another example is dynamic legged robot locomotion, in which impact during walking on uneven terrain or running represent normal operating conditions (Fig. 21).

In order to look more closely into the impact mitigation properties of elastic robot elements, consider again the system depicted in Fig. 20. The impact occurs at the instant of time $t = 0\,\mathrm{s}$.

7.2.1 Impact on a Motor with High Transmission Ratio

In the first scenario, the motor is at rest. It displays a large inertia compared to the impact partner and a strong damping due to a high transmission ratio. As a result,

Fig. 21 Video: Impact mitigation during harsh physical interaction (youtube link). Courtesy of Istituto Italiano di Tecnologia, Humanoids & Human Centred Mechatronics Lab

the impact is assumed to not back-drive the motor and $x_m = \dot{x}_m = \ddot{x}_m = 0$. In the Laplace domain, the equation of motion for the motor mass reduces to:

$$f_m(s) = -kx_l(s) . \tag{100}$$

Correspondingly, the equation of motion for the load mass is:

$$f_l(s) = \left(m_l s^2 + d_k s + k \right) x_l(s) . \tag{101}$$

Combining the two leads to:

$$f_m = -\frac{k}{m_l s^2 + d_k s + k} f_l(s) = -\frac{\omega_0^2}{s^2 + 2\zeta \omega_0 s + \omega_0^2} f_l(s) , \tag{102}$$

where in this case $\omega_0 = \sqrt{k/m_l}$ and $\zeta = d_k/2\sqrt{km_l}$. This transfer function is the force transmissibility from the load side to the motor. In time domain the corresponding response to an impulsive force $f_l(t)$ computes to:

$$g(t) = \frac{1}{\sqrt{1 - \zeta^2}} \sqrt{\frac{k}{m_l}} \, e^{-\zeta \omega_0 t} \sin(\omega_d t), \quad t \geq 0 , \tag{103}$$

where $\omega_d = \omega_0 \sqrt{1 - \zeta^2}$. Interestingly, a heavy load mass reduces the impact force at the motor. This is comparable to an accident between a heavy truck and a light motor bike. The truck drive experiences significantly less severe impact forces compared to the bike driver.

The spring damping directly affects the impact response, as small values of $\sqrt{1 - \zeta^2}$ will rise the peak response of $f_m(t)$. The important effect however is that a lower spring stiffness k directly lowers the impulse peak force response of $f_m(t)$ and moreover slows down the response by reducing ω_0 and consequently ω_d. The peak force occurs at the time $t_{\max}$ after the collision incident:

$$t_{\max} = \frac{\pi}{2\omega_d} = \frac{\pi}{2\sqrt{1 - \zeta^2}} \sqrt{\frac{k}{m_l}} . \tag{104}$$

A sufficiently soft spring can buy a motor controller the time to react, before a harmful force level for the motor is reached. The spring effects an incoherence between the force variables at the link and the at the motor. The same applies to the motion variables on both sides of the spring. In literature this effect is often referred to as "dynamic decoupling of the link and motor".

7.2.2 Impact of the Motor on a Heavy Object

We now consider the case of a motor colliding with a heavy object, such that $m_m \ll m_l$ and $x_l = \dot{x}_l = \ddot{x}_l = 0$. The equations of motion become:

$$f_m(t) = m_m \ddot{x}_m(t) + (d_k + d_b)\dot{x}_m + k\ddot{x}_m ,$$ (105)

$$f_l(t) = -d_k \dot{x}_m - k\ddot{x}_m .$$ (106)

Combining the two equations in the Laplace domain yields:

$$f_l(s) = \frac{\frac{d_k}{m_m}s + \omega_0}{s^2 + 2\zeta\omega_0 s + \omega_0^2}\, f_m .$$ (107)

This time, we have $\omega_0 = \sqrt{k/m_m}$ and $\zeta = (d_k + d_b)/2\sqrt{km_m}$. The response to an impulsive motor force f_m consists of two contributions $g_d(t)$ and $g_k(t)$ defined for $t \geq 0$:

$$g_k(t) = \frac{1}{\sqrt{1-\zeta^2}}\sqrt{\frac{k}{m_m}}\, e^{-\zeta\omega_0 t} \sin(\omega_d t) ,$$ (108)

$$g_d(t) = \frac{d_k}{m_m \omega_d}\, e^{-\zeta\omega_0 t} (\omega_d \cos(\omega_d t) - \zeta\omega_0 \sin(\omega_d t)) .$$ (109)

The total impulse response $g(t)$ "felt" by the object is the superposition of the two functions $g(t) - g_k(t) + g_d(t)$. Qualitatively, the first term $g_k(t)$ has the identical properties to the case of collision with a heavy motor. A softer spring will reduce the force peak seen by the object and prolong the time for the peak to build up. This buys time to react to collisions.

A fundamental difference to the case before lies in the fact that the damping ratio ζ now is the sum of the in practice often relatively low spring damping d_k and the in practice often rather large and not negligible motor damping d_b. The second term $g_d(t)$ is present only, if spring damping d_k cannot be neglected. Then however it produces an instantaneous force response at $t = 0\,$s, that is directly proportional to the damping value d_k.

In summary, the elastic element can contribute to the protection of both, the motor as well as the load side. The evolution of the force response is qualitatively sketched in Table 3.

7.2.3 Stiffness Selection for Impact Resilience

Good knowledge of the impact force magnitudes the robot is expected to experience during its life-time helps to dimension the structurally elastic members for impact resilience. The stiffness k can be dimensioned to downscale the maximally transmitted force magnitude for the highest expected magnitude impact using Table 3.

The maximally transmitted force magnitude along with the stiffness k yield the maximum member deflection. In order to absorb the impact energy without damage, the maximal potential energy transferred into the deflection should remain well below the energy capacity (11) of the elastic member. For a simple member in the linear elastic domain this corresponds to:

$$W_y = \int_0^{x_{max}} f \, dx = \frac{1}{2}kx_{max}^2 \leq \eta_A V \frac{\sigma_y^2}{2E} \tag{110}$$

This allows to select elastic member geometries and materials (see Sect. 2.4) along with the material that yield a robot maintenance interval or life-time expectancy.

Finally, it must be noted that impacts are in fact very complex processes and the principles presented here are approximations that serve for initial conceptual designs. In practice, impact tests are necessary to evaluate the true impact force evolution and transmission over time.

7.3 Bandwidth and Delays

In the same way, as the elastic elements help to mitigate undesired potentially harmful collision forces, they also affect the transmission of desired interaction forces. Some interesting thoughts and analysis on the effect of elastic elements are presented in [9, 10] . Here, we consider again the contact between the motor and a heavy object. For simplicity, we assume the spring damping d_k has been minimised by design and is negligible. The damping ratio is then given by the motor damping $\zeta = d_b/2\sqrt{km_m}$ and the force transmissibility becomes the standard second order transfer function:

$$\frac{f_l(s)}{f_m} = \frac{\omega_0}{s^2 + 2\zeta\omega_0 s + \omega_0^2}, \tag{111}$$

with the frequency dependent magnitude and phase characteristic depicted in the Bode diagram in Fig. 22.

The frequency axis of this diagram is normalised by the natural frequency ω_0. The interpretation is that a harmonic force f_m generated by the motor will be scaled in amplitude and delayed before it arrives at the load. The scaling and delay depends on the damping ratio ζ as well as on the ratio between the desired harmonic frequency ω and the resonance frequency ω_d. In practice, motor damping will be present, but in a reasonable motor design we should expect an underdamped response, so that we can focus on the solid and dashed curves.

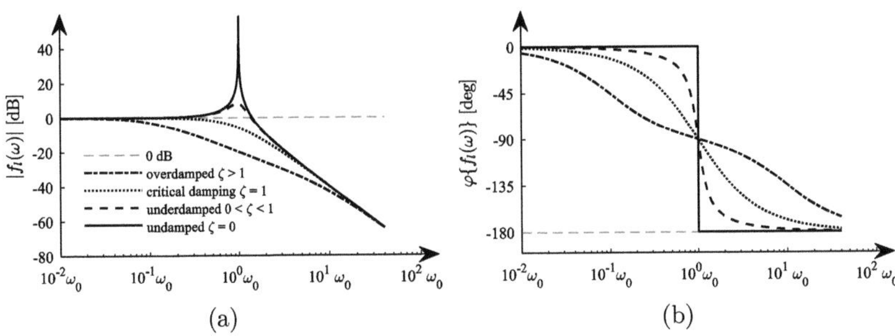

Fig. 22 Frequency magnitude (**a**) and phase (**b**) for a mass-spring-damper element for different damping ratios ζ. The frequency axes are normalised by the system natural frequency ω_0

Barely any magnitude and phase drop occur in the frequency range well below the resonance frequency $\omega \ll \omega_d$. As a rule of thumb, this is the case for up to $\omega \approx \frac{1}{3}\omega_d$.

Any generated harmonic motor force f_m beyond the lowest resonance frequency ω_d will experience a magnitude drop by $-40\,\mathrm{dB}$ per frequency decade.

In addition, the system response to the generated force will be delayed by up to 180°, as visible in Fig. 22b. When the force delivered to the load is actively controlled by a feedback loop, this phase lag can substantially degrade torque and motion tracking accuracy. As it shrinks the stability margin, it can even threaten closed-loop stability.

Controllers that can stably control forces around and beyond the eigenfrequency ω_d exist in literature, but the compensation of the magnitude drop requires disproportionate actuation effort [24, 27]. Given that any actuator can generate only limited power, the lowest resonance frequency therefore fundamentally limits the achievable control bandwidth and is of particular importance for the robot design with respect to a given application.

An often neglected fact is that the bandwidth of a robot's actuation system is not fully characterised by a single number. The actuator power limitations translate into different force limitations at different speeds. As an example, Fig. 23 shows the physically feasible force bandwidth limit ω_c for an arbitrary elastic electric actuator as a function of the normalised force amplitude. The bandwidth axis is normalised by the resonance frequency ω_d. The force axis is normalised by the actuator peak force f_p. The dotted grey line indicates the bandwidth ω_0, achieved if the actuator permanently generates harmonic forces with rated force amplitude f_r. Correspondingly, the dotted black line indicates the bandwidth ω_0 if the actuator permanently generates harmonic forces with peak force amplitude f_p. The dashed black line represents the limit due to the speed generation capability of the motor at different torques. The lower the elastic element's stiffness, the larger the deflection to reach the same torque. With limited speed and lower stiffness, the achievable bandwidth limit has to lower as well. The total physical bandwidth limit

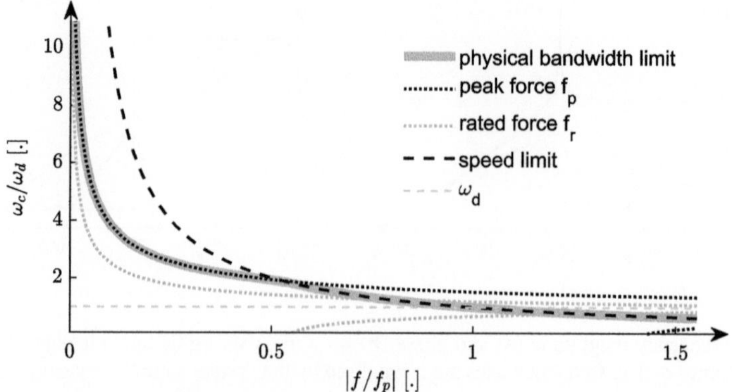

Fig. 23 Achievable force bandwidth per force amplitude

is, therefore, the minimum of the peak torque limit and the speed limit, as indicated by the grey solid line.

From the figure it is also apparent that around the resonance frequency ω_d, the generated torque is significantly amplified, such that force levels higher than the nominal actuator peak force can be attained. Large resonance forces can damage the robot and harm any object or human in the vicinity of the robot. In consequence, operation around resonance has to be handled with caution.

An in depth investigation of the force bandwidth for electric actuators can be found in [24]. In essence, if a robot design shall be optimised for high control bandwidth, it should be made as stiff as possible. However, it must be noted that a better trade-off between bandwidth criterion and other design criteria can be achieved, if the force magnitude dependency of the bandwidth is taken into account.

7.4 Efficiency

7.4.1 Resonance Exploitation

While operating around the system's resonance can be dangerous and harmful, it can also be exploited for energy efficient robot actuation during repetitive motions.

The instantaneous mechanical power $P(t)$ transferred to a mass is the product of its velocity $\dot{x}(t)$ and the force $f(t)$ acting on it. Consider a single linear mass-spring-damper system with vanishing initial conditions. In the Laplace domain we can write:

$$P(\mathrm{s}) = \mathrm{s}x(\mathrm{s})\, f(\mathrm{s})\,. \tag{112}$$

Solving (19) for the force $f(s)$ and resubstituting (17), we obtain:

$$P(s) = Z(s)\, \dot{x}^2(s) = \left(m\,s + d + \frac{k}{s} \right) \dot{x}^2(s) \,. \tag{113}$$

The function $Z(s)$ is called the impedance of the mass-spring-damper system. If translated into the Fourier domain ($s = j\omega$), it represents the force f that the system produces to "impede" a periodic motion $\dot{x}(j\,\omega)$ of frequency ω. The absolute value $|Z(j\,\omega)|$ serves as a scaling factor of the power required to maintain the periodic motion $\dot{x}(j\omega)$. It computes to:

$$|Z(j\,\omega)| = \sqrt{\left(m\,\omega - \frac{k}{\omega} \right)^2 + d^2} \,. \tag{114}$$

Figure 24 displays the absolute value $|Z(j\,\omega)|$ in decibel as a function of the motion frequency normalised by the resonance frequency ω_d.

For frequencies well below and well above the resonance frequency, the power requirement is amplified ($|Z(s)| > 0$ dB). For low frequencies, $\omega \to 0$, the power demand (114) is dominated by the stiffness term $\frac{k}{\omega}$. The power is required to create the spring deflection and thereby to charge and discharge the spring with potential energy at the desired rate. For higher frequencies $\omega \to \infty$, the main contribution in (114) is the inertial term involving $m\,\omega$. The power is required to charge and discharge the inert mass with kinetic energy at the desired rate. Both contributions are indicated by the green and red dashed lines in Fig. 24.

Viscous damping demands constant power, independent of the desired motion frequency. With the help of (17) it can be shown that the mass and stiffness terms cancel at resonance. Around resonance, the power is thus only required to compensate damping induced energy losses.

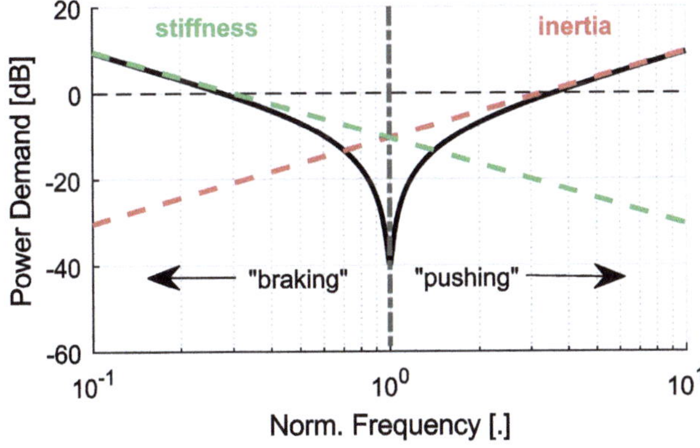

Fig. 24 Power demand

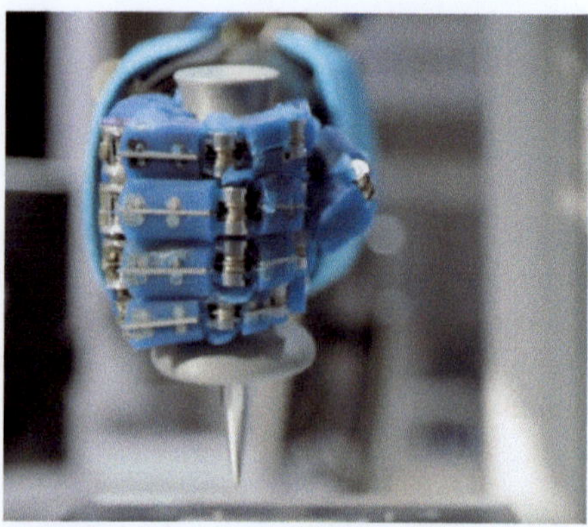

Fig. 25 Video: Exploiting intrinsic compliance for efficient execution of periodic tasks (youtube link). Courtesy of German Aerospace Center (DLR), Institut für Robotik und Mechatronik

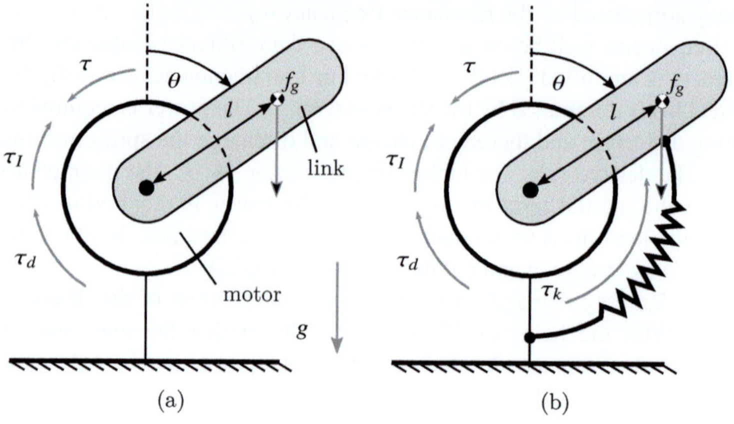

Fig. 26 A single link under gravitational load driven by a motor (**a**) and driven by a motor with parallel spring (**b**)

As the discussion shows, robots intended to perform periodic motions can significantly benefit from intrinsically elastic elements. The design objective for the elastic element is to match the structural member resonances with the desired operation frequencies. Further reading can be found in [12, 15, 20] (Fig. 25).

7.4.2 Intrinsic Dynamics Compensation—A Linearised Example

Elastic structural design elements can be used to fully or partially compensate undesired intrinsic dynamics and thereby unburden the employed actuators. A simple example is intrinsic gravity compensation. Consider a simple rigid link connected to an electrical motor as depicted in Fig. 26a. The motor produces the

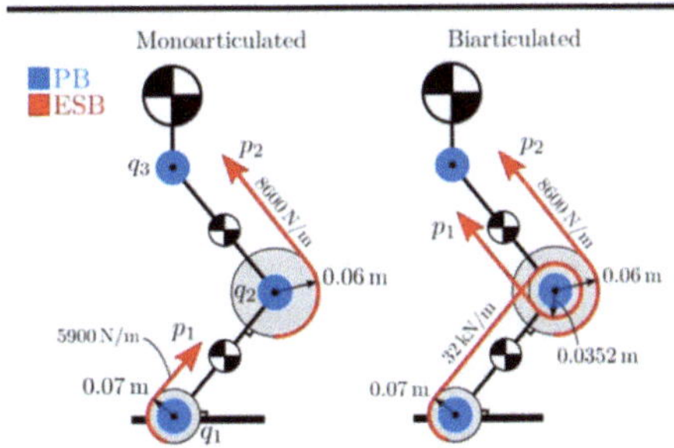

Fig. 27 Video: 3-DoF Leg with parallel compliant actuation for tunable passive load compensation (youtube link). Courtesy of Istituto Italiano di Tecnologia, Humanoids & Human Centred Mechatronics Lab

torque τ to move the link by an angle θ. The motor torque τ is opposed by the inertial torque $\tau_I = I\,\ddot{\theta}$, the frictional damping $\tau_d = d\,\dot{\theta}$ and the gravitational load torque $\tau_g = l\,f_g\sin\theta$. For small displacements around $\theta = 0$ we can linearise $\sin(\theta) \approx \theta$, so that:

$$\tau = I\,\ddot{\theta} + d\,\dot{\theta} + l\,f_g\theta \,. \tag{115}$$

This equation has a mathematical structure similar to the linear simple mass-spring-damper system (15). We introduce an elastic structural element with stiffness k between the mechanical ground of the motor and the link, in parallel to the drive train. In this example, the spring has its equilibrium for $\theta = 0$ and extends the torque balance (115) by the torque $\tau_k = k\theta$:

$$\tau + k\theta = I\,\ddot{\theta} + d\,\dot{\theta} + l\,f_g\theta \,. \tag{116}$$

If the stiffness of the elastic element is designed to be $k = l\,f_g$, it compensates the link weight around the linearisation point $\theta = 0$. The residual frequency dependent power demand for harmonic link motions in the gravity compensated case corresponds to the red dashed curve in Fig. 24. Further reading can be found in [28, 39, 41] (Fig. 27).

8 Variable Impedance Actuation

The torque sensing capabilities of elastic elements allow active impedance control using ongoing active control efforts to reflect actuator impedance in a wide range.

There are three reasons motivating variable impedance actuators: efficient adaptation to different tasks, (modal) dynamics matching, and increasing the total actuator peak power capacity. For the first case, the variable physical impedance allows an efficient adaptation regarding the required contact sensitivity and positioning tolerance as well as accuracy and force level. Therein, only small actuation effort is needed to vary the impedance. In the second case, the system's natural dynamics are matched via control of the actuator's physical impedance with desired cyclic movement patterns, like walking or running. The last case exploits the energy-storing property of the elastic elements to complement the actuator peak power capacity when releasing the stored energy. This is beneficial for impulsive movement tasks, such as running, jumping, throwing, or manipulation of heavy objects.

Multiple works comprehensively survey the principles and hardware of variable impedance actuators [14, 21, 40, 42], almost exclusively utilising equivalent spring assumptions. The following section briefly depicts present approaches to mutually change the reflected stiffness, damping and inertia.

8.1 Variable Stiffness Principles

According to [40], variable stiffness actuators can be divided into three groups: preloading/-tensioning of springs, altering the load-spring transmission, and physically changing the spring structure (Fig. 28). When altering the load-spring transmission, the achievable passive joint range is comparably small. Figure 29 shows examples of all groups.

Additionally, [40] discusses different kinds of nonlinear spring implementations, alterable effective spring lengths or cross sections, nonlinear mechanical interlinks augmenting the cam profile from Fig. 29h, and continuously alterable transmissions

.

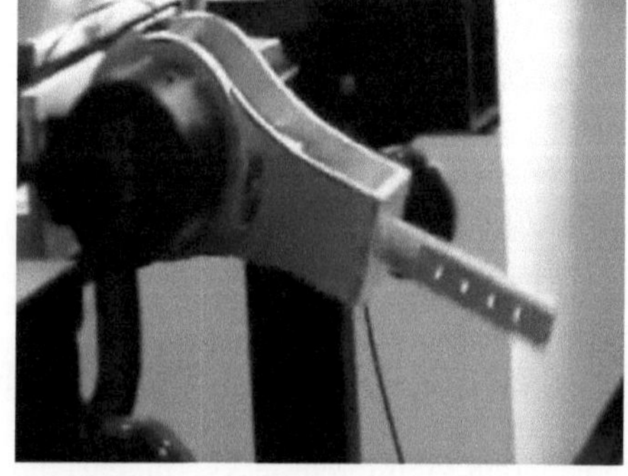

Fig. 28 Video: An example of the "Actuator with Adjustable Stiffness" (AwAS-II) (youtube link). Courtesy of Istituto Italiano di Tecnologia, Humanoids & Human Centred Mechatronics Lab

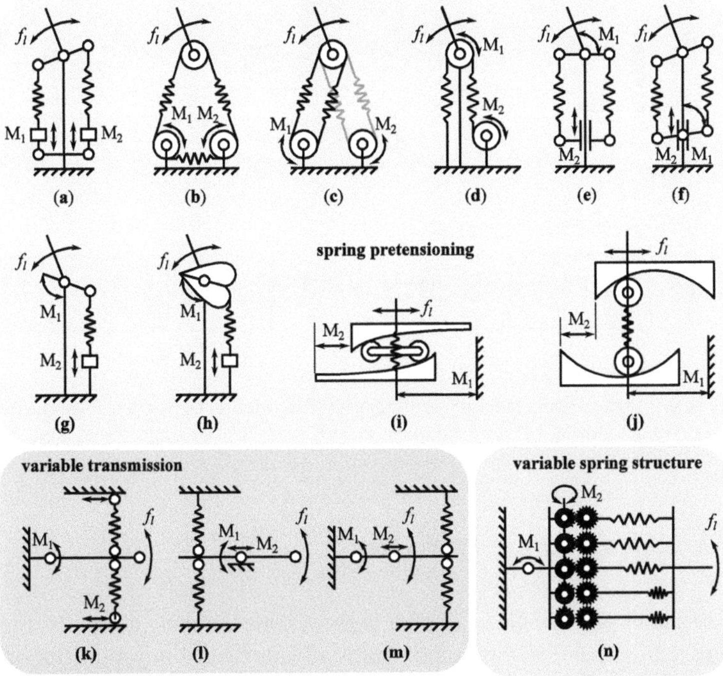

Fig. 29 Principles of variable stiffness mechanisms [25]: unidirectional antagonistic springs (**a**), coupled antagonistic springs (**b**), bi-directional antagonistic springs (**c**), springs in push-pull configuration (**d**), push-pull configured springs with separate position motor M_1 (**e–f**), spring loaded lever arms (**g–h**), spring loaded cam rollers (**i–j**), variable lever-pivot-spring arrangements (**k–m**), and variable spring recruitment (**n**)

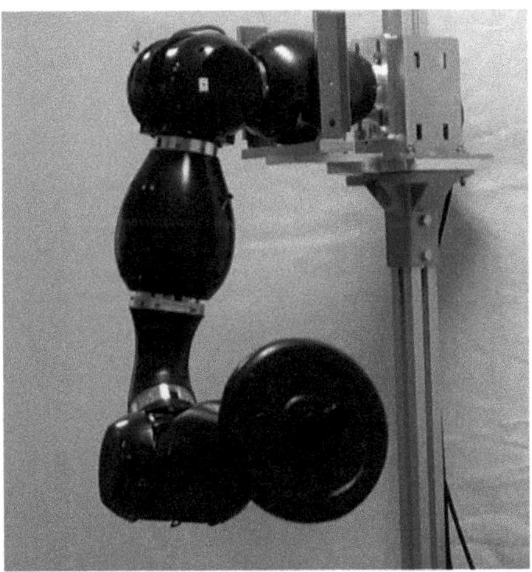

Fig. 30 Video: An example of a robot with variable damping actuators (video link). Courtesy of Istituto Italiano di Tecnologia, Humanoids & Human Centred Mechatronics Lab

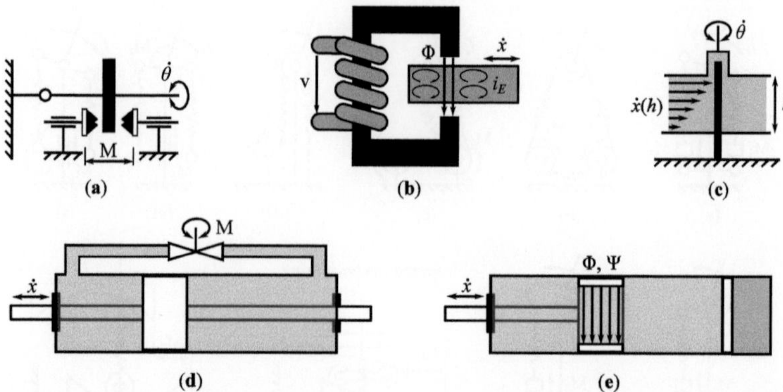

Fig. 31 Principles of variable damping mechanisms [25]: friction breaks (**a**), eddy current breaks (**b**), linear (**c**) and nonlinear (**d**) fluid dampers, as well as electro- or magnetorheological dampers (**e**)

8.2 Variable Damping Principles

Figure 30 shows an example of a robot with variable damping actuators. Figure 31 depicts several variable damping principles also described in [40]. With regard to (93), variable damping concepts introduce an additional control input on the elastic degrees of freedom.

8.3 Variable Inertia Principles

From the actuator output, the motor inertia appears scaled quadratically with the transmission ratio. Therefore, the transmission ratio can be changed in order to adjust the reflected inertia to meet the task requirements. Continuously alterable planetary transmissions, toroidal alterable transmissions, or belt drives with alterable diameter pulleys are examples of variable transmissions.

Video List

Video No.	Image No.	Video-name in App
1	Fig. 1	Elasticity of a supposedly rigid robot.
Filename Video		youtube link
Filename Image		reis.PNG
Video No.	Image No.	Video-name in App
2	Fig. 13	Mode shapes of a cantilever beam.
Filename Video		youtube link
Filename Image		modeShapeCantilever.png
Video No.	Image No.	Video-name in App
3	Fig. 17	An example of an elastic link robot under interaction control.
Filename Video		youtube link
Filename Image		videoTUDOR.png
Video No.	Image No.	Video-name in App
4	Fig. 18	Fast tip positioning of an elastic link robot.
Filename Video		youtube link
Filename Image		fastTipLink.png
Video No.	Image No.	Video-name in App
5	Fig. 21	Impact mitigation during harsh physical interaction.
Filename Video		youtube link
Filename Image		videoCentauro.png
Video No.	Image No.	Video-name in App
6	Fig. 25	Exploiting intrinsic compliance for efficient execution of periodic tasks.
Filename Video		youtube link
Filename Image		videoRepetitive.png
Video No.	Image No.	Video-name in App
7	Fig. 27	3-DoF Leg with parallel compliant actuation for tunable passive load compensation.
Filename Video		youtube link
Filename Image		videoBungeeLeg.png
Video No.	Image No.	Video-name in App
8	Fig. 28	An example of the "Actuator with Adjustable Stiffness" (AwAS-II).
Filename Video		youtube link
Filename Image		videoAwas.png
Video No.	Image No.	Video-name in App
9	Fig. 30	An example of a robot with variable damping actuators.
Filename Video		video link
Filename Image		videoVariableDamper.png

Cross-References

This chapter contains cross-references to content published in other chapters belonging to the four "Robotic Goes MOOC" books: KNO3, KNO4, KNO6, DES5, DES6, DES8, INT2, INT3, INT5, INT6. The full list of chapters abbreviations is available in the Preface.

Acknowledgments We are grateful to the German Research Foundation (DFG, BE 1569/7-1 and BE 1569/12-1) as well as the DAAD-MOET-project 322 and the European Research Council projects WALKMAN (no. 611832), CENTAURO (no. 644839) and CogIMon (no. 644727) for the partial financial support of the work presented in this chapter.

Appendix: Exercises

All provided exercises are classified according to the three objectives of repeating*, applying**, and pursuing/expanding*** the learned content.

Introduction and Elasticity of Objects

Exercise A.1 Knowledge Questions*

This exercise repetitively asks questions on the content of Sect. 2.

(a) Name the six perspectives and criteria which individually lead to considering a robot elastic.
Solution: The application requires a higher accuracy than the rigid-body kinematics and dynamics can achieve. A force/torque control requires the load-dependent deformation measurements. The application exploits structural resonances. The application exploits the intrinsic robot dynamics to save energy. The application exploits the intrinsic robot dynamics to absorb impact energy. The application requires forces equally distributed across a contact area.

(b) Name some general advantages/disadvantages of elastic elements in robot design. What applications are they predestined for with regard to this?
Solution: disadvantages: oscillations, deflections, limit control bandwidth -> prolonged cycling times, loss in precision advantages: lightweight, require simpler support structures, relax precision requirements -> flexible tasks accounting for varieties of positioning, efficient, use as sensors and exploit elastic deformations (e.g. impact mitigation) -> physical contact; compliantly manipulating of objects, e.g. blindly screwing the cap onto a bottle; dynamic legged robot locomotion (walking on uneven terrain, running)

(c) When must the intrinsic elasticity of components be considered?
Solution: Whenever the (robot's) elasticity affects the application.
(d) Name and describe two elastic moduli.
Solution: Young's modulus E, shear modulus G, see Sect. 2
(e) How can the stiffness of an object be increased?
Solution: geometry, support, material

Exercise A.2 Multiple-Choice Questions*

This exercise also repetitively asks questions on the content of Sect. 2. More than one answer might be possible and necessary.

(a) Oscillations occur in elastic robots...

- only due to external forces. *Solution: false*
- due to impulsive loading. *Solution: true*
- due to sine-profile movements. *Solution: false*
- due to step movements. *Solution: true*

Dynamics of Elastic Elements

Exercise A.3 Knowledge Questions*

This exercise repetitively asks questions on the content of Sect. 3.

(a) Which requirements have to be satisfied to model elastic objects as mass-spring-dampers?
Solution: wide linear deflection range, small deformations, or linearised deformations
(b) What does the parameter ω_0 represent and what is its dimension?
Solution: frequency of the harmonic response of the undamped system, natural or characteristic frequency of the system, Hz
(c) What is the fastest way to transfer a system from the initial condition to the equilibrium without oscillations?
Solution: critical damping $\zeta = 1$
(d) Which two parameters represent the i-th natural mode of oscillations of a system?
Solution: $\boldsymbol{\phi}_i$ and ω_i^2
(e) What does *normal mode* mean?
Solution: normalised modal vector, normalised by its length, the first or the largest vector element
(f) What is the main advantage of the modal vectors of a system?
Solution: can be used to diagonalise the system

(g) What is meant by *modal analysis* and what is its use?
    Solution: Derivation of the description of a system's response to initial excitation in terms of eigenvectors and eigen frequencies Usage: e.g. modal matching for efficient gaits?

(h) Name the respective pendant regarding a rotational motion corresponding to these translational physical quantities:

  - position.
  - force.
  - velocity.
  - mass.

Solution: angle, (moment of) torque, angular velocity, torque of inertia

Exercise A.4 Multiple-Choice Questions*

This exercise also repetitively asks questions on the content of Sect. 3. More than one answer might be possible and necessary.

(a) The decay time of oscillations of a mass-spring-damper is proportional to...

  - the initial displacement. *Solution: false*
  - the initial velocity. *Solution: false*
  - the initial acceleration. *Solution: false*
  - non of the aforementioned. *Solution: true*

(b) The spring stiffness k...

  - vanishes at rest position of the spring ($k = 0$). *Solution: false*
  - describes the relation between reset force of the spring and its displacement. *Solution: true*
  - determines the amplitude of the harmonic oscillation of a mass-spring-damper. *Solution: false*
  - determines the frequency of the harmonic oscillation of a mass-spring-damper. *Solution: true*

(c) What do elastic joints and elastic links have in common?

  - Their equations of motion can be derived via Newton's laws or the Lagrangian equations. *Solution: true*
  - Both can get traced back to mass-spring-dampers. *Solution: true*
  - Both have infinite degrees of freedom. *Solution: false*
  - Non of the aforementioned. *Solution: false*

(d) The oscillation duration of a pendulum doubles if...

  - the mass gets divided by 2. *Solution: false*
  - the mass gets multiplied by 2. *Solution: false*
  - the mass gets multiplied by 4. *Solution: false*
  - the length of the pendulum gets multiplied by 2. *Solution: false*
  - the length of the pendulum gets multiplied by 4. *Solution: true*

(e) Which of the following forces are inertial forces that only occur in rotational reference systems?

- Centripetal force. *Solution: false*
- Centrifugal force. *Solution: true*
- Gravitational force. *Solution: false*
- Coriolis force. *Solution: true*

(f) Your pendulum clock runs slow. Which actions will force the clock to run faster?

- Add a mass above the centre of gravity. *Solution: true*
- Add a mass below the centre of gravity. *Solution: true*
- Move the clock to the moon. *Solution: false*
- Increase the pendulum's length. *Solution: false*
- Decrease the pendulum's length. *Solution: true*

(g) An harmonic mass-spring-damper neglecting gravity looses half of its mass at the position of maximum deflection. Which of the following motion quantities will change?

- The position amplitude. *Solution: false*
- The velocity amplitude. *Solution: true*
- The oscillation frequency. *Solution: true*
- The potential energy. *Solution: false*
- The kinetic energy. *Solution: false*
- The total energy. *Solution: false*

(h) Which of the following statements are true with regard to a simple mass-spring-damper neglecting gravity?

- The amplitude of oscillations increases with increasing mass. *Solution: false*
- The rest position changes as the mass changes. *Solution: false*
- The decay time of oscillations decreases with increasing spring stiffness. *Solution: true*
- The oscillation frequency decreases with increasing mass. *Solution: true*

(i) Regarding a harmonic oscillation...

- the amplitude is a function of time. *Solution: false*
- the deflection is a function of time. *Solution: true*
- the kinetic energy is a function of time. *Solution: true*
- the angular frequency is a function of the position. *Solution: false*

(j) Which of the following properties affect the natural frequency of a linear mass-spring-damper?

- The initial deflection. *Solution: false*
- The initial velocity. *Solution: false*
- The rest position. *Solution: false*

- The mass. *Solution: false*
- The damping. *Solution: false*

(k) Which of the following statements are true?

- The decay time of oscillations of a mass-spring-damper only depends on the spring stiffness and the mass.
- The geometry of an elastic object plays no role regarding its oscillations. *Solution: false*
- Reaching any static or dynamic stress limit will result in material failure. *Solution: no, see Sect. 2*
- Massive industrial robots can always be considered rigid. *Solution: no, depending on application e.g. high load*
- Damping and oscillation frequency depend on each other. *Solution: true*

(l) As a fundamental property of oscillating systems (mass positions as coordinates)...

- the mass matrix is diagonal. *Solution: true*
- the damping matrix is diagonal. *Solution: false*
- the stiffness matrix is diagonal. *Solution: false*
- the mass, damping, and stiffness matrices are symmetric. *Solution: true*

Linear-Elastic Joints

Exercise A.5 Knowledge Questions*

This exercise repetitively asks questions on the content of Sect. 4.

(a) Name at least 3 elements that introduce joint elasticity.
Solution: belts, long shafts, cables, harmonic drives, or cycloidal gears
(b) Where do torques about the shaft axis emerge from?
Solution: driving actuator, robot link structure, load and interaction forces
(c) What is the wave speed within an uniform elastic joint shaft with rotation symmetric cross section?
Solution: $c_\theta = \sqrt{\dfrac{G}{\rho}}$
(d) How is the torsional moment τ_x for a uniform, circular cross section with inner radius r_i and outer radius r_o computed?
Solution: (44)
(e) Calculate the inertia per unit length for a uniform cylindrical shaft of density ρ.
Solution: (49)
(f) Note the equivalent stiffness for a number of n series/parallel springs.
Solution: (54), (55)
(g) Name a model reduction approach to simplify single-joint models.
Solution: equivalent lumped spring element

(h) Which are three typical control objectives and how are they related to each other?
   Solution: position regulation, trajectory tracking, force/torque control (ordered)

Exercise A.6 Multiple-Choice Questions*

This exercise also repetitively asks questions on the content of Sect. 4. More than one answer might be possible and necessary.

(a) Which statements about robots with elastic joints are true?

   - The forward kinematics of the arm don't change. *Solution: false*
   - The state vector of the dynamic equations changes. *Solution:*
   - The robot gets redundant. *Solution: false*
   - The principles of elastic effects of rotational joints are different from those of translational ones. *Solution: false*

(b) Which statements about uniform elastic joint shafts are true?

   - Due to symmetry the sheared cross section remains parallel and flat. *Solution: true*
   - The elastic properties are almost continuously distributed along the shaft. *Solution: true*
   - They can be modelled as uniform beams. *Solution: false*
   - The torsion constant T depends on the shear modulus G of the material and the cross section dimensions. *Solution: true*

(c) The mass of the elastic joint shaft is negligible due to...

   - the effective inertia of the remaining transmission. *Solution: true*
   - the actuating torque. *Solution: false*
   - the link and load inertia. *Solution: true*
   - the high torsion wave propagation speed in relation to the joint shaft width. *Solution: false*
   - the mass of the robot's base. *Solution: false*

(d) Which statements about non-uniform elastic joint shafts are true?

   - For accurate torque measurements, deflection sensors should be placed on rather stiff parts. *Solution: false*
   - Shafts with piecewise constant radius can be treated as a parallel connection of multiple springs. *Solution: false*
   - When combining a rather soft and a rather stiff shaft section, the analysis can focus on the soft one. *Solution: true*
   - If subdividing the shaft into uniform sections is not reasonable, the tangential stress σ has to be integrated over the varying the cross section. *Solution: true*

Linear-Elastic Links

Exercise A.7 Knowledge Questions*

This exercise repetitively asks questions on the content of Sect. 5.

(a) Which theory is usually applied to model elastic links? Which assumptions have to hold therefore?
Solution: Euler-Bernoulli beam theory, see Sect. 5.1

(b) What is the meaning of Hooke's law?
Solution: Deflections remain in the linear elastic and reversible material regime. Thus, the object without external forces/torques will turn back to its original geometry.

(c) How to compute the strain ε on a beam surface? On which variables does it depend?
Solution: (57)

(d) Write the Euler-Bernoulli beam equation. What does it describe? In terms of which variables can it be written?
Solution: (67), it describes the bending wave relation in terms of the bending angle θ_b, the bending force f_b or the bending torque τ_b

(e) Name an approach to obtain exact solutions to elastic link bending problems.
Solution: Modal analysis.

(f) What is the main property of a homogeneous and uniform link? Which advantages does that bring along?
Solution: EI_b and ρ_b are constant. $\rightarrow$ the spatial and temporal dependencies are separated in the beam equation, representing independent eigenvalue problems with equal, constant solutions, one of the rare examples that admit a closed-form solution

(g) What is the eigenfrequency of a homogeneous, uniform link?
Solution: (69)

(h) How are the initial value problem and the boundary value problem associated?
Solution: Both are eigenvalue problems; the first with respect to time, the second with respect to space.

(i) How is the wave number defined?
Solution: It is the spatial equivalent to the temporal frequency ω and measures the number of periods per unit length (72).

(j) How many integration constants are necessary to solve the boundary value problem? How are they usually chosen?
Solution: 4 as it is a forth order spatial derivative; two kinematic boundary conditions for each link end

(k) Name the kinematic boundary conditions for a free, clamped and pinned link end.
Solution: see Table 6

(l) Why are kinematic boundary conditions usually preferred over dynamic ones?
Solution: Dynamic boundary conditions mostly occur in the form of masses, lumped springs and dampers connected to elastic link ends. While their effective values strongly depend on the joint configuration, kinematic constraints remain almost constant.

(m) How can dynamic couplings between links, dynamic loads and contacts of arbitrary compliance be considered?
Solution: By incorporating them into the bending problem as external load force

(n) Which parameters describe the i-th bending mode of a link?
Solution: the wave number $k_{\omega,\,i}$, the natural frequency ω_i and the mode shape $\phi_i(x)$

(o) What does the modal expansion theorem for bending beams state?
Solution: Any function $\phi(x)$ satisfying the boundary conditions of the link and ensuring $\frac{\mathrm{d}^2}{\mathrm{d}x^2}\left[E\mathcal{I}_b(x)\frac{\mathrm{d}^2\phi(x)}{\mathrm{d}x^2}\right]$ to be continuous, can be expanded in the absolutely and uniformly convergent series of the eigenfunctions (86).

(p) Which is the main advantage of the modal expansion and the orthogonality of the mode shapes?
Solution: The solution of the initial value problem as a partial differential equation can be converted into a system of mutually independent ordinary modal differential equations.

(q) Why do we need approximate solutions for elastic ink bending?
Solution: Real robots usually show more complex link geometries and non-uniform mass distributions, such that a closed-form solution does not exist. Additionally, there is an infinite number of modes, theoretically.

(r) What are the requirements for successful implementation of control algorithms?
Solution: A finite set of simplified, real-time solvable model equations.

(s) What is the equivalent spring constant for a clamped-free beam under lumped bending force or bending torque?
Solution: Fig. 14a/b

(t) Which is the condition for the approximation of an elastic link as an equivalent massless spring? What does this lead to?
Solution: The mass of the elastic link itself needs to be small compared to the mass of the joints and the payload. This leads to a bending wave propagation speed that is high in the frequency range of interest compared to the elastic link length and the variation of the bending variables along the link becomes negligible.

(u) Name three different equivalent spring approximations to model elastic links. Why are they often only used as initial approximations to design elastic robot components?
Solution: Fig. 14. In practice, the requirements to model the elastic link as an equivalent massless spring are seldom fulfilled.

(v) State at least the five classes of approximation techniques to model elastic link dynamics.

Solution: Fig. 15, series discretisation methods, lumped parameter methods, finite element techniques, frequency domain techniques and wave based approaches.

Exercise A.8 Multiple-Choice Questions*

This exercise also repetitively asks questions on the content of Sect. 5. More than one answer might be possible and necessary.

(a) Which of the following statements are assumptions to regard an elastic link as an Euler-Bernoulli beam?

- The torsion along the neutral fibre is high compared to the beam width. *Solution: false*
- The length of the neutral fibre remains constant. *Solution: true*
- The deflections are small compared to the beam height. *Solution: false*
- Deflections exceed the linear elastic and reversible material regime. *Solution: false*

(b) Which statements about Euler-Bernoulli beams are true?

- The bending torque is anti-proportional to the surface strain. *Solution: false*
- The surface strain depends on the time. *Solution: true*
- The shear is negligible. *Solution: true*
- The bending torque depends on the displacement along the beam length. *Solution: true*
- $\mathcal{I}_b(x)$ incorporates all material and geometry parameters that define the beam's rigidity. *Solution: false*

(c) Which statements about the moment balance about the coordinate origin of a beam element are true?

- The bending torque has to be considered at both ends. *Solution: true*
- Gravitation is negligible. *Solution: false*
- The bending force has to be considered at both ends. *Solution: false*
- External forces might contribute to the moment balance. *Solution: true*
- For an infinitesimal beam element, second order terms of the infinitesimal element length $\mathrm{d}x$ can be neglected. *Solution: true*

(d) Which statements about elastic link bending are true?

- The deflection can be expressed as the product of a shape function ϕ and an amplitude function v. *Solution: true*
- The deflection varies along the link. *Solution: true*
- Typically, the elastic link is modelled as a Timoshenko beam. *Solution: false*
- The shape function ϕ as well as the amplitude function v depend on the location along the link. *Solution: false*

(e) What is the procedure to solve the Euler-Bernoulli equation?

- Solve the initial value problem before the remaining boundary value problem. *Solution: false*
- Solve the boundary value problem before the remaining initial value problem. *Solution: true*
- Solving the boundary value problem is sufficient. *Solution: false*
- Solving the initial value problem is sufficient. *Solution: false*

(f) Which of the following statements about homogeneous, uniform links are wrong?

- The bending wave speed is proportional to the temporal frequency. *Solution: true*
- The so-called wave number is proportional to the bending wave speed. *Solution: false*
- Bending waves in elastic links display dispersive effects. *Solution: true*
- All bending excitations propagate at the same bending wave speed. *Solution: false as the speed depends on the frequency*

(g) How many modal differential equations does the system of equations for the exact solution for elastic link bending contain when exploiting the modal expansion and the orthogonality of the mode shapes?

- None, the system is described by partial differential equations. *Solution: false*
- One modal differential equation per mode. *Solution: true*
- One modal differential equation per link. *Solution: false*
- Two modal differential equation per mode, due to their orthogonality. *Solution: false*

(h) Which statements about an elastic link with a small mass compared to the mass of the joints and the payload are true?

- The variation of the bending variables along the link becomes negligible. *Solution: true*
- The bending wave propagation speed in the frequency range of interest is low compared to the elastic link length. *Solution: false*
- The link's mass needs to be considered. *Solution: false*
- Its dynamics can be approximated by an equivalent massless spring. *Solution: true*

(i) Which statements about modelling the elastic links of a robot as torsional springs are true?

- They can always be modelled as equivalent springs. *Solution: false*
- The dynamics of the elastic links can be approximated analogous to those of elastic joint shafts. *Solution: true*

- The relations presented for combinations of series connections of equivalent torsional springs hold. *Solution: true*
- The relations presented for combinations of parallel connections of equivalent torsional springs do not hold. *Solution: false*

Multi-Degree-of-Freedom Robots

Exercise A.9 Knowledge Questions*

This exercise repetitively asks questions on the content of Sect. 6.

(a) Which two formalism can be applied to derive the equations of motions for elastic joint robots and which for elastic link robots?
Solution: Newton-Euler formalism and the Euler-Langrange formalism can be used to derive the equations of motion for elastic joint robots as well as elastic link robots.

(b) State the general equations of motion of an elastic robot. What is the meaning of the different parts?
Solution: (92), torques = mass matrix times acceleration + Coriolis and Centrifugal terms times velocity + damping times velocity + stiffness times position + gravity terms.

(c) How many degrees of freedom does a conventional robot posses compared to an elastic robot?
Solution: Conventional: motor angles for each of the n_θ robot joints, Elastic: $n_\theta + n_\delta$, number of joints + elastic DOF.

(d) Why are the elastic DOF usually not actuated by their own actuators? How to deal with the resulting underactuated system?
Solution: In order to reduce system complexity and cost, additional actuators are renounced. The controllers exploit the dynamic couplings through inertia, gyroscopic and stiffness matrices to control the elastic degrees of freedom.

(e) Why do engineers usually try to avoid eccentricity in joints?
Solution: In order to decrease joint wear.

(f) Why is the structural damping of elastic joint robots usually neglected? What is the problem with that? How do elastic links behave comparably?
Solution: The engineering materials that form the elastic joint component are typically metals, that exhibit very low structural damping. Many control problems of elastic joint robots emerge from the very low intrinsic joint damping. This is the same for elastic links, usually. Elastic link designs based on composite materials represent an exception.

(g) State the reduced equations of motion of an elastic joint robot.
Solution: (93)

(h) Often a high gain is chosen for joint actuation controllers of elastic link robots. Why?
Solution: To regard the link on the joint actuator side as clamped.

(i) Write the reduced equations of motion of an elastic link robot.
Solution: (98)

(j) How can the problem of underactuation of an elastic link robot be faced?
Solution: Besides controllers exploiting dynamic couplings, additional actuators can be used, e.g. piezoelectric patches.

(k) What is meant by controllability and modal accessibility of oscillations? In which cases are they questionable?
Solution: In the sense of control theory, controllability refers to the overall feasibility to control the robot motion and to dampen undesired structural oscillation modes with the available actuators. Modal accessibility quantifies the potential of the present actuators to dampen individual oscillation modes. Both become questionable when considering an elastic link robot with link bending occurring outside the plane of rigid motion and no mounted patch actuators.

(l) Why can elastic joint robots be expected to show complete controllability?
Solution: The torsional deflection axis always coincides with the joint actuation axis.

(m) Name three metrics to compute the controllability of elastic robots.
Solution: Binary controllability indicator, modal controllability metric, and geometric controllability metric.

(n) How can the problem of controllability be circumvented?
Solution: By carefully designing the robot, such that oscillations only occur in planes which can be governed by the present actuators.

Exercise A.10 Multiple-Choice Questions*

This exercise also repetitively asks questions on the content of Sect. 6. More than one answer might be possible and necessary.

(a) Which of the following statements are true?

- An elastic joint model assumes the predominant elasticity to occur in the mechanical interface between motor and rigid robot links. *Solution: true*
- An elastic joint model assumes the predominant elasticity to occur in the links. *Solution: false*
- An elastic link model assumes the predominant elasticity to occur in the mechanical interface between motor and the rigid robot links. *Solution: false*
- An elastic link model assumes the predominant elasticity to occur in the links. *Solution: true*

(b) Which statements about the Newton-Euler and Euler-Lagrange formalism are true?

- Newton-Euler formalism can be applied to revolute joints. *Solution: true*
- Euler-Langrange formalism can be applied to revolute joints. *Solution: true*

- The principal mathematical structure of the resulting equations of motion for a multi-degree-of-freedom elastic robot depend on the formalism. *Solution: false*
- Newton-Euler formalism can be applied to prismatic joints. *Solution: true*
- Euler-Langrange formalism can be applied to prismatic joints. *Solution: true*
- Both formalisms are not applicable to revolute nor to prismatic joints. *Solution: false*

(c) Which statements about the general equations of motion of elastic robots are true?

- The mass matrix depends on the joint angles solely. *Solution: false*
- The stiffness matrix depends on the elastic displacements solely. *Solution: false*
- The stiffness matrix is a square matrix. *Solution: true*
- The mass matrix' elements are linear functions of the coordinates $\boldsymbol{\theta}_m$ and $\boldsymbol{\delta}$. *Solution: false*
- The damping terms are independent of joint angles and elastic displacements. *Solution: true*
- All coordinates and their derivatives are time-dependent. *Solution: true*
- $\mathbf{D}$ is the diagonal dissipative damping matrix. *Solution: true*
- $\mathbf{C}$ is the diagonal dissipative matrix collecting the gyroscopic terms. *Solution: no, the states of multi-degree-of-freedom robots are mutually coupled through non-dissipative velocity dependent gyroscopic terms*
- The gravitational forces might be non-linear functions of the coordinate vector. *Solution: true*

(d) Which simplifying assumptions are usually made for elastic joint robots?

- The joint rotor and gear inertia is asymmetric about the joint's axis of rotation. *Solution: false*
- The joint elasticity is modelled as an equivalent spring. *Solution: true*
- The damping is considered to emerge from gearing and air friction. *Solution: no, only gearing friction.*
- The gear inertia is symmetric about the joint's axis of rotation. *Solution: true*
- The kinetic energy of the motor shaft is mainly due to its own rotation. *Solution: true*
- Joint axes are considered as non-uniform cylinders. *Solution: false*

(e) Which statements about elastic joint robots are true?

- The only rotor contributions to the robot inertia matrix $\mathbf{M}$ are the off-diagonal block matrices $\mathbf{M}_{m\delta}(\boldsymbol{\theta}_m, \boldsymbol{\theta}_l)$ and $\mathbf{M}_{m\delta}^{\top}(\boldsymbol{\theta}_m, \boldsymbol{\theta}_l)$. *Solution: no, those matrices vanish.*
- Typically high transmission ratios are used. *Solution: true*

- Any gyroscopic terms $\mathbf{C}_m$, $\mathbf{C}_{m\delta}$ and $\mathbf{C}_{\delta m}$ related to the rotor mass and inertia in $\mathbf{C}$ vanish. *Solution: true*
- A low transmission ratio scales the apparent rotor inertia as well as the rotational speed about the joint axis up. *Solution: false*
- The stiffness Matrix $\mathbf{K}$ is considered independent of the coordinates. *Solution: true*

(f) Which simplifying assumptions are usually made for elastic link robots?

- The inertia matrix $\mathbf{M}$ is assumed to be a function of all coordinates. *Solution: no, a function of the joint angles $\boldsymbol{\theta}_m$ only.*
- The bending deformations are assumed to be limited to the plane orthogonal to the rigid motion. *Solution: no, in the same plane*
- The link deflections stay small. *Solution: true*
- The stiffness matrix $\mathbf{K}$ only contains the diagonal modal stiffness terms for each link in the bottom right block. *Solution: true*
- The inertia sub-matrix $\mathbf{M}_\nu$, associated with the modal coordinates, is constant. *Solution: true*

(g) Which statements about elastic link robots are true?

- The boundary conditions on the joint actuator side are usually chosen to be clamped boundary conditions. *Solution: true*
- Link elasticity may occur as an unexpected design issue. *Solution: true*
- The elements of $\mathbf{C}$ thus depend on the joint angles as well as joint accelerations. *Solution: no, angles and velocities.*
- Small link deflections reduce the complexity of the multi-link equations. *Solution: true*
- The Euler-Bernoulli equation has to be solved for each link approximately. *Solution: true*
- Many multi-elastic-link robots posses extra actuators. *Solution: false*

(h) Which statements comparing elastic joint and elastic link robots are true?

- As the stiffness increases, the dynamics model of elastic joint robots converges to the one of elastic link robots. *Solution: false*
- The dynamics of elastic link robots are easier to compute that those of elastic joint robots. *Solution: false*
- The key difference to elastic joint robots is that with elastic links the off diagonal blocks of the inertia matrix can be neglected.*Solution: false*
- The experienced damping of elastic joint robots is analogue to the one of elastic link robots. *Solution: true*
- The degree of underactuation of elastic joint robots might be higher than the one of an equivalent robot with elastic links instead. *Solution: false*

Design Considerations for Elastic Elements

Exercise A.11 Knowledge Questions*

This exercise repetitively asks questions on the content of Sect. 7.

(a) What is the purpose of the intentional integration of an elastic element as a sensor in a robotic actuator?
Solution: It serves for the purpose of calculating the load force based on the spring deflection measurements

(b) How can the system pictured in Fig. 20 mathematically be described?
Solution: Eq. (99)

(c) Why is the spring design in elastic joints a challenging task? Which aspects need to be considered?
Solution: minimise spring damping to use as sensor, desired stiffness, strength, resolution, and linearity, suppressing parasitic effects such as ripples, cross-talk to non-axial torque loads, noise, thermal and temporal drift

(d) Name the two most common ways to sense deflections.
Solution: strain gauge based and encoder based

(e) How can the usage of elasticity in robot design be helpful?
Solution: Elasticity can help mitigate the forces building up during contacts between a robot and a work-piece or/and a robot and a human

(f) How can the relationship between the force against ground and the load force be mathematically described in the Laplace domain?
Solution: Eq. (102)

(g) In which cases would which of the deflection sensing techniques be preferred?
Solution: strain gauge based: higher stiffness, no thermal changes encoder based: lower stiffness, thermal changes

(h) Which location is best to mount the sensing device respectively?
Solution: strain gauges: location of highest strain encoder: location of highest displacement

(i) What is the relation between the impulse peak force response at the motor and the characteristic frequency?
Solution: $t_{\max} = \frac{\pi}{2\omega_d} = \frac{\pi}{2\omega_0\sqrt{1-\zeta^2}}$

(j) Consider a motor colliding with a heavy object. How can the motions be mathematically described in such a case?
Solution: (105)

(k) How is a *soft joint* able to protect the motor?
Solution: softer spring -> lower impulse peak force response at the motor and more time to react via motor controller, before a harmful force level is reached

(l) Do the effects of an elastic joint occur both ways (regarding a collision either on motor or joint side)? Reason your answer and describe differences if applicable.
Solution: Sect. 7.2.2

(m) Why are impact tests necessary?
Solution: impact tests are necessary to evaluate the true impact force evolution and transmission over time

(n) How do elastic elements also help (besides the fact that they have an impact on mitigating undesired potentially harmful collision forces)?
Solution: They also affect the transmission of desired interaction forces

(o) Which are the total physical bandwidth limits?
Solution: the minimum of the peak torque limit and the speed limit

(p) How should the robot be designed, if it shall be optimised for high control bandwidth?
Solution: it should be made as stiff as possible

(q) What do the protecting properties of elastic elements mean with regard to potentially harmful forces when it comes to *normal operation mode*?
Solution: desired interaction forces are also delayed

(r) How does an elastic joint influence a load side force control?
Solution: phase lag decreases accuracy and can threaten stability; magnitude drop -> disproportionate actuation effort

(s) How is the impedance of the mass-spring-damper system mathematically defined?
Solution: $ms + d + \frac{k}{s}$, see term in brackets in Eq. (113)

(t) What is the power around resonance only required for? Why is that?
Solution: around resonance, the power is only required to compensate damping induced energy losses. Because the mass and stiffness terms cancel at resonance.

(u) How is the torque τ in Fig. 26a defined for small displacements around $\theta = 0$?
Solution: Eq. (115)

Exercise A.12 Multiple-Choice Questions*

This exercise also repetitively asks questions on the content of Sect. 7. More than one answer might be possible and necessary.

(a) Which of the following statements are true for a collision on the load side of an elastic joint?

- Heavy load masses reduce the impact force at the motor. *Solution: true*
- High transmission rates lead to zero force transmissibility from load side to motor. *Solution: false*
- Higher spring stiffness fastens the impulse peak force response at the motor. *Solution: true*
- A high transmission ratio reduces the damping of impacts on the motor. *Solution: false*
- Higher spring damping increases the impulse peak force response at the motor. *Solution: false*
- Higher spring stiffness leads to higher impulse peak force response at the motor. *Solution: true*

(b) Which of the following statements about elastic joints are true?

- A harmonic force f_m generated by the motor will be scaled and delayed while transferred to the load side. *Solution: true*
- In a reasonable motor design, we should expect an overdamped response. *Solution: false*
- The scaling and delay of a transferred harmonic force only depends on the damping ratio. *Solution: no, also on the ratio between desired harmonic frequency and resonance frequency*
- The generated torque is independent of the resonance frequency. *Solution: no, Around the resonance frequency ω_d, the generated torque is significantly amplified.*
- Any torque applied via an elastic joint leads to safe operation. *Solution: no, Accurate torque tracking in the vicinity of the resonance peak requires a controller to actively damp the system, to avoid damages of the robot actuation and harm of objects or humans in contact with the robot.*
- A lower stiffness leads to a higher control bandwidth. *Solution: false*
- The control bandwidth is limited by different effects of the actuator power limitations. *Solution: yes, as the actuator power limitations translate into different force limitations at different speeds*

Variable Impedance Actuation

Exercise A.13 Knowledge Questions*

This exercise repetitively asks questions on the content of Sect. 8.

(a) Which are the key parameters variable physical impedance allows efficient adaption to?
Solution: variable physical impedance allows an efficient adaption regarding the required contact sensitivity and positioning tolerance as well as accuracy and force level
(b) In which groups are variable stiffness actuators divided into?
Solution: preloading/-tensioning of springs, altering the load-spring transmission, and physically changing the spring structure
(c) What does variable damping concepts introduce regarding to [40]?
Solution: variable damping concepts introduce an additional control input on the elastic degrees of freedom

Exercise A.14 Multiple-Choice Questions*

This exercise also repetitively asks questions on the content of Sect. 8. More than one answer might be possible and necessary.

(a) Which of the following reasons are motivations for variable impedance actuators?

> - non-adaption to different tasks *Solution: false, the right answer would be: adaption to different tasks*
> - (modal) dynamics matching *Solution: true*
> - increasing the total actuator peak power capacity *Solution: true*
> - decreasing the total actuator peak power capacity *Solution: false*

(b) How does the motor inertia appear to scale with the transmission ratio viewed from the actuator output?

> - linearly *Solution: false*
> - quadratically *Solution: true*
> - exponentially *Solution: false*
> - there is no relationship between inertia and transmission ratio *Solution: false*

References

1. L. Bascetta, G. Ferretti, B. Scaglioni, Closed form Newton–Euler dynamic model of flexible manipulators. Robotica, 1–25 (2015)
2. W.J. Book, Modeling, design and control of flexible manipulator arms. PhD thesis, Massachusetts Institute of Technology (1974)
3. W.J. Book, Recursive Lagrangian dynamics of flexible manipulator arms. Int. J. Robot. Res. **3**(3), 87 (1984)
4. W.J. Book, S.L. Dickerson, G.G. Hastings, S. Cetinkunt, T.E. Alberts, Combined approaches to lightweight arm utilization. Georgia Institute of Technology (1985)
5. G. Buondonno, A. De Luca, A recursive Newton-Euler algorithm for robots with elastic joints and its application to control, in *IEEE/RSJ International Conference On Intelligent Robots and Systems (IROS), 2015* (IEEE, 2015), pp. 5526–5532
6. A. de Luca, B. Siciliano, Closed-form dynamic model of planar multilink lightweight robots. IEEE Trans. Syst. Man Cybern. **21**(4), 826–839 (1991)
7. F. Duarte, P. Ballesteros, F. Ullah, C. Bohn, Modeling and sliding mode control of a single-link flexible robot to reduce the transient response, in *20th International Conference On Methods and Models in Automation and Robotics (MMAR), 2015* (IEEE, 2015), pp. 235–240
8. S. Dwivedy, P. Eberhard, Dynamic analysis of flexible manipulators, a literature review. Mech. Mach. Theory **41**(7), 749–777 (2006)
9. S. Eppinger, W. Seering, Understanding bandwidth limitations in robot force control, in *IEEE International Conference On Robotics and Automation. Proceedings. 1987*, vol. 4 (IEEE, 1987), pp. 904–909
10. S.D. Eppinger, W.P. Seering, Three dynamic problems in robot force control, in *IEEE International Conference on Robotics and Automation, 1989. Proceedings, 1989* (IEEE, 1989), pp. 392–397
11. A. Ghazavi, F. Gordaninejad, N. Chalhoub, Dynamic analysis of a composite-material flexible robot arm. Comput. Struct. **49**(2), 315–327 (1993)
12. J.W. Grizzle, J. Hurst, B. Morris, H.-W. Park, K. Sreenath, Mabel, a new robotic bipedal walker and runner, in *American Control Conference, 2009. ACC'09.* (IEEE, 2009), pp. 2030–2036

13. K.-H. Grote, J. Feldhusen, *DUBBEL: Taschenbuch Für Den Maschinenbau* (Springer-Verlag, Berlin, 2007)
14. R. Ham, T. Sugar, B. Vanderborght, K. Hollander, D. Lefeber, Compliant actuator designs. IEEE Robot. Autom. Mag. **16**(3), 81–94 (2009)
15. C. Hubicki, J. Grimes, M. Jones, D. Renjewski, A. Spröwitz, A. Abate, J. Hurst, Atrias: design and validation of a tether-free 3d-capable spring-mass bipedal robot. Int. J. Robot. Res. **35**(12), 1497–1521 (2016)
16. F. Irgens, *Continuum Mechanics* (Springer Science & Business Media, Berlin, 2008)
17. F.I. John, J. Malzahn, T. Bertram, Controllability and accessibility of vibrations in multiple planes on link-elastic robot arms, in *2017 IEEE International Conference on Robotics and Biomimetics (ROBIO)* (IEEE, 2017), pp. 1491–1496
18. A. Konno, M. Uchiyama, Y. Kito, M. Murakami, Configuration-dependent controllability of flexible manipulators, in *Experimental Robotics III* (Springer, Berlin, 1994), pp. 529–544
19. A. Konno, M. Uchiyama, M. Murakami, Configuration-dependent vibration controllability of flexible-link manipulators. Int. J. Robot. Res. **16**(4), 567–576 (1997)
20. D. Lakatos, W. Friedl, A. Albu-Schäffer, Eigenmodes of nonlinear dynamics: definition, existence, and embodiment into legged robots with elastic elements. IEEE Robot. Autom. Lett. **2**(2), 1062–1069 (2017)
21. K.F. Laurin-Kovitz, J.E. Colgate, S.D. Carnes, Design of components for programmable passive impedance, in *IEEE International Conference on Robotics and Automation, 1991. Proceedings* (IEEE, 1991), pp. 1476–1481
22. A. Luca, B. Siciliano, Regulation of flexible arms under gravity. IEEE Trans. Robot. Autom. **9**(4), 463–467 (1993)
23. J. Malzahn, T. Bertram, Video: Exploiting Link Elasticity in a Conventional Industrial Robot Arm (2014). https://youtu.be/cAk5CPEWRMk
24. J. Malzahn, N. Kashiri, W. Roozing, N. Tsagarakis, D. Caldwell, What is the torque bandwidth of this actuator?, in *2017 IEEE/RSJ International Conference on Intelligent Robots and Systems (IROS)* (IEEE, 2017), pp. 4762–4768
25. J. Malzahn, V. Barasuol, K. Janschek, Actuator modeling and simulation, in *Humanoid Robotics: A Reference*, ed. by A. Goswami, P. Vadakkepat Humanoid robotics: a reference. Springer Publishing Company Incorporated 2018
26. L. Meirovitch, *Fundamentals of Vibrations* (McGraw-Hill, Boston, 2001)
27. N. Paine, J.S. Mehling, J. Holley, N.A. Radford, G. Johnson, C.-L. Fok, L. Sentis, Actuator control for the NASA-JSC Valkyrie humanoid robot: a decoupled dynamics approach for torque control of series elastic robots: actuator control for the NASA-JSC Valkyrie humanoid robot. J. Field Robot. **32**(3), 378–396 (2015)
28. D. Paluska, H. Herr, The effect of series elasticity on actuator power and work output: implications for robotic and prosthetic joint design. Robot. Auton. Syst. **54**(8), 667–673 (2006)
29. G.A. Pratt, M.M. Williamson, Series elastic actuators, in *IEEE/RSJ International Conference On Intelligent Robots and Systems 95.'Human Robot Interaction and Cooperative Robots', Proceedings. 1995*, vol. 1 (1995), pp. 399–406
30. G. Robinson, J. Davies, Continuum robots - a state of the art, in *1999 IEEE International Conference on Robotics and Automation, 1999. Proceedings*, vol. 4 (IEEE, 1999), pp. 2849–2854
31. J. Rösler, H. Harders, M. Bäker, *Mechanisches Verhalten der Werkstoffe.* Lehrbuch, 5., aktualisierte und erweiterte auflage edition (Springer Vieweg, Wiesbaden, 2016). OCLC: 951092036
32. J.E. Shigley, *Shigley's Mechanical Engineering Design* (Tata McGraw-Hill Education, New York, 2011)
33. H.-C. Shin, S.-B. Choi, Position control of a two-link flexible manipulator featuring piezoelectric actuators and sensors. Mechatronics **11**(6), 707–729 (2001)
34. M.W. Spong, Modeling and control of elastic joint robots. J. Dyn. Syst. Meas. Control **109**(4), 310 (1987)

35. M.W. Spong, Control of flexible joint robots: a survey. Coordinated Science Laboratory Report no. DC-116 (UILU-ENG 90-2203) (1990)
36. L. Sweet, M. Good, Re-definition of the robot motion control problem: effects of plant dynamics, drive system constraints, and user requirements, in *The 23rd IEEE Conference on Decision and Control, 1984* (IEEE, 1984), pp. 724–732
37. S. Tosunoglu, S.-H. Lin, D. Tesar, Identification of inaccessible oscillations in n-link flexible robotic systems, in *Proceedings of the 28th IEEE Conference on Decision and Control, 1989* (IEEE, 1989), pp. 2512–2518
38. S. Tosunoglu, S.-H. Lin, D. Tesar, Accessibility and controllability of flexible robotic manipulators. J. Dyn. Syst. Meas. Control **114**(1), 50–58 (1992)
39. N.G. Tsagarakis, S. Morfey, H. Dallali, G.A. Medrano-Cerda, D.G. Caldwell, An asymmetric compliant antagonistic joint design for high performance mobility, in *IEEE/RSJ International Conference On Intelligent Robots and Systems (IROS), 2013* (IEEE, 2013), pp. 5512–5517
40. B. Vanderborght, A. Albu-Schaeffer, A. Bicchi, E. Burdet, D. Caldwell, R. Carloni, M. Catalano, O. Eiberger, W. Friedl, G. Ganesh, M. Garabini, M. Grebenstein, G. Grioli, S. Haddadin, H. Hoppner, A. Jafari, M. Laffranchi, D. Lefeber, F. Petit, S. Stramigioli, N. Tsagarakis, M. Van Damme, R. Van Ham, L. Visser, S. Wolf, Variable impedance actuators: a review. Robot. Auton. Syst. **61**(12), 1601–1614 (2013)
41. T. Verstraten, P. Beckerle, R. Furnémont, G. Mathijssen, B. Vanderborght, D. Lefeber, Series and parallel elastic actuation: impact of natural dynamics on power and energy consumption. Mech. Mach. Theory **102**, 232–246 (2016)
42. S. Wolf, G. Grioli, O. Eiberger, W. Friedl, M. Grebenstein, H. Hoppner, E. Burdet, D.G. Caldwell, R. Carloni, M.G. Catalano, D. Lefeber, S. Stramigioli, N. Tsagarakis, M. Van Damme, R. Van Ham, B. Vanderborght, L.C. Visser, A. Bicchi, A. Albu-Schaffer, Variable stiffness actuators: review on design and components. IEEE/ASME Trans. Mechatron. **21**(5), 2418–2430 (2016)
43. M. Zinn, O. Khatib, B. Roth, J.K. Salisbury, Playing it safe [human-friendly robots]. IEEE Robot. Autom. Mag. **11**(2), 12–21 (2004)

Soft Robotics

Cecilia Laschi

Abstract Should robots be stiff to resist external mechanical perturbations or should they adapt to and maybe exploit the interaction with their environments? Soft robotics is an approach to design and develop deformable robots that can comply with, and even exploit, interactions for accomplishing their tasks. Soft robotics is then the use of soft materials or deformable structures to build robots. Though a young field, a wide range of techniques and technologies exist today for designing and building soft robots, including methods for fabrication, technologies for actuation and sensing, as well as control techniques. Together with such technological development, soft robotics technologies enable robot abilities that were not possible before, like squeezing, stretching, morphing, stiffening, growing, self-healing, evolving. They open up new scenarios for robotics, leading towards robots that can effectively and efficiently adapt to their environments and tasks. It becomes possible to pursue applications of soft robots in several fields that range from explorations in unstructured environments to biomedical applications, where a soft interaction with human patients is required.

1 A Rationale for Soft Robotics: Embodied Intelligence and Morphological Computation

> Many tasks become much easier if morphological computation is taken into account.
>   Rolf Pfeifer

Morphology and body mechanics determine the way we move more than we think. Movements and sensory-motor coordination emerge from the complex interaction of our body mechanical properties, shape and arrangement of muscles and receptors with our physical environment [1]. The role of body in shaping our behavior is named *Embodied Intelligence* [2] and the part of computing done by the

---

C. Laschi (✉)
National University of Singapore, College of Design and Engineering, Singapore, Singapore
e-mail: mpcclc@nus.edu.sg

© Springer Nature Switzerland AG 2025
B. Siciliano (ed.), *Robotics Goes MOOC*,
https://doi.org/10.1007/978-3-319-75823-7_5

physical body in motor control is referred to as *Morphological Computation* [3, 4]. This is true in living beings, and in robots as well [5]. If properly designed, robots can take advantage of the physical interaction with the environment to perform the desired movements and reduce the computational burden on their control system. To some extent, this may make a difference in the success of robots negotiating complex natural environments, enabling them to perform tasks and ultimately supporting the materialization of the huge growth of service robotics market that is often forecast.

In order to take advantage of robot-environment interaction, the robot bodyware needs to make use of compliance, instead of fighting it. Properly designed compliance produces deformations that help motor behaviour. A simple example is the compensation of impacts on feet and legs while walking, done by the compliant knee joints; and compliance is actually used in many models for locomotion [6]. But compliance has an important role in manipulation, too [7, 8].

1.1 Soft Robotics Definitions

Soft Robotics is about introducing compliance in robot bodyware. It can be done by actuators with variable stiffness or with compliant control techniques ([7]; also see Chapter 4). But it can even be done by using soft materials for building soft robots. In this second case soft robots are simply defined as "robots built with soft materials" [9], or as "soft-bodied robots", in analogy with "soft-bodied animals. In living beings, we observe that the vast majority of animals are soft-bodied: even animals with stiff skeletons are mainly composed of soft tissues and liquids [10]. Some definitions focus on the materials: "systems that are capable of autonomous behaviour, and that are primarily composed of materials with moduli in the range of that of soft biological materials" [11] "soft-matter robotics", based on the well-known concept of "soft matter" used for materials [12].

More precisely, soft robotics can be defined as:

> Soft robots/devices that can actively interact with the environment and can undergo 'large' deformations relying on inherent or structural compliance.
> RoboSoft Community[1]

The definition given by the RoboSoft community is the most comprehensive, as it outlines the deformability of the robot bodyware, either by intrinsic material compliance, i.e. low Young's modulus, or by extrinsic morphology that magnifies strains of rigid-material structural elements (i.e. plastic or metallic thin layers or fibers).

---

[1] RoboSoft is a Coordination Action on Soft Robotics funded by the European Commission in 2013–2016. The RoboSoft Community accounted for 34 member institutions for a total of 100+ scientists.

2 How to Build Soft Robots: Soft Robotics Actuation Technologies

How to make a piece of soft material deform and move? Biology suggests muscle-like contractions, but artificial muscles are still an unmatched goal. Some smart materials can approximate muscle-like contractions and electrical or fluidic actuation can also fit the purpose to some extent and can produce controlled deformations of soft structures. In this chapter, we analyse a few possible approaches to contraction, deformation and stiffening, as summarised in Figs. 1 and 6. For each, we describe the working principle and the main mathematical relations governing it, where relevant.

2.1 Electro-active Polymers—EAPs

EAPs promisingly show a possibility to fill the gap between artificial and natural muscles [13] especially for their high deformability and energy developed. EAPs change their shape and dimension upon electrical stimulation [14]. They are usually classified in two groups: electronic and ionic (see Table 1).

For the simplicity of their working principle and for their versatility, the most widely investigated are Dielectric Elastomers—DEAs, belonging to the first group. DEAs consist of a very compliant, electrically insulator, elastomer, generally planar (one dimension much smaller than the other two), combined with very thin layers of highly compliant, electrically conductive, material on the two main areas (e.g.,

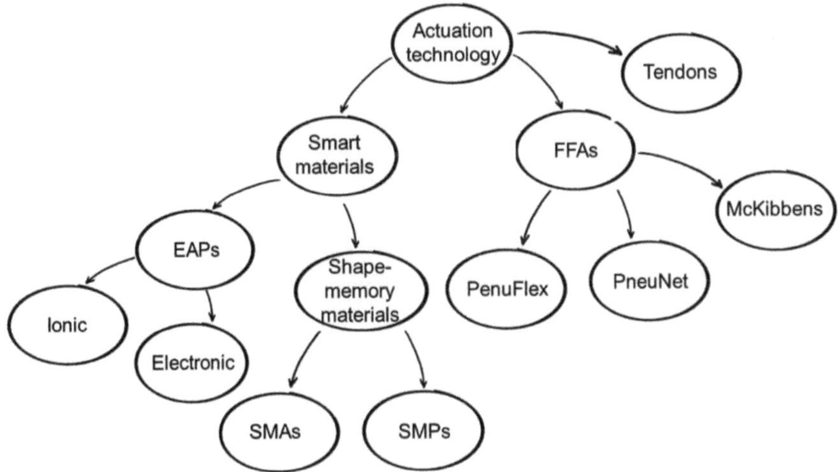

Fig. 1 A taxonomy of the technological approaches to soft robot actuation described in this section

Table 1 Main EAP classes in the two groups of electronic and ionic EAPs

Electronic EAPs	Ionic EAPs
Dielectric Elastomers (DEA)	Carbon NanoTubes (CNT)
Electrostrictive Graft Elastomers	Ionic Polymers Gel (IPG)
Ferroelectric Polymers	Ionomeric Polymer-Metal Composite (IPMC)
Liquid Crystal Elastomer (LCE) Materials	Conductive Polymers (CP)

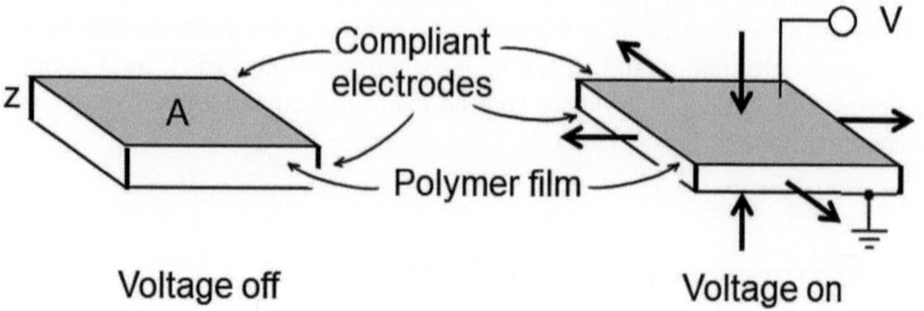

Fig. 2 Scheme of a DEA, consisting of a layer of a compliant insulator with conductive thin layers on both surfaces; the contraction obtained by applying a voltage is shown on the right

spray graphite, carbon grease, graphite powder, very thin metallic layers) (see Fig. 2 (left)). Such two layers work as electrodes and when connected to an electric voltage they generate an electric field through the non-conductive, dielectric, material. This is the phenomenon occurring in a capacitor, but here the electric charge generates electrostatic attraction between the two electrodes, winning over the resistance force of the compliant dielectric material. In other words, the charges on the electrodes generate an attractive force squeezing the material in-between (Maxwell stress) (see Fig. 2 (right)).

With a sufficiently high field, the elastomer can be significantly deformed. Considerations on the energetic aspects can start from the energy density of a capacitor:

$$u = \frac{1}{2}\varepsilon_r\varepsilon_0 E^2$$

where ε_r and ε_0 are the dielectric constants and E is the electric field. Transforming this equation by using the definition of capacitance (C) and the relation between the applied voltage V and E gives the energy stored in a capacitor:

$$U = uAz = \frac{1}{2}CV^2$$

with A the electrode areas and z the distance between electrodes. With constant voltage, the energy differential becomes:

$$dU = \frac{\varepsilon_r \varepsilon_0 V^2}{2} d\left(\frac{A}{z}\right) = \frac{\varepsilon_r \varepsilon_0 V^2}{2}\left(\frac{1}{z}dA - \frac{A}{z^2}dz\right)$$

The differential of A/z can be found by considering that elastomers usually have Poisson's coefficients very close to 0.5, meaning that their deformations occur at quasi-constant volumes:

$$d\left(Az\right) = zdA + Adz = 0 \iff \frac{dA}{A} = -\frac{dz}{z}$$

which can be inserted above to obtain dU:

$$dU = -\frac{\varepsilon_r \varepsilon_0 V^2}{2} 2\frac{A}{z^2}dz$$

Force and stress between the two electrodes are, respectively:

$$F = -\frac{dU}{dz} = \frac{\varepsilon_r \varepsilon_0 V^2}{z^2}A \qquad S = \frac{\varepsilon_r \varepsilon_0 V^2}{z^2}$$

Notwithstanding a quadratic relation between forces obtained and voltage applied, forces are usually low, unless high voltages are used. Reducing polymer thickness is not enough to compensate the high negative powers of dielectric constants (typically ~10–12 F/m). Thus, high voltages in the order of kV are necessary.

2.2 Shape-Memory Materials

Shape Memory Alloys—SMAs are metal alloys that can undergo large plastic deformations and go back to their original shape by just an increase of temperature [15]. They are produced in wires that can in turn be shaped in springs, which contracts when heated, typically by applying an electrical current [16]. Actuators built with SMAs can dramatically reduce size, weight and complexity of robotic systems. Their high force/weight ratio, the large lifecycle, the negligible volume, the capability to work as sensors as well, and the complete absence of noise support the use of SMA-based actuators in soft robotics [17]. On the other side, SMAs require high currents and the transduction process is inefficient. In addition, the material activation is highly non-linear and presents high hysteresis, making SMAs difficult to control accurately. Their thermo-mechanical behaviour is characterized by a series of interdependent parameters, difficult to control and model.

A simple analytical model often used for preliminary analysis is based on the power transformations and the relations between the variables which play a relevant role in heat development, transformation and dissipation:

$$\rho c V \frac{d T(t)}{dt} + \rho V \Delta H \frac{d\xi}{dt} = Ri^2(t) - hA(T(t) - T_e)$$

where:

- $\rho c V \frac{dT(t)}{dt} + \rho V \Delta H \frac{d\xi}{dt}$ is the power absorbed and used for increasing alloy temperature (in addition to most common terms like SMA density ρ and its specific heat c, the SMA material volume V and temperature T(t) at a time t, it is worth noting that ΔH stands for the latent phase-transformation heat and ξ is a measure of variation of the crystal structure at time instant t),
- $Ri^2(t)$ is the electrical power provided to the actuator,
- $hA(T(t) - T_e)$ is the heat dissipated in the environment (with h thermal exchange coefficient, A thermal exchange surface, T_e environmental temperature).

A part of the SMA drawbacks is solved by using Shape Memory Polymers—SMPs, which use the same principle as SMAs but other stimuli than electricity. Chemical or thermal stimuli, light or magnetic fields are the most used ones and they show a high transduction efficiency, at the cost of a higher response time. SMPs belong to the smart polymer class, attracting attention recently for their applicability in Micro-Electro-Mechanical Systems—MEMS and in biomedical devices. In many applications they result suitable to replace metallic ones, thanks to their flexibility, biocompatibility and wide modification range. An exhaustive analysis of such materials is given in Ratna & Karger-Kocsis [18].

2.3 *Fluidic Actuation*

While pneumatic actuators have been well-known in robotics for a long while, like the McKibben artificial muscles [19], more forms of fluidic actuation are widely used in soft robotics today. A very popular configuration consists of chambers molded in a soft material, usually silicone, which are pressurized with air [20]. So-called *Flexible Fluidic Actuators—FFAs*—include a wide range of systems that include an expansion chamber defined by an internal wall of containment connected to at least two anchoring points, that can translate the chamber expansion into deformations along preferential directions.

Different solutions can be adopted to guide the balloon-like deformation given by inflation: the internal or external surface of chambers can be patterned in order to have preferential inflation areas; materials with different Young's modulus can be combined, in order to constrain the deformation in the desired direction; external fibers can be wrapped around the soft body in order to constrain the deformation along an axis, for instance for obtaining elongation out of inflation (see Fig. 3).

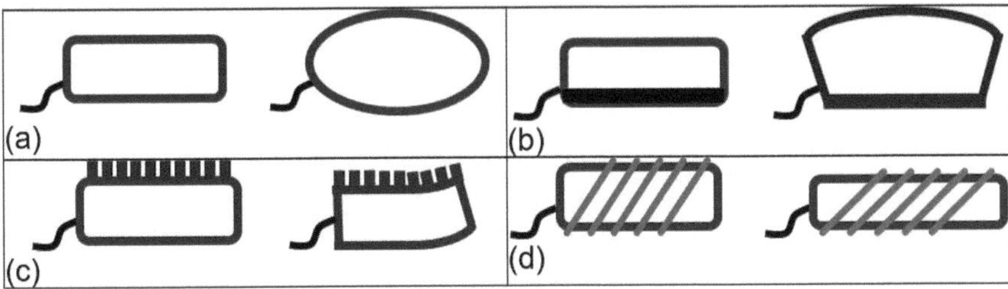

Fig. 3 Basic deformations in FFAs. Each picture shows the shapes at rest (left) and when pressurized (right): (**a**) a box-shape chamber, (**b**) patterning on the upper surface, (**c**) a stiffer material for the chamber base, (**d**) fibers wrapped around the chamber

Such actuators can deform and transform the force produced by the fluid pressure on the internal wall into a traction/compression force or a bending. This general principle can be used in several ways and several configurations are reported in literature: when the internal elastomeric chamber is coupled with a braid, we have McKibben actuators [19]; when the containment system is composed of elicoidal wires around the chamber, we often refer to PneuFlex (Fig. 3d); when the actuator geometry, shape, size and chamber arrangement are designed to have a preferential deformation direction, we refer to PneuNets (Fig. 3b, c).

For designing McKibben actuators, analytic relations can be used:

$$L = b\cos\theta$$

$$D = \frac{b\sin\theta}{n\pi}$$

$$b = \sqrt{L^2 + D^2 n^2 \pi^2}$$

$$V = \frac{b^3 \cos\theta \sin^2\theta}{4n^2\pi}$$

where, in addition to the parameters described in Fig. 4, V is the volume of the cylinder representing the actuator. This geometric approach gives a first approximated description of the relation between the braid angle and the actuator length and diameter.

Fig. 4 Geometry of a
McKibben actuator

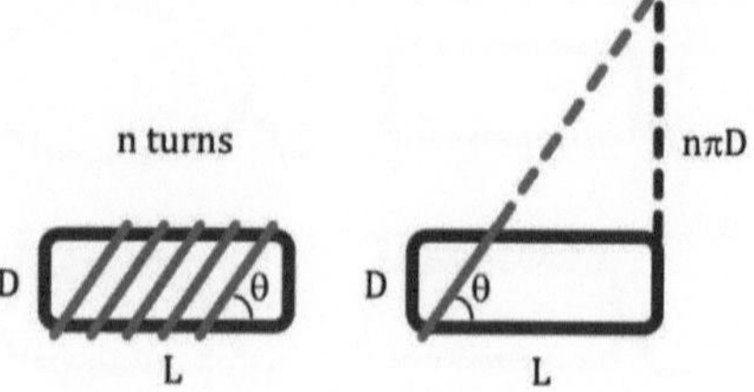

From this relation, by using the principle of virtual works, we have:

$$dW_{in} = \int_{S_i} (P - P_0)\, dl_i \bullet dS_i = (P - P_0) \int_{S_i} dl_i \bullet dS_i = P'dV$$

$$dW_{out} = -FdL$$

$$dW_{out} = dW_{in} \Rightarrow F = -P'\frac{dV}{dL} = \frac{P'b^2 \left(3\cos^2\theta - 1\right)}{4\pi n^2}$$

where P represents the absolute internal pressure, P_0 the environmental pressure, P' the relative pressure, S_i the internal surface, dS_i the area vector, dl_i the internal surface displacement, dV the volume change, F the axial force and dL the axial displacement.

FFAs usually show high power density, but they need cumbersome sources of pressurised fluids, which can hardly be miniaturised. An analysis of such technologies is given in De Greef et al. [21].

2.4 Tendons

A very effective way to actuate soft robot structures is by using tendon-like embedded cables. An advantage is obtaining a distributed and continuous action. Cables can be easily embedded in soft robots, while other actuators might not. In fact, motors for cable tension can be hosted outside of the soft robot structure, keeping it compliant and deformable. Control is also simplified by the action being distributed, but cable friction may reduce the system controllability. Compared to other actuation methods, tendons have low inertia, small size and weight, fast response time and wide-range transmission of forces and power [22]. In the example of Fig. 5, a cable is embedded into a silicone arm and make it curl on one side when pulled by an external motor.

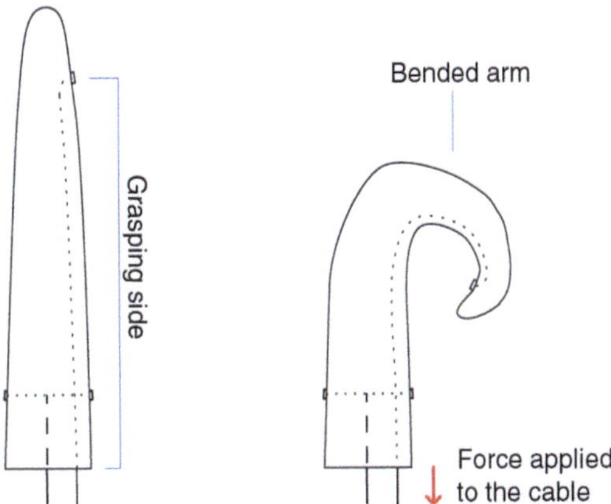

Fig. 5 (Left) Concept of a tendon embedded into a soft material and connected to an external motor; (right) when the actuator pulls the tendon, a deformation of the soft structure is obtained. Return movement is given by the material elasticity. The simplicity of this solution comes with the well-known drawbacks of tendon systems when used for mechanical transmission, such as the inaccuracy of positioning, the possible modification of cable length and tension with time, the need for proper routing channels

2.5 *Stiffening*

What's most interesting in soft robots is not just their compliance, but especially their possibility to tune their mechanical stiffness. In order to perform effective manipulation and locomotion, soft robots need to generate forces and to apply them on their environment. That's why they need to stiffen their body parts as required. A direct control of the level of stiffness of structural components is an effective way to adapt the body to tasks and to variable external conditions, especially when it allows decoupling stiffening from motion. Stiffening while keeping the shape is an important ability, which improves the functionality of a soft robotic system and enriches its behaviour.

Two main approaches can be adopted for this purpose: exploiting the antagonistic arrangement of active actuators (like those described in the previous sections) and using semi-active actuators (Fig. 6).

2.5.1 Active Actuators in Antagonistic Arrangement

The principle used in this first approach is based on exploiting the contrasting forces that two or more actuators can apply to each other, or on coupling active actuators with non-deformable structures. The equilibrium position depends on the combination of the equilibrium positions of the parts. So it is possible, in some

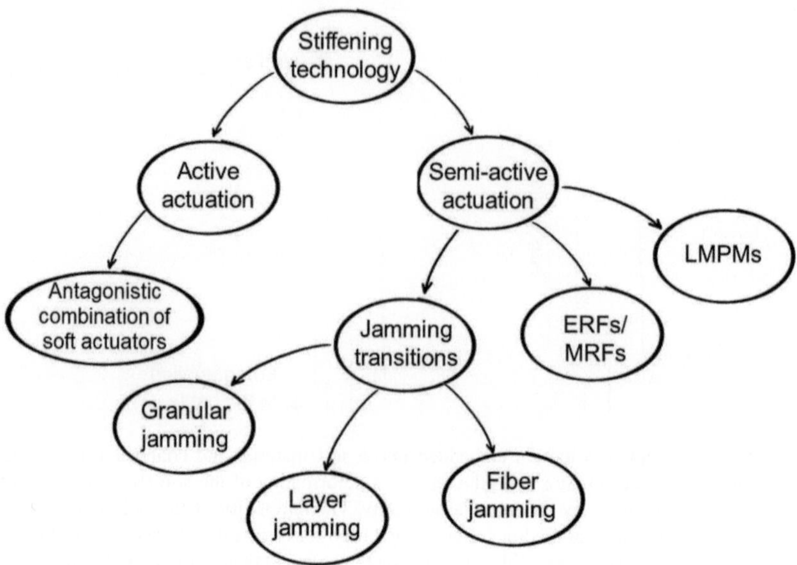

Fig. 6 A taxonomy of the technological approaches to soft robot stiffening described in this section

cases, to activate single parts and keep the other parts of the system at rest. This allows to independently set the equilibrium position and the stiffness of the system.

An example of coupling antagonistic active technologies is the combination of McKibben actuators designed and arranged in a purposive way [23]. As mentioned, such actuators can deform when connected to a pressurised air supply, but they can both elongate and shorten. The direction of deformation is given by one, important, parameter: the initial angle of fibers θ_0 (Fig. 4). For purely geometrical reasons (given the inextensibility of the single fibers composing the braid), an initial angle smaller than 54.7° makes the system a contractile unit, while a larger initial angle makes it elongate (Fig. 4). At this point, it is easy to see how the combination of more units, with same length, but with different braid angles, gives an antagonistic action that translates into an isometric stress (without deformation). The result is purely an overall stiffness increase.

An example of coupling active actuators with non-deformable structures is given by the OCTOPUS robot, inspired by the morphology and functionality of *Octopus vulgaris*. It has eight arms, two of which devoted to manipulation. Their actuation system is based on SMA springs arranged like the muscles of a *muscular hydrostat*, a special muscular structure found in the animal world (in octopus arms, in elephant trunks, in squid tentacles). The octopus *muscular hydrostat* includes longitudinal and transverse muscles [24] with an antagonistic action on each other (the longitudinal ones allow contractions, while the transverse ones allow elongation). The muscles are kept together by connective tissue and, since the structure is composed of uncompressible tissues, the co-activation of the two

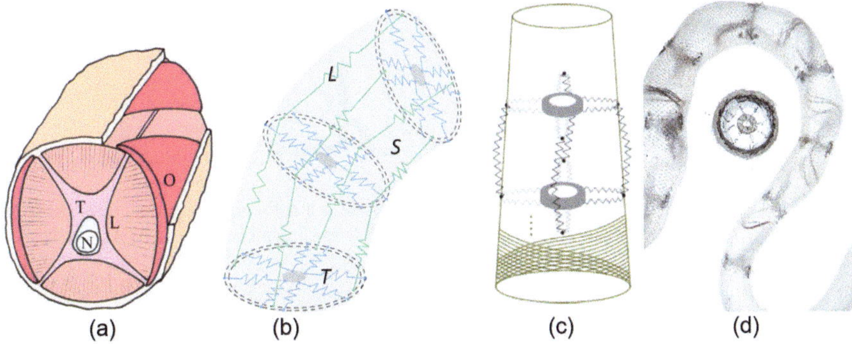

Fig. 7 (**a**) Scheme of muscular hydrostat, showing longitudinal (L), transverse (T) and oblique (O) muscles; (**b**) concept of soft robot arm with longitudinal (L) and transverse (T) SMA springs; (**c**) design of longitudinal and transverse SMA springs inside a braided support structure; (**d**) picture of the soft arm prototype. (Reprinted with permission from [25])

kinds of muscles stiffens the system without deformation (see Fig. 7a). According to this principle, SMA actuators have been inserted longitudinally and transversally (see Fig. 7b) inside a braid (similar to the McKibben one). The braid works as a support of the actuation system, but it also transmits the deformations given locally by the SMA springs and it enables the antagonistic effect on two normal planes on which the SMA spring lays (see the design in Fig. 7c) and a photo in Fig. 7d).

Despite the existence of several unexplored possibilities (basically, all the combinations of active actuation technologies are good candidates), coupling heterogeneous technologies can bring some risks (like scalability, for instance). On the other side, combining active and passive structures in a smart arrangement can bring new concepts for robot design, where support materials can have both structural and functional roles.

2.5.2 Semi-active Actuators

Differently from active actuators, some smart materials are used as semi-active actuators, meaning that they only dissipate energy during mechanical interactions. This special class of materials offers the possibility of changing the intrinsic mechanical properties in response to controlled physical stimuli, without movement or deformations (isometric stiffness variation).

Among the strategies for changing the stiffness of a soft robot, the jamming-based technologies are emerging as a new set of possibilities: creating vacuum inside a membrane containing granular material, or thin layers, or fibers, produces a significant stiffening of the overall structure, which goes back to complete compliance when air is put back in. Despite being well-known since long ago (in the food industry and in agriculture), so-called granular jamming has recently been used

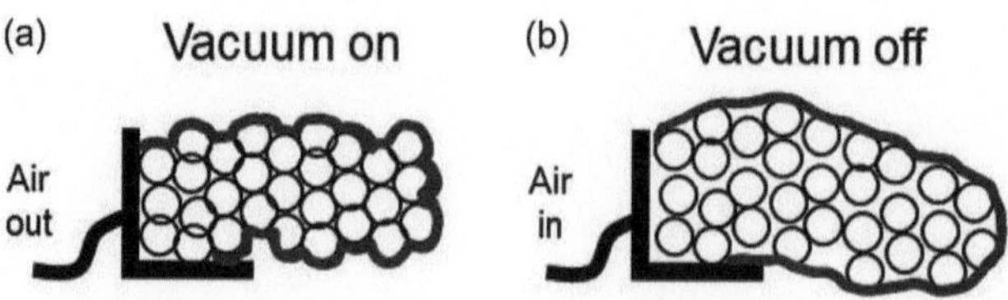

Fig. 8 (**a**) A granular material inside a flexible membrane is a soft and floppy structure; (**b**) when air is removed, particle contacts increase and the overall structure becomes stiff. The process is completely reversible

in robotics for its capability to enable a reversible transition between a quasi-fluid and a quasi-solid state, with a limited volume change [26].

There are three systematic approaches for using the jamming phenomenon, as mentioned: membranes with granular material or thin layers or fibers. All use the same activation principle: vacuum applied inside the membrane generates a phase change, increasing the relative shear force between the particles or the layers or the fibers (see Fig. 8 for the granular jamming case). The principle is still being studied under the theoretical point of view, in order to derive a deeper understanding and knowledge (especially about the relations between the parameters that govern the interaction, like shape, size, granular material stiffness, superficial texture and thickness, membrane stiffness). Nevertheless, several applications already exist in soft robotics which show its effectiveness [27]. One of them is the Universal Gripper [28, 29], a tool that can grasp objects of diverse shapes and sizes, because it can adapt its shape to the object, when pressed against it in soft state, and it can keep it stably when stiffened.

Another semi-active technology for stiffness tuning are electro- or magneto-rheological fluids (ERFs and MRFs, respectively). These materials can change their rheological properties when an electric or magnetic field is applied: sensitive particles align themselves along the field lines and link themselves in response to particle interaction; this translates into an increase of resistance to deformation. They are usually employed in the automotive industry, for adaptive bumpers or dampers. If constrained inside an elastomeric material, this principle can be used for varying the stiffness of soft structures as well, but the embedding, contamination and possible losses limit a large-scale use. More successfully, materials based on MR or ER particles can be employed: elastomers and foams. Their sensitiveness to an electrical or magnetic field can be embedded in a soft material by enclosing the particles during the polymerization process [30].

The most effective approach, in terms of range of stiffness variation, is based on a further technology, using low melting (or phase-change) point materials (LMPMs). LMPMs melt at a low temperature and can move from solid to liquid state by an electrical stimulus, thus greatly varying their stiffness. A simple example to show this principle is wax: a reticular structure composed of this material shows a stiffness range covering three orders of magnitude [31]. The same kind of activation gives more limited results with some thermoplastic materials, but with more structural stability. This is the case of polyvinyl alcohol—PVA, which undergoes a glass transition that significantly modifies the material elastic modulus [32]. A disadvantage with this technology is the nature of its activation: all mechanisms based on thermal phenomena generally have a low efficiency in energy transduction, they often require additional heating elements and their velocity is limited by thermal parameters, like material conductivity and thermal exchange coefficient, as well as environmental temperature.

3 How to Make Soft Robots Move: Soft Robot Control

Once found an actuator for building a soft robot, how to control the deformations it can produce for obtaining a desired robot movement? In a soft robot, first of all, the relation between the actuator space and the joint space is not as direct as usual: signals to an actuator do not always correspond to one joint motion, and the concept of joint itself is ambiguous; most importantly, the relation between the joint space and the task space passes through the robot configuration space: the joints deform the robot and the deformation corresponds to a position in task space [33]. See Fig. 9 for a representation of the transformations involved, (see also [34]).

Two classes of approaches can be taken to describe and use such transformations among operating spaces in order to control soft robots: (1) well-known model-based approaches, where a model of the robot and its operating space transformations is used, directly and inversely, to derive the motor commands for reaching a desired position and orientation in the task space; (2) model-free approaches, based on learning techniques that can encode the operating space transformations.

3.1 *Model-Based Approaches to Soft Robot Control*

It is straightforward that in order to use a model-based control approach, a model of the soft robot and its deformations, as given by its actuators, has to be built first.

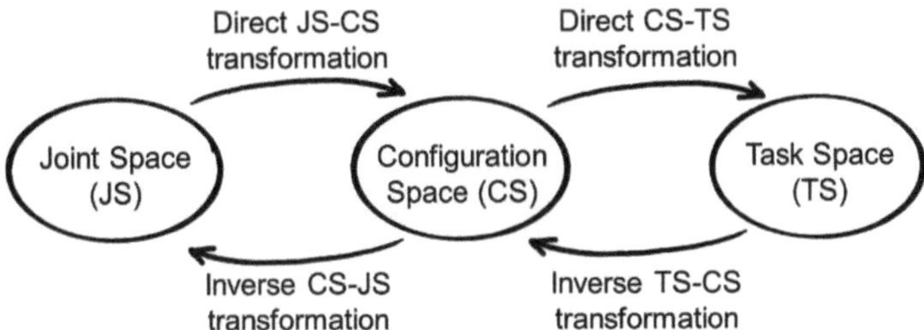

Fig. 9 Operating spaces in soft robots

The high dimensionality gives interesting challenges, as continuum soft robots have virtually infinite joints. Let's take the steady state assumption first, considering static conditions where under force equilibrium the soft robot configuration is given by a low-dimensional state space representation. In this context, with the constant curvature (CC) approximation, a three-dimensional (3D) soft robot can be considered as consisting of a finite number of curved links. These links are described by a finite set of arc parameters. The configuration space of the soft robot can be parameterized by three variables. For multi-section soft arms, more CC sections can be concatenated to obtain a piecewise-constant-curvature (PCC) model [35–37]. The CC model can be extended to the variable constant curvature (VCC) approximation, where a single module is modeled as n segments modeled as CC [38]. Continuum geometrically exact approaches can also be adopted to model a soft arm. As an example, using beam theory, a Cosserat geometrically exact model has been built for a tendon-driven continuum robot [39, 40]. In this approach, the body is considered as a Cosserat continuum medium, i.e. a continuum set of infinitesimal rigid micro-bodies. In this way, it is possible to properly simulate the behaviour of a soft robot arm with non-constant curvature, as it considers the arm torsion.

The inverse relations can be found starting from these direct models, so as to derive the variables in the actuator space for a desired position in task space. Both approaches of closed-loop control in joint space and in task space can be adopted and common PID controllers can be used to close the loop. Some examples are reviewed in Thuruthel et al. [34]. See Fig. 10 for a general scheme. The well-known Jacobian method can be used to invert the model, as shown in the case of a non-constant curvature tendon-driven soft arm [41], inverting the Cosserat direct model from Renda et al. [39].

Like for the case of traditional robots, model-based approaches ensure accurate and stable control. Difficulties in applying these approaches with soft robots may arise from the complexity of models and related transformations.

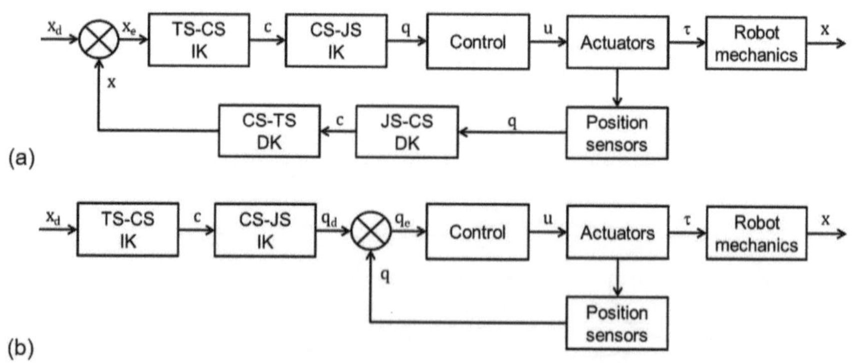

Fig. 10 General schemes of model-based closed-loop control systems for soft robots: (**a**) in task-space; (**b**) in joint-space. Variables are: $x \in TS$, $c \in CS$, $q \in JS$

3.2 Model-Free Approaches to Soft Robot Control

Learning-based approaches have been used in robot control, by developing so-called neurocontrollers, e.g. neural networks that solve robot kinematics and dynamics transformations by learning. They do not need robot models, as the controller is built through a learning phase based on robot movements: in the learning phase, a robot can associate the motor commands with the effect of movements, in terms of end effector position and orientation, i.e. it builds the actuator space to task space transformation; the networks finally encode the robot kinematics and dynamics. In closed-loop control systems, learning-based modules can replace the inverse modules, like those implementing the transformation between task and joint. Some examples are reviewed in Thuruthel et al. [34].

Interestingly, Giorelli et al. [42] presented a comparison of the performance of a soft arm controller developed with a neural approach and with an inverse Jacobian approach, showing how the accuracy of the Jacobian method is higher, but strongly dependent on the accuracy of the model. The works shows experimentally how inaccuracies in the model, due for instance to fabrication inaccuracies, bring a growing error in the control system. On the contrary, a simple neural control system can take into account the inaccuracies of the arm model with no effect on the performance.

An analysis of current approaches for controlling soft robots and a survey of current achievements are reported in Thuruthel et al. [34].

Soft robotics is very much motivated by the implementation of embodied intelligence (as explained in the beginning of this chapter) and morphological computation is expected to reduce the computational burden on the control system: the forces generated by the interaction of robot body and environment can be used for obtaining the desired motor behavior [2, 5]. In this sense, modelling a soft robot in order to control its position and orientation in task space, without proper account of the interactions, can be limitative, with respect to the larger set of goals that soft robot control can have: we might be interested not necessarily in the end effector final position, but in the arm posture, in relation to external objects (for instance, for grasping by whole arm object wrapping). The interaction between robot and environment, and task, is complex to model, with the approaches described above. Such models can be built with learning techniques, based on direct experience of interaction and its physical consequences. In the learning phase, a soft robot can associate the motor commands with the effect of movements, in terms of deformations and interaction with the environment. In the end, internal models encode the part of control which is given by the interaction of the physical body with the environment, as in the morphological computation paradigm. Learning approaches can help using morphological computation in soft robots.

Dynamic controllers have been proposed based on model-free, learning-based methods. Thuruthel et al. [43] used a recurrent neural network for learning the forward dynamic model in conjunction with a trajectory optimization method. The proposed controller, implemented on a soft robot arm, was able to follow predefined trajectory with high speed and accuracy. Interestingly, the system tends to attractors states, representing self-stabilizing trajectories, as in the paradigm of morphological computation ([2], Ch. 4).

Learning-based approaches to control are generally less accurate and require relatively long learning phases. They present some advantages when used with soft robots, as models and transformations are built by learning and they intrinsically encode morphological computation.

4 What Soft Robots Can Do: Soft Robot Abilities

In addition to the disruptive impact of soft robotics on robotics technologies and on robot modelling and control, a significant impact is evident in what soft robots can do, in terms soft robot *abilities*. Abilities are intended here as additional to usual robot functions of manipulation and locomotion and are not necessarily related to an application field [44].

The following sections report some of the main ones, outlining the principles behind each of them and the way they can be achieved with soft robotics technologies. We consider some of them as 'basic' abilities, while others are 'composed', i.e. a combination of basic ones.

4.1 Basic Soft Robot Abilities

4.1.1 Bending

Bending is the ability of a (slender) structure to change shape and assume a curved configuration. It refers to structures where at least one of their dimensions is significantly smaller than the others, e.g. 1 order of magnitude. A rod is a structure with length significantly bigger than the other two dimensions.

In such cases, bending is given by an asymmetric change in length of the two sides of the structure, while keeping the section area constant (see Fig. 11a). For bending in elastic robots, please refer to Chapter 4. In soft robotics, bending can be obtained by independently actuating the two sides, with the technologies described in previous sections. The section on soft robot control mainly refers to bending of soft robot arms.

4.1.2 Elongation/Shortening

Elongation and shortening are the abilities of increasing or reducing the length of a structure, respectively. In the case of a rod, it corresponds to uniformly changing the length.

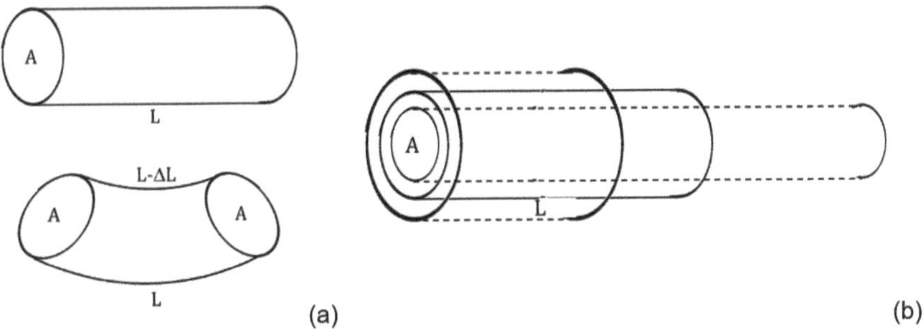

Fig. 11 (**a**) How bending can be obtained in a rod-like structure by contracting one side and keeping the section area constant. (**b**) Elongation and shortening at constant-volume, corresponding to diameter reduction and increase, respectively. Composed soft robot abilities

If the structure volume can change, elongation and shortening can be obtained by increasing or reducing the volume, respectively. This is the case of fluidic actuation, for instance, as described in a previous section. Pneumatic actuators like the McKibben artificial muscles can elongate or shorten by air supply.

In hydrostatic structures, where the volume is constant, elongation and shortening corresponds to diameter reduction or increase, respectively (see Fig. 11b).

4.1.3 Peristalsis

Peristalsis is a wave-like movement given by an alternate radially symmetrical contraction and relaxation of a tubular structure. It is very typical of the gastrointestinal tract, where the wave-like movement generates a one-direction displacement of the contents. Several worms use peristalsis for locomotion: a wave of radial contractions travels backward along the body, pushing the animal forward (see Fig. 12).

Peristaltic locomotion is based on the friction that the body is subject to: radial contractions of a body segment correspond to elongations of that segment; other segments are not active and are subject to static friction; in order to elongate the active segment in the forward direction, the friction force in the backward direction of the body segments that are not active has to be larger than the friction force in the forward direction. In other terms, the maximum body mass that can be contracted radially and elongated is less than the ratio between the backward friction and the sum of backward and forward friction. For instance, if they are equal, than the maximum body mass that can elongate is less than half that friction value.

4.1.4 Jumping

Jumping is an effective locomotion consisting of a resting phase followed by a flying phase. The resting phase is necessary to store potential energy in elastic elements.

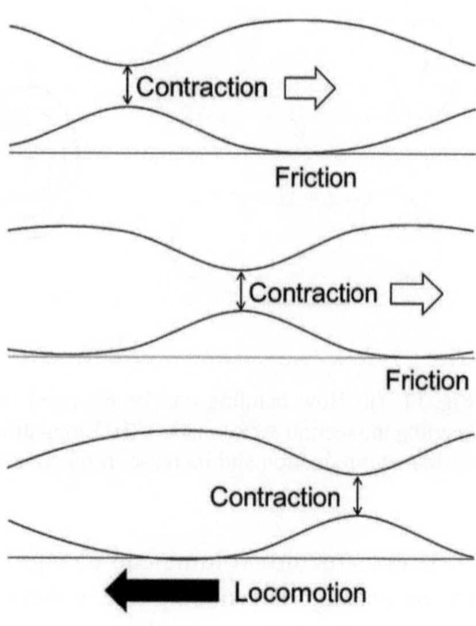

Fig. 12 In peristaltic locomotion, a contraction wave propagates backwards, translating the friction area and generating a forward push

Such energy is then released at once in a jump, consisting of a quasi-ballistic flying phase.

The flying phase starts with a v_0 velocity, but the horizontal speed of jumping is defined as the length of the jump divided by the overall time needed for energy storing and for jumping, i.e. the resting phase plus the flying phase.

4.1.5 Swimming

Swimming is the ability of generating propulsion of own body by hydrodynamic forces, in water. It can be achieved by diverse body movements, most of which require bending. Most common swimming movements are oscillatory or undulating. Jet propulsion can also be achieved in water by expelling water out of a shell-like structure, through a funnel.

5 How to Use Soft Robots: Soft Robotics Applications

The abilities that soft robots possess open up new scenarios for robotics applications, where interaction is important in the task, especially in the biomedical field when a soft interaction with a patient is preferable.

Few cases are described in the following as examples of how soft robots can be designed in response to application needs and how the soft robotics technologies described before can be integrated into a usable soft robot.

5.1 Soft Robots for Personal Assistance: The Case of the I-SUPPORT[2] Arm for Personal Hygiene

5.1.1 Application Purposes and Requirements

Personal assistance stands as an ideal case for application of soft robots, thanks to the safer and more effective interaction that they offer. Soft robotics technologies allow to build robots that can increase independent ageing, even in areas related to privacy, like personal care and hygiene. In this context, the idea has been pursued of developing a robotic arm for personal care of elderly people, especially in those personal hygiene tasks that are more dangerous, like showering, in case they are no longer fully autonomous. In showering, reaching and washing some body parts can get difficult and requires assistance even in case the elderly is still independent in most of his/her daily tasks.

The I-SUPPORT arm is a soft arm conceived to be installed in a regular bathroom in the shower area, close to usual water faucet and water supply tubes (see Fig. 10). It can spray water and detergent and gently wipe the user's skin. The main requirements have been given by the scenario of use, usually containing water and vapour, and by the workspace defined by the anthropometric dimensions of a typical user. The user profile is a still-autonomous elderly person, with some difficulties in moving in the shower environment and accomplishing washing tasks, who may wish to keep autonomous in showering, in a seated position. The study of the workspace has been based on the volumes occupied by a sitting person and on the corresponding positions of the parts to wash.

5.1.2 Technical Design and Development

In order to comply with the requirements of the intended application, i.e. safe interaction in personal assistance, and of the scenario of use, i.e. limited space, with presence of water and humidity, the actuation technologies chosen in the design of the I-SUPPORT arm are a combination of two active actuation technologies: FFAs and tendons [45]. Such a hybrid system combines a soft interaction given by the intrinsic compliance of FFAs with the stability and force given by tendons. From the functional point of view, the two technologies are complementary: while the FFAs provide elongation and bending, tendons provide shortening and they help keeping curved postures. The combination of the two does not give a combination of their functions: with a proper arrangement of actuators, co-activation generates an agonistic-antagonistic couple, resulting in an increase of stiffness, and thus provides the possibility to tune and control the overall system stiffness.

The arm is composed of two modules. Each module has three FFAs arranged longitudinally and angularly equidistant around the central longitudinal axis of the

---

[2] I-SUPPORT is a project funded by the European Commission, under contract #643666.

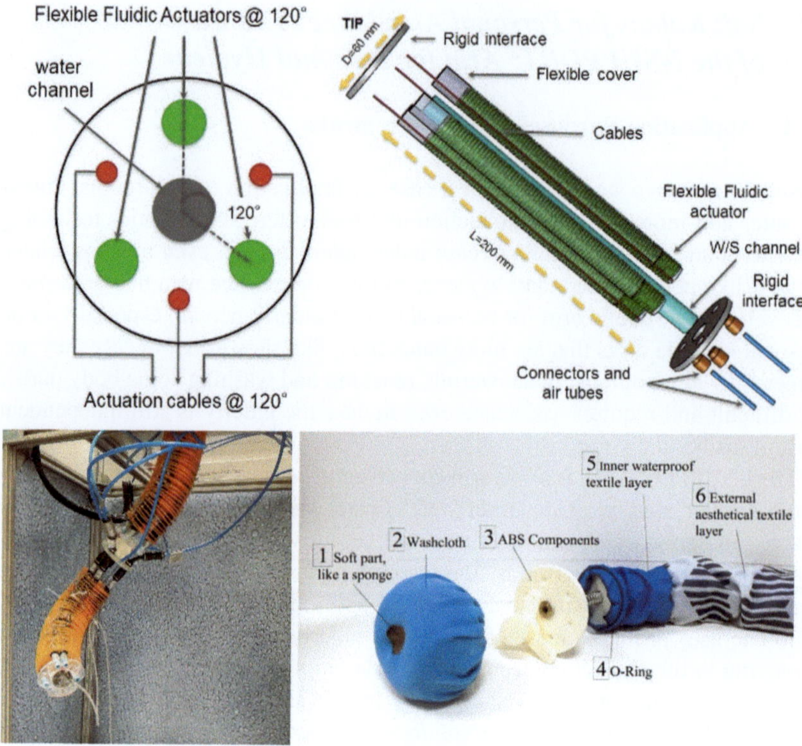

Fig. 13 (Top) CAD model of one module of the I-SUPPORT arm, showing its main components, FFAs and tendons, and their arrangement in the section. (Bottom) Two-module arm and prototype for experimental trials and clinical validation

arm. Between each FFA couple, tendons are placed longitudinally, at equal distance from each FFA, and are connected to electric motors contained in the arm base (see Fig. 13 for details of the arm components).

A first experimental characterization of the module has shown that the prototype can elongate of about 68% (from 205 mm to 345 mm) by the activation of the FFAs and it can reach a 25% shortening (from 205 mm to 155 mm) by pulling the tendons. The module has also demonstrated to cover a workspace that is appropriate to the intended application. The workspace is maximized by the combination of the two actuation technologies.

5.2 Soft Robotics Technologies for Minimally Invasive Surgery: The Case of the STIFF-FLOP[3] Endoscope

5.2.1 Application Purposes and Requirements

Minimally-Invasive Surgery—MIS—is today the most promising approach to several surgical procedures, for its advantages in reducing patient's distress and hospitalization time and costs. From the patient's point of view, the advantage of operating through a few-mm aperture is evident, while from the technical point of view this scenario is challenging. Surgical robotics provides sophisticated devices that help 'virtually' bring the surgeon into the abdominal cavity. They are often based on rigid materials and parts, in order to effectively transmit forces from the surgeon outside to the surgical site inside the patient's body. Limited visibility is also an issue which introduces risks for involuntary interactions with nearby tissues.

5.2.2 Technical Design and Development

The requirements for intrinsic safety, dexterity and stiffness variation make soft robotics technologies ideal candidates in this application scenario. In this case, the environment is very delicate and the technological choices play a fundamental role. High electric fields, high currents or high temperatures cannot be used, so the technological choices are reduced. Among remaining technologies, an interesting synergy of fluidic actuation can be adopted: FFAs can provide high dexterity and granular jamming can provide stiffness variation by using negative pressures.

This is the approach used in the STIFF-FLOP project, aiming at a new surgical device, able to elongate, bend in all directions, and stiffen parts of its body. Its body is composed of several modules [46]. Each module is made of silicone and hosts four chambers: three chambers are empty and represent FFAs. They are arranged longitudinally, at equal distance from the central longitudinal axis. They provide elongation and all-direction bending. An external containment constrains the radial expansion of chambers and optimizes their deformation directionality (see Fig. 14). Dexterity and safe interaction are ensured but, once reached the surgical site, the system needs to stiffen in order to transmit higher forces. This is obtained by granular jamming, integrated in a central longitudinal chamber.

By using FFAs and granular jamming, several versions of the endoscope have been built and tested, showing a high potential in surgery. The endoscope can double its length and cover a wider workspace and it can increase its stiffness by 36%, thus increasing the force applied and opening new possibilities and operational capabilities to surgeons.

---

[3] STIFF-FLOP is a project funded by the European Commission, under contract #287728.

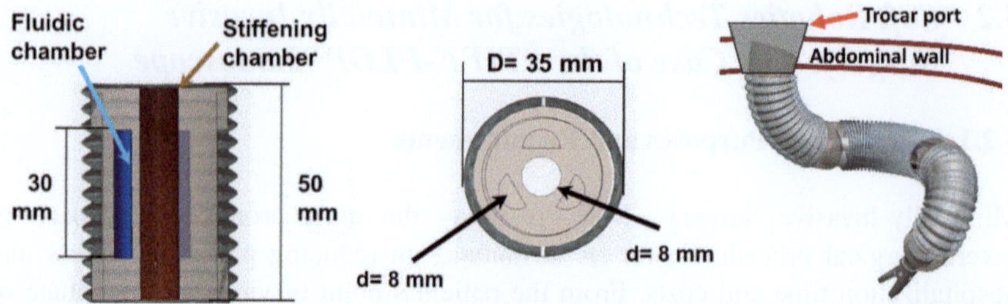

Fig. 14 The STIFF-FLOP endoscope: (Left) longitudinal view of fluidic and granular jamming chambers; (Center) section view of chamber arrangement; (Right) three-module version of the integrated endoscope

5.3 Soft Robotics Technologies for Organ Simulators: The Case of the Biorobotic Larynx

5.3.1 Application Purposes and Requirements

Organ simulators are important in medicine for training and for research. Soft robotics technologies can be used to build realistic simulators that match the mechanical properties and the movements of target organs.

The human larynx is a complex organ, consisting of many parts of small dimensions, composed of diverse tissues (cartilages, muscles, articulations and fibrous tissue). They have different mechanical properties, like compliance and elasticity, and they can perform small movements, functional to voice production. In vivo studies of the larynx are not easy to perform and a realistic simulator can serve the purpose with important outcomes, scientifically and clinically. The core element of the larynx, vocal cords, can vibrate and produce sounds thanks to the combination of tissues with different compliances, arranged in three layers: body, ligament and cover. Reproducing such layers with the same mechanical properties is the main requirement for building vibrating vocal cords.

5.3.2 Technical Design and Development

From the functional synthesis of the mechanical properties of the three layers composing the vocal cords, three different materials have been realized by using commercial silicones: bi-component EcoFlex 0030 combined with Silicone Thinner, both by Smooth-On Inc. Silicone Thinner was added in different ratios to obtain specific elastic and rheological property values for the cover, body and ligament layers. Our aim was to obtain silicone samples with elastic modulus of 15 kPa, 20 kPa and 30 kPa for cover, body and ligament, respectively, as in the human vocal cord layers [47].

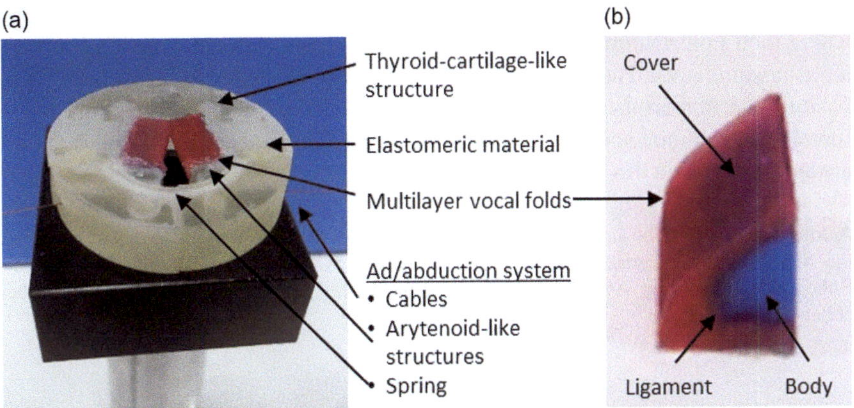

Fig. 15 (**a**) Prototype with two vocal cords for experimental testing of vibration; (**b**) prototype of one vocal cord. (Reproduced with permission from [48])

A CAD model of the vocal cords allowed a finite element analysis (FE) showing the deformations when air flows through the cords. A prototype built from the CAD model, by using the three silicone materials (Fig. 15b) allowed experimental testing (Fig. 15a). The experimental results obtained with a controlled air flow, measuring the cord vibrations with a high-speed camera (NAC HotShot 512 SC) show that the combination of the cord morphology and mechanical properties generates a vibration frequency (average value f = 154 Hz) and a vertical displacement in the physiological range of human vocal cords [49].

6 Conclusions

Soft robotics is a young yet promising field of research and development, contributing to a further progress of robotics and providing robots with further abilities and application perspectives.

Many challenges are still open in soft robotics: the fabrication of components and systems will benefit from established materials, methods and processes; structured methodologies can be defined for soft robot design; solid theories are being defined for soft robot modelling and control. Despite such challenges and open research questions, soft robotics technologies reached a sufficiently high level of systematization to enable the definition and teaching of principles and and main tools.

As a contribution to future developments of robotics, soft robotics technologies especially allow thinking of further robot abilities, given by body deformability. The use of soft materials also opens perspective for the use of materials that embed abilities themselves, like self-healing or biodegradability.

Soft robots are finding applications in diverse fields, like the biomedical one, where their contribution spans from personal assistance to endoscopy, to building realistic simulators. They bring benefits in the way robots interact physically with patients and in reproducing human physiology realistically. Good trade-offs can be found with common soft robots limitations in payload and accuracy, which are not always priorities in this field.

Acknowledgments The authors acknowledge the support of the European Commission through the STIFF-FLOP Integrated Project (#287728), the I-SUPPORT project (#643666) and the RoboSoft Coordination Action (#619319) and thank the partners of those projects for their collaboration.

Self-Assessment

Questions

1. Which sentence better defines soft robotics?
   1. Software robots
   2. Robots built with soft materials
   3. Highly deformable robots and structures
2. Which sentence better defines morphological computation?
   1. The part of control outsourced to the physical body
   2. The part of control delegated to the peripheral nervous systems
   3. The part of control required for morphological deformation
3. The working principle of EAP is based on:
   1. Newton forces
   2. Maxwell effect
   3. Bernoulli effect
4. Mechanisms for guiding deformation of a fluidic-actuated beam to bend (more than one answer is correct):
   1. Materials with different Young's modulus on one side
   2. Materials with different Poisson ratio on one side
   3. Elicoidal wrapping of inextensible fibers around the beam
   4. Symmetric cables embedded in the beam
   5. Asymmetric chambers inside the beam
   6. Preshaping the desired curvature
5. Is the stiffening process of granular jamming reversible?
   1. Yes
   2. No
6. Which of the following sentences is true:
   1. Model-based approaches cannot be used for controlling soft robots
   2. Learning-based approaches can be used for controlling soft robots
   3. Model-based approaches need to be complemented with learning-based approaches for controlling soft robots

7. Tick abilities that soft robots possess (more than one answer is correct):
   1. Bending
   2. Volume reduction
   3. Elongation
   4. Regeneration
   5. Jumping
   6. Crawling
8. Advantages of the use of soft robots in biomedical applications (more than one answer is correct):
   1. Safe interaction with patients
   2. High payload in personal assistance
   3. High accuracy in surgery
   4. Realistic stiffness in body part simulators
   5. Safe navigation inside the human body
   6. Biocompatibility

Answers

1. 3
2. 1
3. 2
4. 1, 5
5. 1
6. 2
7. 1, 3, 5, 6
8. 1, 4, 5

References

1. Pfeifer, R., Lungarella, M., Iida, F.: Self-organization, embodiment, and biologically inspired robotics. Science. **318**(1088) (2007)
2. Pfeifer, R., Bongard, J.: How the body shapes the way we think: a new view of intelligence. The MIT Press, Cambridge, MA (2007)
3. Müller, V.C., Hoffmann, M.: What is morphological computation? On how the body contributes to cognition and control. Artif. Life. **23**(1), 1–24 (2017)
4. Zambrano, D., Cianchetti, M., Laschi, C.: The morphological computation principles as a new paradigm for robotic design. In: Hauser, H., Füchslin, R.M., Pfeifer, R. (eds.) Opinions and Outlooks on Morphological Computation, pp. 214–225 (2014)
5. Laschi, C., Mazzolai, B.: Lessons from animals and plants: the symbiosis of morphological computation and soft robotics. IEEE Robot. Autom. Mag. **23**(3), 107–114 (2016)
6. Calisti, M., Picardi, G., Laschi, C.: Fundamentals of soft robot locomotion. J. R. Soc. Interface. **14**(130), 20170101 (2017)
7. Albu-Schaffer, A., Eiberger, O., Grebenstein, M., Haddadin, S., Ott, C., Wimbock, T., Wolf, S., Hirzinger, G.: Soft robotics. IEEE Robot. Autom. Mag. **15**(3), 20–30 (2008)
8. Deimel, R., Brock, O.: A novel type of compliant and underactuated robotic hand for dexterous grasping. Int. J. Robot. Res. **35**(1–3), 161–185 (2016)

9. Laschi, C., Cianchetti, M.: Soft Robotics: new perspectives for robot bodyware and control. Front. Bioeng. Biotechnol. **2**(3) (2014)

10. Kim, S., Laschi, C., Trimmer, B.: Soft robotics: a bioinspired evolution in robotics. Trends Biotechnol. **31**(5), 287–294 (2013)

11. Rus, D., Tolley, M. T.: Design, fabrication and control of soft robots. Nature. 521(7553), 467–475 (2015)

12. Wang, L., Iida, F.: Deformation in Soft-Matter Robotics: A Categorization and Quantitative Characterization. IEEE Robotics & Automation Magazine, 22(3), 125–139 (2015)

13. Carpi, F., Kornbluh, R., Sommer-Larsen, P., Alici, G.: Electroactive polymer actuators as artificial muscles: are they ready for bioinspired applications? Bioinspir. Biomim. **6**(4), 045006 (2011)

14. Mirfakhrai, T., Madden, J.D., Baughman, R.H.: Polymer artificial muscles. Mater. Today. **10**, 30–38 (2007)

15. Funakubo, H., Kennedy, J.B.: Shape memory alloys. Gordon and Breach Science Publishers (1987)

16. Follador, M., Cianchetti, M., Arienti, A., Laschi, C.: A general method for the design and fabrication of shape memory alloy active spring actuators. Smart Mater. Struct. **21**(11), 115029 (2012)

17. Cianchetti, M.: Fundamentals on the use of shape memory alloys in *soft* robotics. In: Habib, M.K., Davim, J.P. (eds.) Interdisciplinary Mechatronics: Engineering Science and Research Development, pp. 227–254. Wiley-ISTE (2013)

18. Ratna, D., Karger-Kocsis, J.: Recent advances in shape memory polymers and composites: a review. J. Mater. Sci. **43**, 254–269 (2008)

19. Chou, C.-P., Hannaford, B.: Measurement and modeling of McKibben pneumatic artificial muscles. IEEE Trans. Robot. Autom. **12**(1), 90–102 (1996)

20. Shepherd, R.F., Ilievski, F., Choi, W., Morin, S.A., Stokes, A.A., Mazzeo, A.D., Chen, X., Wang, M., Whitesides, G.M.: Multigait soft robot. Proc. Natl. Acad. Sci. USA. **108**, 20400–20403 (2011)

21. De Greef, A., Lambert, P., Delchambre, A.: Towards flexible medical instruments: review of flexible fluidic actuators. Precis. Eng. **33**(4), 311–321 (2009)

22. M. Calisti, A. Arienti, M.E. Giannaccini, M. Follador, M. Giorelli, M. Cianchetti, B. Mazzolai, C. Laschi, P. Dario, Study and fabrication of bioinspired Octopus arm mockups tested on a multipurpose platform, in *IEEE RAS/EMBS Int. Conf. on Biomedical Robotics and Biomechatronics – BioRob*, Tokyo, Japan, 26–29 September (2010), pp. 461–466

23. K. Suzumori, S. Wakimoto, K. Miyoshi, K. Iwata, Long bending rubber mechanism combined contracting and extending fluidic actuators, in *IEEE/RSJ Int. Conf. in Intelligent Robots and Systems – IROS*, Tokyo, Japan, 3–7 November (2013), pp. 4454–4459

24. Kier, W.M., Stella, M.P.: The arrangement and function of octopus arm musculature and connective tissue. J. Morphol. **268**, 831–843 (2007)

25. M. Follador, M. Cianchetti, C. Laschi, Development of the functional unit of a completely soft octopus-like robotic arm, in *IEEE RAS-EMBS International Conference on Biomedical Robotics and Biomechatronics (BioRob)*, Rome, Italy (2012), pp. 640–645

26. Liu, A.J., Nagel, S.R.: Nonlinear dynamics: jamming is not just cool anymore. Nature. **396**, 21–22 (1998)

27. Fitzgerald, S.G., Delaney, G.W., Howard, D.: A review of jamming actuation in soft robotics. Actuators. **9**(104) (2020)

28. Amend Jr., J.R., Brown, E., Rodenberg, N., Jaeger, H., Lipson, H.: A positive pressure universal gripper based on the jamming of granular material. IEEE Trans. Robot. **28**, 341–350 (2012)

29. Brown, E., Rodenberg, N., Amend, J., Mozeika, A., Steltz, E., Zakin, M.R., Jaeger, H.M.: Universal robotic gripper based on the jamming of granular material. Proc. Natl. Acad. Sci. **107**(44), 18809–18814 (2010)

30. Carlson, J.D., Jolly, M.R.: MR fluid, foam and elastomer devices. Mechatronics. **10**(4–5), 555–569 (2000)

31. Cheng, N.G., Gopinath, A., Wang, L., Iagnemma, K., Hosoi, A.E.: Thermally tunable, self-healing composites for *soft* robotic applications. Macromol. Mater. Eng. **299**(11), 1279–1284 (2014)
32. Capadona, J.R., Shanmuganathan, K., Tyler, D.J., Rowan, S.J., Weder, C.: Stimuli-responsive polymer nanocomposites inspired by the sea cucumber dermis. Science. **319**(5868), 1370–1374 (2008)
33. Webster, R., Jones, B.: Design and kinematic modeling of constant curvature continuum robots: a review. Int. J. Robot. Res. **29**, 1661–1683 (2010)
34. Thuruthel, T.G., Ansari, Y., Falotico, E., Laschi, C.: Control strategies for soft robotic manipulators: a survey. Soft Robot. **5**(2), 149–163 (2018)
35. Boyer, F., Porez, M., Khalil, W.: Macro-continuous computed torque algorithm for a three-dimensional eel-like robot. IEEE Trans. Robot. **22**(4), 763–775 (2006)
36. Camarillo, D.B., Carlson, C.R., Salisbury, J.K.: Configuration tracking for continuum manipulators with coupled tendon drive. IEEE Trans. Robot. **25**(4), 798–808 (2009)
37. B.A. Jones, R.L. Gray, K. Turlapati, Three dimensional statics for continuum robotics, in *IEEE/RSJ Int. Conf. on Intelligent Robots and Systems*, St. Louis, MO, USA, 11–15 October (2009), pp. 2659–2664
38. Mahl, T., Hildebrandt, A., Sawodny, O.: A variable curvature continuum kinematics for kinematic control of the bionic handling assistant. IEEE Trans. Robot. **30**, 935–949 (2014)
39. Renda, F., Cianchetti, M., Giorelli, M., Arienti, A., Laschi, C.: A 3D steady-state model of a tendon-driven continuum soft manipulator inspired by the octopus arm. Bioinspir. Biomim. **7**(2), 025006 (2012)
40. Renda, F., Giorelli, M., Calisti, M., Cianchetti, M., Laschi, C.: Dynamic model of a multi-bending soft robot arm driven by cables. IEEE Trans. Robot. **30**, 1109–1122 (2014)
41. Giorelli, M., Renda, F., Calisti, M., Arienti, A., Ferri, G., Laschi, C.: A two dimensional inverse kinetics model of a cable driven manipulator inspired by the octopus arm. IEEE Int. Conf. Robot. Autom., 3819–3824 (2012)
42. Giorelli, M., Renda, F., Calisti, M., Arienti, A., Ferri, G., Laschi, C.: Neural network and Jacobian method for solving the inverse statics of a cable-driven *soft* arm with nonconstant curvature. IEEE Trans. Robot. **31**, 823–834 (2015)
43. Thuruthel, T.G., Falotico, E., Renda, F., Laschi, C.: Learning dynamics and trajectory optimization for octopus inspired soft robotic manipulators. Bioinspir. Biomim. **12**, 066003 (2017)
44. Laschi, C., Mazzolai, B., Cianchetti, M.: Soft robotics: technologies and systems pushing the boundaries of robot abilities. Sci. Robot. **1**(1) (2016)
45. Ansari, Y., Manti, M., Falotico, E., Mollard, Y., Cianchetti, M., Laschi, C.: Towards the development of a soft manipulator as an assistive robot for personal care of elderly people. Int. J. Adv. Robot. Syst. **14**(2) (2017)
46. Ranzani, T., Gerboni, G., Cianchetti, M., Menciassi, A.: A bioinspired *soft* manipulator for minimally invasive surgery. Bioinspir. Biomim. **10**, 035008 (2015)
47. Alipour, F., Vigmostad, S.: Measurement of vocal folds elastic properties for continuum modeling. J. Voice. **26**(6), 816.e21–816.e29 (2012)
48. Cianchetti, M., Laschi, C.: Pleasant to the touch. IEEE Pulse Mag. **3**, 34–37 (2016)
49. M. Manti, M. Cianchetti, A. Nacci, F. Ursino, C. Laschi, A biorobotic model of the human larynx, in *IEEE Int. Conf. of the Engineering in Medicine and Biology Society – EMBC*, Milan, Italy, 25–29 August (2015), pp. 3623–3626

Bioinspired Robot Design

Mark R. Cutkosky

Abstract As robots move beyond relatively controlled environments and into the world at large, they increasingly can benefit by taking lessons from nature. This realization has lead to a proliferation of bioinspired and soft robots. The trend is supported by advances in our understanding of how biological systems work, at scales ranging from proteins to entire organisms. Among the recurring lessons are: how to manage the flow of energy, how to interact effectively with the environment, how to integrate passive mechanical properties and structures with sensing and control, and how to be robust—both physically and operationally—in the face of disturbances. Bioinspired robots have also promoted advances in fabrication and materials to create compliant, multi-material and multi-functional structures like those that organisms exploit. This chapter begins with the background and rationale behind bioinspired robotics and then considers specific topics in greater detail, drawing upon two case studies. The emphasis is on the design, construction, actuation and control of robots rather than robot swarms or bioinspired computation. The chapter also reviews trends in materials and fabrication that support bioinspired robotics. Each section concludes with design questions to provoke further thought and exploration. A final section introduces bioinspired design methods and exercises to help the reader explore opportunities in bioinspired robotics.

1 Introduction

Bioinspiration is as old as design itself. We live a world surrounded by plant and animal examples and cannot help but be influenced by them. As engineering design has become recognized and understood as a discipline, the role of nature in informing and inspiring our designs has similarly evolved. Some early examples of this recognition come from the Renaissance. In Leonardo da Vinci's notebooks,

M. R. Cutkosky (✉)
Stanford University, Stanford, CA, USA
e-mail: cutkosky@stanford.edu

B. Siciliano (ed.), *Robotics Goes MOOC*,
https://doi.org/10.1007/978-3-319-75823-7_6

where sketches of birds or human cadavers sometimes occupy the same pages as sketches of machines, as well as other writings from the period (e.g. Alberti), we find articulated two ideas of bioinspired design:

- The bodies of animals, including humans, can be viewed as machines with levers, tendons, tubes, pumps, etc. that allow for daily functions such as locomotion and physical work.
- The machines that we design are reflections of what we see in nature. Thus we may talk about the feet or the skeleton of a machine.

The design of robots, in particular, is unavoidably bioinspired—robots, after all, are attempts to create autonomous beings like those we find in nature. This too is an old idea, with many examples of animal and human *automata* created over the years [1]. It has also lead to debate about the extent to which animals are models for intelligent and autonomous behavior. Descartes infamously thought not [2] but today we recognize diverse animals ranging from crows to porpoises as intelligent and potentially self-aware [3] and, at least in science fiction, we assume that robots will be too.

Increasingly, we find bioinspiration also in the design and operation of specialized subsystems, e.g. for propulsion, or sensing, or physical interaction with the environment. This trend has grown more important as robots transition from collections of stiff, metal links driven by gears and electric motors to adaptive systems fabricated from combinations of hard and soft materials, often powered by artificial muscle actuators, with motion constrained by tendons, membranes or exoskeletal shells.

Supporting the trend toward bioinspired design are two additional trends from science and engineering. On the one hand we have much better tools for investigating and understanding in detail how biological systems work, from the level of cells, chemical processes and microstructured surfaces at the bottom end, to entire animals and even swarms of organisms at the upper end. These discoveries produce ideas for new designs over a wide range of applications, from synthetic gecko-inspired adhesives and swimming microrobots to control and planning for swarms of autonomous air vehicles.

Learning about how natural organisms work is inspirational, but it is only an enabling strategy if we also have the technology to implement what we discover. Thus, an equal contributor to bioinspired robotics has been a recent proliferation of new materials and manufacturing processes. Examples range from soft lithography and multimaterial 3D printing, to kirigami and origami robotics. These techniques enable us to create systems that, like their natural counterparts, are compliant, heterogeneous, robust, and based on integrated actuation, sensing and control. Even so, bioinspired robots present numerous practical challenges. Unlike natural organisms, our synthetic creations typically require different bulk manufacturing processes and specialized equipment for different materials and feature length scales. Thus each new element of integration and mechanical complexity is comparatively expensive.

A related challenge, which is only starting to be addressed, is how to model and control bioinspired designs. Traditional robotic analysis methods, based on stiff links, precise revolute or prismatic joints, and separate subsystems for kinematics, actuation, sensing and control, are no longer appropriate. Bioinspired robots may have many degrees of freedom, with combinations of stiff and soft materials and with coupled sensing, actuation and control. These factors are leading to new analysis techniques that aim to model and control their behavior without excessive computational complexity [4]. These and related issues are covered in a thought-provoking discussion of "grand challenges" in robots [5].

1.1 Motivation and Themes

Even if the superiority of nature is a myth, such a vastly different technology can only be a gold mine of stimulation, even inspiration. (S. Vogel: "Nature's swell but is it worth copying?") [6]

Animals and plants have always had to contend with both expected and unexpected changes in their environments. As robots move out of highly controlled environments like manufacturing facilities, they too will need to tolerate changing conditions and unexpected events. Accordingly, bioinspiration is a useful design tool for increasing robots' robustness and versatility. Across the wide range of examples that nature provides, there are a number of recurring themes. We review these briefly in the following paragraphs because they either influence or motivate many bioinspired designs.

1.1.1 Robustness

In comparison to robots, animals and plants are remarkably robust, both physically and in terms of being able to function despite adverse circumstances. For survival in an unpredictable world, robustness is a necessity. As an illustrative example, cockroaches can withstand compressive forces of $\approx$300 times their weight and compress to a fraction of their standing height to slip through a crack [7]. Moreover, they can run surprisingly well with one, two, or even more of their six legs missing [8, 9]. This kind of physical and operational robustness is clearly beyond the capabilities of robots. From the standpoint of design, an interesting question is: *What is responsible for the robustness that we observe?* Partial answers lie in some of the next paragraphs. However, a more comprehensive answer requires diving into the biology literature and talking with experts. The bioinspired design case study in Sect. 2.1 provides an example of this approach, which has lead to improvements in the locomotion of small robots.

A second point worth remembering is that impressive and robust performance does not imply optimality. Evolution works on the principle of being "good enough" to survive and proliferate. The right solution is furthermore not a static condition—animals are continually responding to changing circumstances and pressures, even as we study them.

1.1.2 Exploiting Complexity

One reason for the versatility of biological systems is that they are far more complex than robots and can have many layers of behavior to draw upon. They can call upon latent behaviors (e.g. to reduce the damage from an unexpected fall or to mitigate the effects of an injury on the speed of locomotion) that are not seen in normal operation. They can adjust their performance and characteristics through physical changes (e.g. increasing blood flow or sweating). They can even self-repair when damage occurs. Returning to the cockroach example in the previous paragraph, the locomotion patterns seen when cockroaches lose multiple legs are different from those used in normal running. Nonetheless, they are available when needed.

The complexity of natural systems is evident at all levels. From a design standpoint, even a single-cell paramecium is formidably complex with macrophages, sodium channels, protein folding and numerous other elements of electrochemical "machinery" for energy acquisition conversion, locomotion, and control [10]. Part of the reason for this complexity is that the cost is low. Plants and animals grow and differentiate cell by cell, responding uniquely to environmental factors as they do. In comparison, virtually all man-made systems are assembled from components, each of which is made using specialized manufacturing techniques (machining, forging, casting, lithography); consequently every local change in properties requires new manufacturing processes and adds to the cost and difficulty of fabrication. As noted in Sect. 3, the situation is beginning to change with the advent of 3D printers that can print multiple materials (e.g. hard and soft, or conductive and insulating) but remains a long way from the variation that we encounter in biology.

1.1.3 Simplifying Principles

Fortunately, as we study biological systems we also discover simplifying principles. Often complex systems with many degrees of freedom are controlled for a certain task as though they had far fewer. A good example is the leg of a running or hopping animal which can often be approximated as a spring-loaded inverted pendulum [11] for the purposes of dynamic modeling and control. This simplification is examined further in Sect. 2.1; analogous principles have been defined for the case of small animals such as lizards that use their tail to maintain orientation and stability while jumping [12].

Other simplifying principles have been identified as "synergies" that reduce the apparent dimensionality of complex systems in stereotypical behaviors. As an example, consider how people grasp everyday objects. The hand is remarkably complex with some 27 degrees of freedom controlled by numerous muscles [13, 14]. Moreover, people's hands vary substantially in terms of size, etc. However, a principal components analysis of the geometry and kinematics of human subjects grasping common objects revealed that most grasps could be captured fairly accurately using just a few principal components [15]. This realization has lead to the design of underactuated bioinspired robotic and prosthetic hands that are actuated and controlled in terms of first principal components observed, thereby greatly reducing the number of motors required and reducing the control complexity [16, 17].

1.1.4 Sensing and Control Integrated with Mechanical Properties

Whereas in robotics the structural elements, actuators, sensors, perception, and control are typically separate systems, assembled in a modular fashion, these functions are intertwined in nature. Muscles are often structural elements and part of the sensory and control system as well as being actuators. From a controls perspective, the actions of reflexes and voluntary movements are wrapped around mechanical systems for energy storage and release, or damping, that provide an immediate passive response to external stimuli [11] (Fig. 1).

The end result of this combination of sensing and control wrapped around passive response is remarkable robustness. A good example, which we will examine in more detail in Sect. 2.1, is the locomotion of a cockroach—despite large variations in terrain and even when multiple limbs are damaged or lost.

1.1.5 Managing Interactions with the Environment

Although biological organisms are often not optimal in terms of their performance for any single task such as walking or flying (recall that nature works on the principle of "good enough" to compete and reproduce) they nonetheless exhibit impressive efficiency in terms of managing the expenditure of energy in interactions with the environment. Given that food is always a limited resource this result is not surprising. The fitness of a species depends on its ability to utilize limited food resources efficiently. As a result, the designers of robotic systems benefit

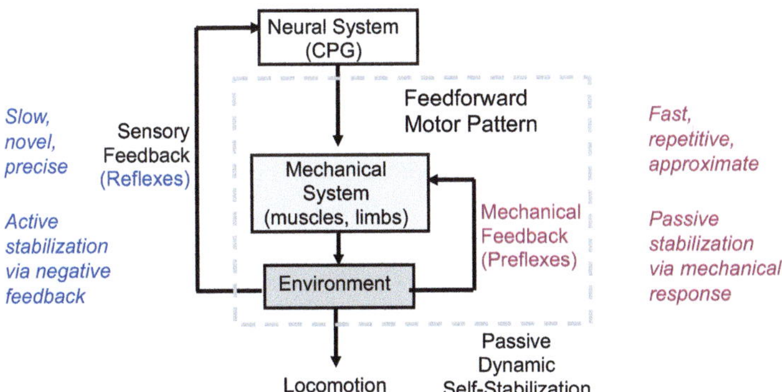

Fig. 1 Integrated sensing, actuation and control in nature arise from a combination of passive response (preflexes) and reactions to sensor feedback (reflexes). Content adapted from [11]

from studying how animals locomote and otherwise interact energetically with their environments for insight on improving efficiency. However, once we understand how to accomplish a new task such as flying, walking, jumping or climbing vertical surfaces we can create robotic systems that outperform their natural counterparts in terms of efficiency. In part this is because the robots have fewer competing agenda— they do not need to eat, reproduce, escape from predators, etc.

1.2 Open Questions for Discussion

There is not a single right answer to the questions in this section. Rather, they can provide a point of departure for delving into some of the cited references and thinking about under what circumstances a bioinspired design approach is most likely to be useful. An ideal way to use these questions might be to set up a small online discussion group with participants taking intentionally disparate views.

Modularity Versus Integration What are the tradeoffs of modular versus integrated design? Modularity confers an ability to combine and decompose systems from components or subsystems—each of which can be specified and analyzed in detail. Modularity thereby facilitates the design, analysis and control of complex systems and allows one to construct proofs regarding stability or performance. Systems Engineering is built on this tradition [18]. In nature, systems for structural support, actuation, sensing, planning etc. are highly integrated and not easily decomposed. This approach leads to complex, layered behaviors that are not easy to analyze but may also underlie the robustness and versatility we observe. A few efforts have been undertaken to produce similarly layered behaviors in robotics [19] but they have not progressed apace with more traditional approaches to motion planning and control. Arguably this is in part because they too are difficult to analyze. Nonetheless, as we develop increasingly bioinspired robots—in which passive mechanical properties, sensing, and actuation are not separate systems—we will need solutions for predicting and analyzing their performance.

Efficient Energy Management How important is it for robots to manage energy consumption? Under what circumstances is managing energy important or relatively unimportant? For example, a robot working in a factory with a continuous power connection does not need to conserve power; maintaining speed and precision are more important. Even mobile robots, if they have access to hydrocarbon fuels or frequent access to recharging stations, can afford to be relatively inefficient. However, in other applications, such as robots for planetary exploration or flying drones, the need to prolong mission life has prompted concerted efforts to increase efficiency. One such program was the DARPA Maximum Mobility and Manipulation (M3) program (https://www.darpa.mil/program/maximum-mobility-and-manipulation) which, among other developments, supported the decidedly bioinspired Cheetah robot from MIT [20].

When to Avoid Bioinspiration When does bioinspiration not make sense? What are some good counter examples? One obvious example is that bioinspiration would not have lead to the invention of the wheel, or indeed anything that involves continuous rotation, like turbine engines or electric motors. A fundamental issue is the need in nature to supply blood to moving parts, so that even hip joints that would otherwise be capable of continuous rotation are constrained by blood vessels and nerves to a kind of bending rotation.

Another area in which bioinspiration can constrain the design space concerns materials. Although, as noted earlier, nature has many examples of remarkable composite materials and structures with high complexity, nature also has a very limited repertoire of raw materials to work with. Metals, crystalline materials and ceramics are used very little. Any material that requires high temperatures and large energy expenditure to extract, purify and shape it is typically excluded. In contrast, going back to the Bronze age, humans have used energy-intensive processes to extract and work with strong metals, ceramics and crystals (e.g. sapphire, diamond). As a consequence, self-repair for objects like the hardened steel gears in a transmission is not an issue because they last almost indefinitely.

Other thought-provoking examples and counter-examples are found in [6], quoted at the top of this section.

2 Design Case Studies

In this section we explore a couple of bioinspired robot designs in more detail to put the general ideas outlined in the previous section in more concrete form.

2.1 *Enhancing Robustness and Stability in Running*

In the 1990s several robotics groups around the world became interested in small, mobile robots that drew inspiration from the locomotion of insects to increase their performance and robustness in navigating rough terrain. An important source of information arose from studies of running cockroaches and other insects and the subsequent examination of insect locomotion as a stable dynamic system. In particular, the cockroach shows a remarkable ability to run over very rough terrain (see Fig. 2, left) without slowing down. Interestingly, the leg motions are too rapid for sensor-based reaction to obstacles, instead the stabilization and ability to reject disturbances must arise from open-loop properties including the materials and kinematic arrangement of the legs.

At normal running speeds, the cockroach, like other insects, uses an alternating tripod gait in which the front and rear leg on one side and the middle leg on the opposite side are actuated in phase, keeping the center of mass over a triangular

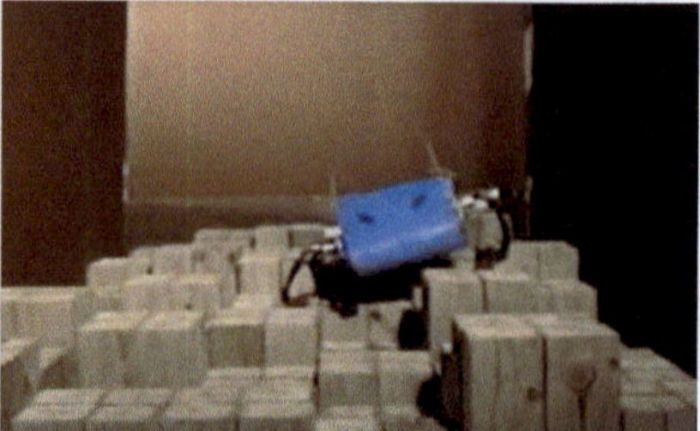

Fig. 2 (Left) A cockroach can run over belly-height obstacles without appreciably slowing down or getting knocked off course due to a combination of stabilizing features [21]. (Right) The Rhex robot running on similarly-scaled rough terrain, thanks to incorporation of principles from the cockroach [22]. Left image provided by S.N. Sponberg, Georgia Tech. with permission; right image by D. Koditschek, U. Penn, with permission

polygon of support. This strategy provides inherent stability as compared to bipedal or quadrupedal locomotion; the center of mass can be kept over the polygon of support formed by the feet in contact with the ground [23].

In addition, in comparison to larger animals, insects have a sprawled posture in which the legs stick out to the sides and in the fore-aft direction. The wide stance further increases stability and results in contact forces that are primarily horizontal rather than vertical. The reaction forces are transmitted mainly along the legs, as opposed to perpendicular to the legs, so that the mode of forward propulsion is more like bouncing on a Pogo stick than rowing a boat. The bouncing locomotion pattern, with energy stored elastically in the limbs and returned at the end of each step, is termed SLIP, or spring-loaded inverted pendulum [23].

Early studies of cockroach locomotion confirmed that recovery from perturbations occurs too rapidly to be explained entirely by response to sensory stimuli [24]. Numerical evidence of the stabilizing effect was presented in [25] with detailed models and proof of asymptotic stability in [26]. A detailed dynamic analysis can be found in [27]. The dynamic behavior is described by an oscillating system with a large "basin of attraction" in phase space for the periodic motions associated with running. Unlike early hopping robots, which also exhibited the SLIP locomotion pattern [28, 29] the insect's alternating tripod running pattern is stable over a wide range of conditions even when driven open-loop, i.e., with actuation controlled by a fixed timer rather than triggered by ground contact sensors [27].

The insights obtained from studying insect running lead to a number of small robotic hexapods [22, 30–33]. Each of these emphasizes a subset of the insect-inspired principles (alternating tripod gait, sprawled posture, contact forces directed primarily along the legs, compliant legs that store and return energy, passive damping that helps to stabilize the system in response to perturbations) while also making compromises to maintain simplicity. For example, each cockroach leg has seven primary degrees of freedom [34]. However, recalling that forces are directed mainly along the legs and that the hip or trochanter femur joint is mostly passive, it is possible to make a running robot with just one actuated degree of freedom per

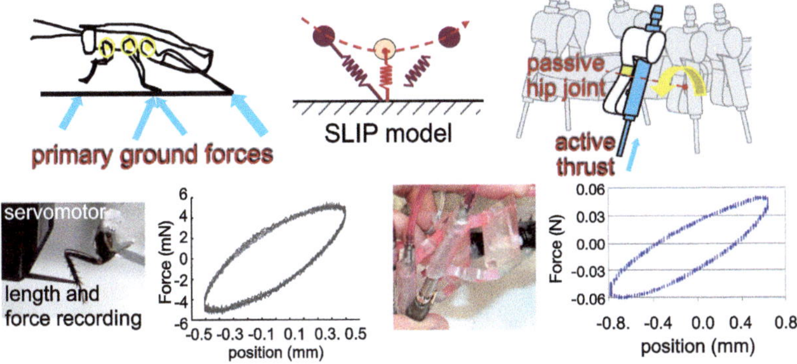

Fig. 3 *Left:* The running cockroach is propelled mainly by thrusting from the legs (light blue arrows). Hip torques are smaller and partly passive (yellow). A single leg shows significant damping, which helps to stabilize locomotion. *Center:* Three legs work together in a tripod, which can be approximated as a single spring so that the cockroach approximates a spring-loaded inverted pendulum (SLIP). *Right:* The Sprawl family of robots embodies this simplification with legs that thrust with one degree of freedom and elastic hip joints. The soft urethane is intentionally hysteretic, which is stabilizing. Adapted from [31, 39]

leg and with passive, damped hip joints. This is the approach taken in the Sprawl robots [31, 33]. Conversely, one can actuate only the hip joint and rely on passive compliance in the radial direction to achieve a bouncing gait as in RHex [22]. In both cases, the RHex and Sprawl exhibit similar open-loop stability. They settle into a regular "period-one" trot with oscillations of the center of mass in the vertical and horizontal directions that match the stepping frequency [35]. Interestingly, damping at the hip increases the range of frequencies for which stable, period-one trotting is possible [36, 37]. However, it does this at the expense of some reduction in the efficiency of locomotion. To achieve damping, the Sprawl robots have hip joints fabricated from hysteretic urethane, increasing the range of stable running speeds [38] (Fig. 3).

2.1.1 Design Discussion Question

Which of the principles from insect locomotion are specific to small robots and which apply broadly over a range of length scales? As a point of reference, hexapedal robots have been produced at everything from insect scales (similar to the robots depicted in Fig. 7) to the OSU Adaptive Suspension Vehicle, which was the size of a truck and had a mass of 2700 kg [40, 41]. All of these robots use the alternating tripod gait, and some relatively large hexapods have used a sprawled insect-like posture [42]. In nature, however, there are no large hexapods. In addition, the sprawled posture seen in insects generally disappears at larger scales. One can hypothesize dynamic, metabolic and structural explanations for this trend.

2.2 Exploiting Interaction with Surfaces in the Environment

For a second case study we consider the theme introduced earlier of exploiting physical interaction with the environment. A particularly compelling example of this phenomenon is the ability of geckos to run across walls and ceilings by exploiting van der Waals forces at the interface between their feet and the surfaces on which they cling.

Gecko-inspired adhesion is also good example of a bioinspired technology made possible by new tools to explore a biological system in detail. Although people have been interested in how geckos stick from at least the time of Aristotle [43], a true understanding of their adhesive system awaited the availability of scanning electron microscopes and extremely sensitive two-axis micro-electromechanical force sensors to measure the forces produced by setal stalks under different loading scenarios [44, 45]. These discoveries paved the way for numerous efforts to create synthetic dry adhesives. Reviews include [46–52].

From a functional perspective, particularly useful properties of the gecko's adhesive apparatus are as follows [53, 54]:

- conformation—Dry adhesion uses van der Waals forces, which only work over molecular distances; hence a dry adhesive system must conform intimately to a surface. The gecko's hierarchical system of lamellae, setal stalks and spatular tips conforms to both rough and smooth surfaces producing enough adhesion that the gecko can hang from a single toe.
- directional adhesion—The gecko's adhesive system adheres only when pulled from the palm toward the tips of the toes (as in climbing a wall); otherwise it is not sticky. Moreover, the amount of adhesion varies with the applied tangential force. Reproducing this proportionality has been instrumental for robotic applications.
- ease of attachment and detachment—The gecko runs up and down walls at speeds on the order of 1 m/s, taking many steps per second. A low effort of attachment and detachment prevents it from wasting energy with each step.
- high cycle rate—Given the number of steps the gecko takes per second, the adhesive must attach and release rapidly. This requirement precludes viscoelastic materials as found in sticky tape.
- self-cleaning—The gecko's feet stay clean even on dusty surfaces [55]. Dirt particles are relatively large compared to the terminal spatulae of the gecko's adhesive system and are more strongly attracted to most wall surfaces than they are to the gecko.

Although there are many synthetic gecko-inspired adhesives that approximate or exceed certain characteristics of gecko adhesion, none matches the collection of desirable capabilities listed above. Nonetheless, they are useful for applications including climbing robots, grasping surfaces in manufacturing processes, and even grasping objects such as solar panels and fuel tanks in space [56, 57].

2.2.1 Controllable Adhesion

From the perspective of robotics, a particularly important characteristic of gecko-inspired adhesion is that, unlike sticky tape, it is *controllable*, in the same way that Coulomb friction is controllable. We take advantage of this controllability in the brakes of automobiles: the harder one presses on the brakes, the greater the friction and the faster the deceleration. Similarly when grasping objects we rely on controllable friction to prevent slipping by regulating the grasp force. The gecko's adhesive system has an analogous property: the more it is loaded tangentially, i.e. in shear, the greater the adhesion (up to a limit). Hence it too is controllable.

For additional insight, it is useful to consider the implications of directional adhesion in terms of robot force and motion planning. For control, it is useful to think of constraints and regions of allowable forces in a multidimensional force space. The objective is to plan force trajectories for the robot foot so that contact forces remain in a safe region. This view of the gecko adhesive forces is summarized in Fig. 4, which shows limiting normal and tangential forces for three different levels of preload, corresponding to between 500 and 700 μm of elastic compression. Increasing the preload slightly increases the maximum adhesive force; a more interesting result is that the adhesion varies almost linearly with tangential force.

When the normal force, F_n, in Fig. 4 is positive, the foot is pressing into the wall and Coulomb friction pertains: $|F_t| \leq \mu F_n$. However, when a positive shear force is applied, pulling from the palm toward the tips of the toes, some adhesion is available. Moreover, the magnitude of the adhesion is in proportion to the shear force, up to a limit: $-F_n \leq \alpha F_t$ for $0 \leq T_t \leq T_{tmax}$ where α is a constant of proportionality, somewhat like μ for Coulomb friction. Thus, the gecko can adjust the amount of adhesion at any instant by controlling the shear force that it applies

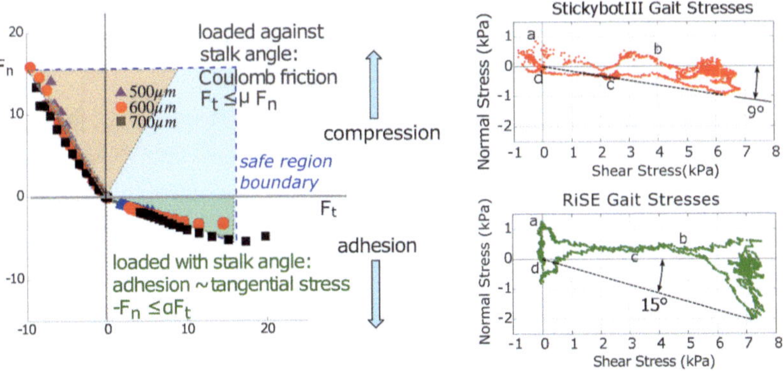

Fig. 4 (Left) Vertical axis shows normal forces (compressive or adhesive), horizontal axis shows tangential forces, safe region encloses forces that will not produce sliding or detachment (adapted from [58]). (Right) Normal and shear (tangential) stresses produced by ≈1 kg Stickybot and ≈4 kg RiSE robots stay within their respective safe regions except when detaching. Labels a–d correspond to labels in the accompanying text (adapted from [59])

with its feet. When taking a step up a wall, it applies a large shear force for maximum adhesion. When ready to detach its foot at the end of a step it relaxes the shear force, bringing the combined normal and shear force toward the origin of the plot and allowing it to detach its foot with almost no detachment force.

The plots at the right of Fig. 4 show how the limit curve is used by a couple of climbing robots, here showing stresses (force/area) for comparison although the RiSE robot is 4 times heavier than StickyBot. The robots assume a diagonal stride gait, with some overlap between pairs of legs. As the foot makes contact with the wall (a) the normal force is initially positive. As the robot loads the foot parallel to the wall, the shear stress first increases (b), which establishes adhesion. At (c) the robot is near the end of its stride as the opposite pair of legs assumes the load and the shear stress continues to diminish until (d) at which point there is no longer adhesion and the left detaches. The different values of α for the two robots correspond to different adhesives used.

The proportional nature of the adhesion also leads to some possibly nonintuitive results. Figure 5 shows a small gecko or robot climbing a vertical wall using a diagonal gait with one upper and one lower foot in contact at each step. Because the center of mass is located a small distance away from the wall surface, the upper foot, shown in green, must produce adhesion ($F_n < 0$) to keep the gecko from falling backward off the wall. The blue lower limb, in contrast, is pressed gently into the wall ($F_n > 0$). With the weight split evenly between the front and rear feet,

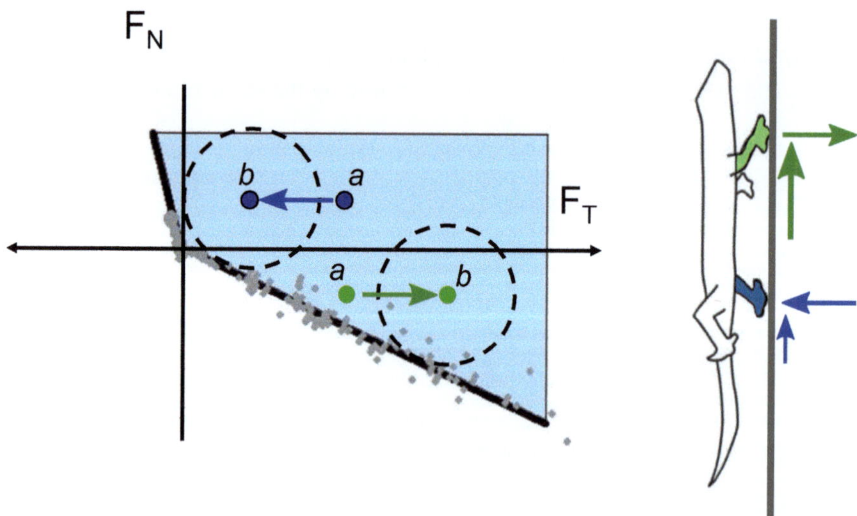

Fig. 5 When climbing a wall, the forelimbs require adhesion but the rear limbs are under compression. Hence, if weight is evenly distributed (a) the forelegs are most in danger of failing (i.e. are nearest to the edge of the safe region). To optimize for adhesion, the forelimbs should be controlled to assume more than 1/2 the weight, as shown in (b). Geckos do this as well [60] (figure adapted from [61])

the green dot corresponding to the upper limb, at position (a) in force space, is closer to the edge of the safe region than the blue dot associated with the lower limb. This situation matches our intuition that the upper limb is more likely to fail and may suggest a control approach that tends to "favor" the upper limb by loading it gently and supporting most of the weight with the lower limb.

However, this is the wrong strategy. Instead, the gecko or robot should pull *harder* with its front limbs, so that it has more adhesion available. The result is shown by moving the forces from (a) to (b) in the figure so that both feet have an equal safety margin.

The same lesson must be adapted for human-scale climbing, as seen in Fig. 6. Adhesion is needed at the hands to prevent pitching back. Hence, like a gecko or the Stickybot robot climbing a vertical wall, we should support the majority of our weight with our front limbs. However, people have greatest strength in their legs and use their legs for upward propulsion when climbing. Therefore it was necessary to develop a system of stirrups to transfer shear load from hands to feet [62]. Simply equipping a person with "gecko gloves" is not useful. Instead, as is often the case in bioinspired design, it is necessary not only to draw inspiration from biology but also to learn the underlying principles and adapt them to different application scenarios while accounting for differences in scale and implementation.

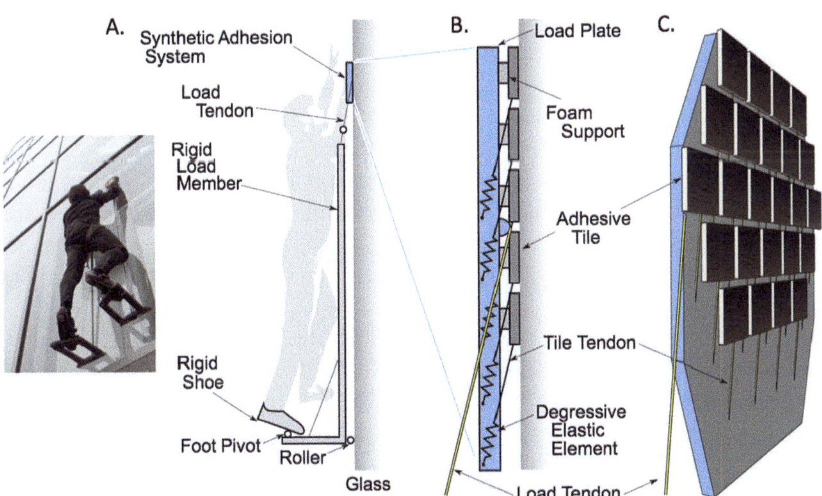

Fig. 6 Climbing a glass wall and view of apparatus for loading, which evenly distributes the adhesive stress over an array of small tiles and then redirects the resultant force to stirrups at the feet (adapted from [62])

3   Fabrication Challenges and Opportunities

A point made in the introduction is that the proliferation of bioinspired designs has proceeded apace with advances in materials and fabrication. For the insect-inspired robots, the running behavior depended on making robust multimaterial legs with flexible, damped hip joints at the proximal ends of comparatively stiff limbs. The method used to create these multimaterial limbs was Shape Deposition Manufacturing (SDM) [63–65]. Today one can also 3D print multimaterial legs with similar properties. Hence SDM, which is a relatively slow process that requires casting and curing polymers in multiple cycles to make a part, is being supplanted by multimaterial 3D printing.

However one interesting advantage of SDM is that it can create a very smooth surface finish. This is because it relies on material removal (e.g. CNC machining) to create mold cavities. Material removal processes like machining and grinding have an inherent smoothing property, which is why they are used for components like shafts, bearings, etc. where surface finish is critically important. In the case of the cockroach robots, SDM provided smooth finished hip joints that could survive millions of bending cycles without fatigue failure.

A different process is typically needed for creating the adhesive microstructure, taking place in a clean room and using lithographic methods [66]. For practical reasons, the facilities used to make microscopic and macroscopic features are typically different; the microscopic processes are too slow for macroscopic parts and the macroscopic machines are not accurate enough for microscopic features. As noted earlier this limitation sets bioinspired robots aside from organisms that grow large and small structures, in parallel, cell by cell.

With the rising popularity of soft robotics, soft lithography [67] and molding have become popular techniques for body fabrication [68]. These processes share some characteristics with SDM because of their basis in a casting workflow but they do not include material removal steps. They are often used to make highly compliant structures.

At a smaller scale, such as that of the miniature robots shown in Fig. 7, alignment and assembly operations become increasingly challenging. The Smart Composite Microstructures (SCM) (Fig. 7) process provides a solution to this difficulty by creating laminated structures in two and a half dimensions before folding into three dimensions [70]. The process involves patterning shapes and components on a series of layers of flexible and hard materials, which can contain reinforcing fibers. The shapes are cut typically with an ultraviolet laser, folded and bonded to create an "origami" structure with precise tolerances. Integration of actuation is an advantage of such a technique; piezoelectric actuators can be tailored for different applications and embedded as seen in Fig. 7b–d. Shape Memory Alloy (SMA) actuators can also be integrated into a structure to create a self-folding reconfigurable origami sheet [71]. Additionally, sensing can be integrated during the SCM process [72]. Emerging CAD and process planning tools simplify the process of converting a solid model to a sequence of processing steps [73]. To further reduce the need for manual

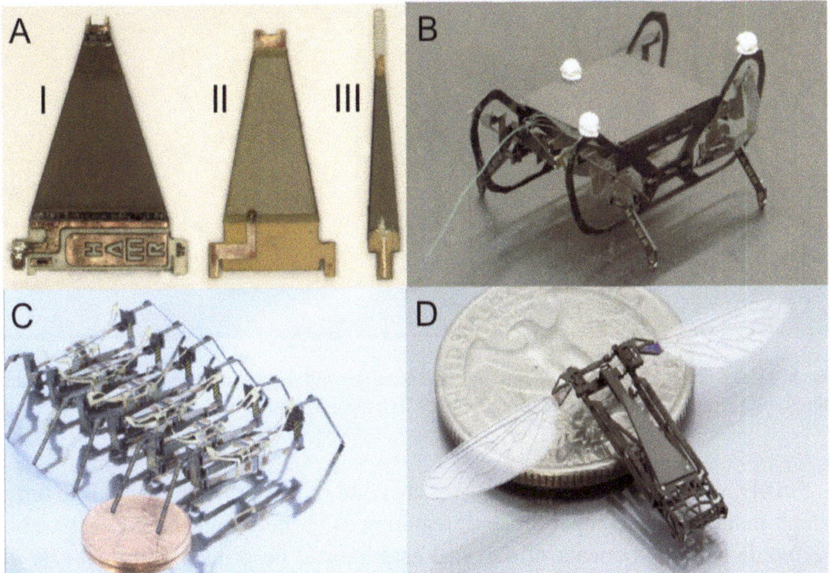

Fig. 7 Smart Composite Microstructure manufacturing is a layered technique for creating multi-material robot bodies with integrated actuation, sensing and control [69]: (**a**) Piezo electric actuators and their control circuitry, (**b**) an assembled quadruped, (**c**) centipede, and (**d**) hoverfly robots (Images from R. Wood, Harvard University), with permission

assembly or other interactions, it is also possible to make structures self-folding [74], or spring-loaded so that they automatically pop-up into their three-dimensional shape [75]. Although we are still a long way from the levels of complexity and integration found in natural organisms, robots like those in Fig. 7 show what is becoming possible with emerging fabrication techniques.

Requiring even less intervention during the manufacturing process are 3D printing or Additive Manufacturing (AM) technologies [76]. 3D printing is becoming ubiquitous as printers grow more capable and inexpensive, even for hobbyist use. It is becoming possible to print soft and multi-material structures with good surface finish and fatigue life directly from CAD models. The resolution of some technologies has reached the sub-micron level (e.g., Fig. 8), and new materials, such as printed matrices with embedded composites or printed electronics [77], are greatly increasing the functionality of printed parts. Biocompatible materials and designs are opening up new avenues for 3D printing applications in biotechnology and biohybrid (combined biological and robotic) machines [78, 79]. 3D printers can also produce materials with functionally graded moduli, allowing designers to create robot structures with spatially varying mechanical properties (as commonly found in plants and animals) and with smooth transitions to prevent the stress concentrations that otherwise arise at discrete material boundaries [80, 81].

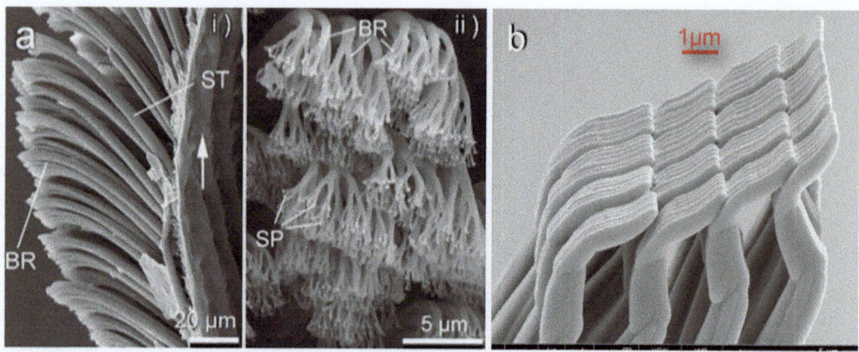

Fig. 8 Gecko setal array (**a**) (from [85] with permission) and approximation (**b**) using two-photon lithography with sub-micron resolution (from O. Tricini, IIT, with permission)

Finally, various new hybrid processes are combining the above methods to further extend manufacturing capabilities. Integrating 3D printing and SDM enables both the molds and components of the final assembly to be printed and then cast into and around discrete components to make monolithic structures [82, 83]. Another hybrid process integrates 3D printing with soft lithography to create an entirely soft device—even the energy source and logic are made of compliant materials [84].

3.1 Remaining Fabrication Challenges

Despite the rapidly advancing capabilities of these technologies, it remains that the emerging fabrication processes typically work on materials in bulk form. Therefore, each new variation in material adds complexity. In addition, most equipment for depositing or shaping materials works over a limited range of length scales [81]. For example, micromachining techniques are impractical for depositing and shaping material at the scale of tens of centimeters. The synthetic gecko setae depicted in Fig. 8 are impressive, but it takes many hours to produce less than 1×1 mm worth of stalks, which is impractically slow [86]. In contrast, the cells of biological organisms grow and differentiate in parallel. The price for complexity in nature is therefore much lower. This realization is one of the motivations behind biohybrid systems that aim to combine the best attributes of robotic and biological components [87, 88], which also have the potential to be self-repairing. Other emerging methods such as block copolymer self-assembly [89] and various techniques based on folding, printing, and assembly at a scale ranging from tens of nanometers to hundreds of micrometers [90] show promise to create macroscopic samples with micro or nanopatterning. In other work, a precision CNC was used to both create microscopic adhesive features and macroscopic support frames for a gecko-inspired gripper [91].

4   Bioinspired Design Process

What is a design process that leads to the development of robots like the cockroach-inspired runners and the gecko-inspired climbers explored in the two previous examples?

A common process lead to the Rhex [22], Sprawl [31, 33], Rise [92], and StickyBot [93] robots. In each case, the design grew from a collaboration among roboticists and biologists, in response to a stated challenge (e.g. "Run rapidly over rough terrain" or "Climb walls quietly and without consuming lots of energy").

Figure 9 illustrates the process, using StickyBot as an example. It begins with an exploration of animals that can achieve a desired capability. In this case we are interested in animals that climb smooth vertical surfaces with agility. Possible exemplars include geckos, lizards, spiders and insects. Among these animals, geckos stand out for their ability to run up both rough and smooth surfaces using dry adhesion.

The exploration continues by exploring the biology literature, a process made easier with resources like Google Scholar (e.g. search "cockroach vertical climbing"

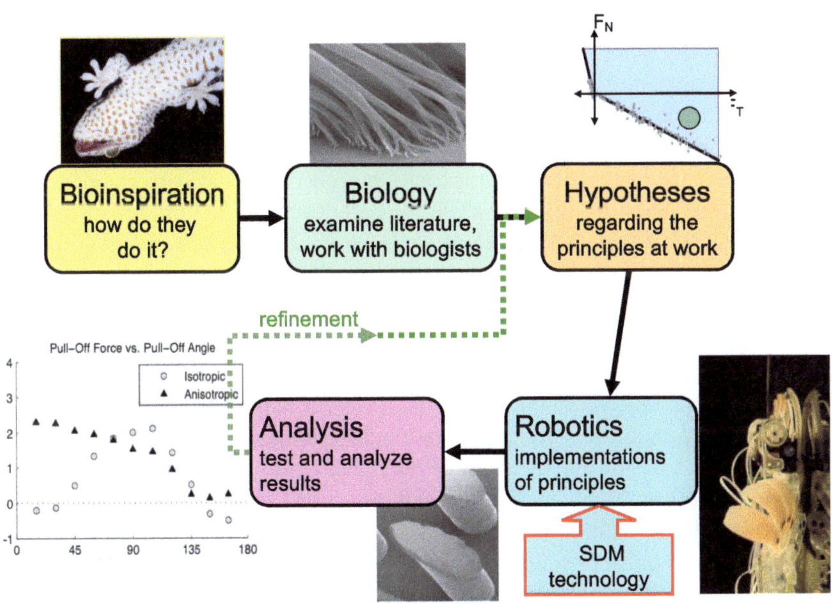

Fig. 9 Bioinspired design starts with a challenge, followed by exploration of possible plant or animal exemplars and examination of the relevant literature. The next step is to formulate hypotheses about the most important principles, which can be adapted to robots. Robot prototypes often use multimaterial manufacturing processes (e.g. SDM as discussed in Sect. 3) and are tested to evaluate the hypotheses. Analysis typically results in refinement and iteration (adapted from [61])

or "spider adhesion"). However, the biological literature is often not written with design inspiration as a goal. Moreover, it is not always obvious where to start (e.g. which animals to examine). Hence, it is useful to make contact with biologists who study topics like animal locomotion, manipulation, or sensing. Recognizing the challenge of mapping from biological literature to engineering objectives, there are also specialized databases such as AskNature (https://asknature.org/) and design research tools, a number of which are described and reviewed in [94–96]. In Sect. 4.1 we introduce one relatively simple technique for articulating functional requirements in a form that can help when reviewing the literature.

The next step of the bio-inspired design process involves creating hypotheses about principles that underlie the animals' success. As noted earlier, natural examples are formidably complex, the more so as we examine them in detail. Because we cannot exactly reproduce the biological systems and structures, we attempt to identify the most important effects so that we can incorporate them into simplified approximations of what we observe in nature.

We fabricate robotic systems or structures that embody those principles, using the best multimaterial methods at our disposal, and test them. It is at this stage that robotics can provide useful information for biologists as well as engineers, because it is much easier to conduct comprehensive tests—to support or refute our hypotheses—on robots than on animals.

We then analyze the results and invariably wind up refining our hypotheses and robotic implementations, and so the cycle repeats.

This process leads to new knowledge as well as new robotic capabilities. Indeed, as we learn more about what is important and how to optimize performance we can deviate from biological examples and exceed biological performance. As an example, it has been estimated that the gecko is the largest animal that can climb using van der Waals forces [97]. However, by understanding in detail the principles of load sharing and controllable adhesion it is possible to design a system that allows a human to climb, despite being $100\times$ heavier.

More broadly, bioinspired design has been the topic of numerous presentations, workshops and even courses (e.g. at U.C. Berkeley's CiBER Institute https://www.cibercenter.net/). Different courses follow different approaches, for example considering bioinspiration as a design tool to employ alongside others or, conversely, to start with bioinspiration and then ask "What is this good for?". The former approach is sometimes called "needs pull" (start with needs and then look for tools to satisfy the corresponding requirements), the latter is closer to "technology push" or, in this case, biology-push (a solution looking for a problem). Both approaches can be successful. In the design exercises below we take a needs-pull approach, following a methodology used in the Stanford engineering design curriculum.

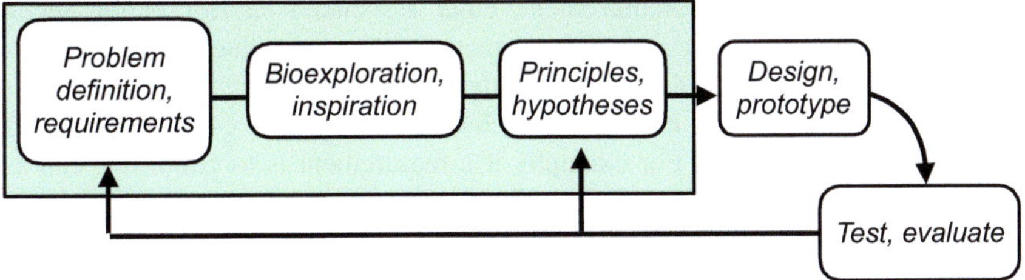

Fig. 10 Design process steps generalized from Fig. 9. Those in the shaded box can be performed online

Table 1 Requirements table using 3-column template from Stanford ME310 [98]

Requirement	Metrics	Rationale
Acquire small, thin items	Min. diameter: 4 mm; Min sheet thickness: 1 mm	Need to pick up pencils, coins, credit cards
Grasp large items with low grasp force	Max. diameter: 20 cm; mass = 1 kg with grasp force < 1 N	Grasp ripe fruit (e.g. pear, large tomato)
Acquire and grasp slippery items	Min. coefficient of friction: $\mu = 0.1$	
Stably hold long tools, poles	Hold cylinder 1 cm $<$ dia $<$ 10 cm $\times$ 1 m long with moments of 2 N m	Able to wield a hammer or staff
Gentle touch for probing loose objects	Endpoint inertia < 100g, low initial contact force (<0.1 N) when moving at 10 cm/s	Able to explore in a cluttered environment without knocking things over
...	...	...

4.1 Design Exercises

Although bioinspired robot design as depicted in Fig. 9 ultimately requires access to fabrication and testing facilities, the first few steps can be undertaken by anybody with access to the internet. These steps occupy the shaded rectangle in Fig. 10. We explore them in a bit more detail in this section with some suggested exercises.

4.1.1 Defining Functional Requirements

In order to examine the biology literature it is valuable to have some functional requirements in mind. We are not looking for biomimicry, but rather to be inspired and informed by findings from biological examples. As a starting point, we need to articulate the objectives that our design should fulfill. Although not particularly bioinspired, the format in Table 1, taken from Stanford's Design Thinking curriculum [98–100], has been found useful. The idea is to list requirements from a functional (what it must do) standpoint. List each requirement in the left column, initially without too much concern for how the requirements are articulated or organized. Then, for each requirement, enter associated *metrics* in the second column. True requirements always have metrics, whether values or ranges. How can we tell if a particular design fulfills the requirements? Thinking about these metrics enforces specificity and may often lead the designer to go back and modify the requirement definition. For example, if a requirement is to climb, we can ask how fast, and what should be the maximum slope (e.g., what about overhanging surfaces?). The table provides a few examples for the case of a bioinspired gripper or hand.

As with the requirements definition in the left column, the idea is not to become unduly concerned about what the metrics and values are but rather to use them as a way to clarify one's thinking. Finally, the third column is used to provide some rationale for the requirement: Why is this requirement sufficiently important to be included in the table?

After articulating a couple of pages full of such requirements the usual process is to cluster them, grouping those that are related, eliminating redundancies and also denoting sub-requirements that lie under a common higher level requirements.

Note also that requirements are not static. As the design evolves and as we learn more information, through early design prototyping and testing, we commonly go back and revisit and modify the requirements—perhaps adding new "discovered requirements" and deprecating ones that we find less important. As an example, the importance of *controllable* adhesion was not fully appreciated in the early days of the project that lead to the climbing robots discussed in Sect. 2.2.1.

4.1.2 Finding Exemplars

With requirements and objectives defined we are ready for step 2, finding examples. As noted in Sect. 4, searching the biology literature has become easier with tools like Google Scholar. However, as also noted in that section, the literature is typically not written with design inspiration in mind, therefore it is useful to be able to communicate with biologists. In addition, there are specialized databases and techniques, some of which were cited in that section.

Note that we are not necessarily looking for a complex behavior (e.g. legged locomotion) but may be looking for a specific function such as storing elastic energy, adhering to wet surfaces, or exploring surfaces with tactile sensors or whiskers—indeed, all of these examples have lead to bioinspired robotic designs. Still another strategy is to focus on particular environments or conditions and then to study those animals that are at home in similar environments.

As robots work outdoors and in other unstructured environments (e.g. a messy home) the list of desired functional capabilities becomes longer and the opportunities for bioinspiration increase. The robots no longer just need to assemble parts, they now need to communicate and respond to actions from other agents, to plan paths, to collaborate, etc. Nature provides countless examples of these capabilities. Recognize that even single cell animals perform sensing and locomotion and colonies of such cells can explore environments for resources.

4.1.3 Extracting Principles and Hypotheses

We need to extract principles that can adapt to robot prototypes. While this activity is necessarily application-specific, there are a few strategies that can be useful:

- *Look for patterns across species.* If something is found widely across many species it may be especially worth looking at. Convergent evolution is often instructive: e.g. cactus and euphorbia for surviving in desert conditions, or common approaches to swimming across fish, pinnipeds and other very different aquatic species.
- *Look for ranges of solutions and consider how functions and metrics vary.* For example, many animals can manipulate. How do insects, squirrels, birds, apes, octopuses and elephants manipulate items like food? What different strategies do they use? What is similar? One can do the same for flight (insects, bats, birds) and swimming, etc.
- *Look for unusual solutions.* We gravitate toward familiar animal examples, but it is also useful to look at plants and even single-cell animals. Plants can explore (e.g., find pockets of soil with desirable moisture or nutrients, and seek light), attach to surfaces, sense chemicals even actuate. The time scale is typically slow (think of the roots of a tree splitting a rock) but can also be fast as in the case of a Venus flytrap.
- *Look for differences with respect to human engineering.* Recognize what is inherently different. As noted in one of the open questions in Sect. 1.2, nature has a limited repertoire of materials, and uses nothing that requires high temperatures or energy to process. In addition, nature needs to keep most of the components alive, i.e. perfused and oxygenated with blood vessels, etc. This makes it hard to support functions like continuous rotation, which in turn makes it hard to have wheels, propellers, etc. Along similar lines, recognize that precision and repeatability are unique attributes of modern engineering. If we grew gears or turbine blades from bioinspired processes, accreting gradually through actions or many small agents (e.g. insects or cells) they would look more like termite mounds or bones. They might be functional, but no two would be exactly alike. Depending on the context this might or might not be a problem.

4.1.4 A Design Challenge

Presumably readers will want to find their own bioinspired robotic design challenges on which to apply some of the ideas and methods presented in this chapter. We part with a challenge used as a warm-up exercise in previous design courses and which prompts an interesting exploration into the literature.

*Design a robotic hand that is **not** inspired by the human hand.* There are a couple of reasons behind this example. Although the human hand is arguably the most sophisticated and readily available one in our experience it is likely more dexterous and surely more complex than most robots require. In addition, there are many other animals that can grasp and manipulate. Some interesting examples from the literature include frogs [101] and insects, and even the octopus. Examples from insects and related hand design principles are reviewed in [102].

Cross-References

This chapter contains cross-references to content published in other chapters belonging to the four "Robotic Goes MOOC" books: DES5, DES8, INT1, IMP3, IMP5, IMP7. The full list of chapters abbreviations is available in the Preface.

References

1. S. Cave, K. Dihal, Ancient dreams of intelligent machines: 3,000 years of robots. Nature **559**(7715), 473–475 (2018)
2. P. Harrison, Descartes on animals. Philos. Q. (1950–) **42**(167), 219–227 (1992)
3. J.L. Gould, Animal cognition. Curr. Biol. **14**(10), R372–R375 (2004)
4. O. Goury, C. Duriez, Fast, generic, and reliable control and simulation of soft robots using model order reduction. IEEE Trans. Robot. **34**(6), 1565–1576 (2018)
5. G.-Z. Yang, J. Bellingham, P.E. Dupont, P. Fischer, L. Floridi, R. Full, N. Jacobstein, V. Kumar, M. McNutt, R. Merrifield et al., The grand challenges of science robotics. Sci. Robot. **3**(14), eaar7650 (2018)
6. S. Vogel, Nature's swell, but is it worth copying? MRS Bull. **28**(6), 404–408 (2003)
7. K. Jayaram, R.J. Full, Cockroaches traverse crevices, crawl rapidly in confined spaces, and inspire a soft, legged robot. Proc. Natl. Acad. Sci. **113**(8), E950–E957 (2016)
8. F. Delcomyn, The effect of limb amputation on locomotion in the cockroach periplaneta americana. J. Exp. Biol. **54**(2), 453–469 (1971)
9. K. Jayaram. Robustness of Biological and Bio-inspired Exoskeletons. PhD thesis, UC Berkeley (2015)
10. H. Machemer, J.W. Deitmer, Mechanoreception in ciliates, in *Progress in Sensory Physiology* (Springer, Berlin, 1985), pp. 81–118
11. R.J. Full, D.E. Koditschek, Templates and anchors: neuromechanical hypotheses of legged locomotion on land. J. Exp. Biol. **202**(23), 3325–3332 (1999)
12. T. Libby, A.M. Johnson, E. Chang-Siu, R.J. Full, D.E. Koditschek, Comparative design, scaling, and control of appendages for inertial reorientation. IEEE Trans. Robot. **32**(6), 1380–1398 (2016)
13. U. Castiello, The neuroscience of grasping. Nat. Rev. Neurosci. **6**(9), 726–736 (2005)
14. G. ElKoura, K. Singh, Handrix: animating the human hand, in *Proceedings of the 2003 ACM SIGGRAPH/Eurographics symposium on Computer animation* (2003), pp. 110–119
15. M. Santello, M. Flanders, J.F. Soechting, Postural hand synergies for tool use. J. Neurosci. **18**(23), 10105–10115 (1998)
16. G. Palli, C. Melchiorri, G. Vassura, U. Scarcia, L. Moriello, G. Berselli, A. Cavallo, G. De Maria, C. Natale, S. Pirozzi, The dexmart hand: mechatronic design and experimental evaluation of synergy-based control for human-like grasping. Int. J. Robot. Res. **33**(5), 799–824 (2014)
17. M.G. Catalano, G. Grioli, E. Farnioli, A. Serio, C. Piazza, A. Bicchi, Adaptive synergies for the design and control of the pisa/iit softhand. Int. J. Robot. Res. **33**(5), 768–782 (2014)

18. B.S. Blanchard, W.J. Fabrycky, W.J. Fabrycky, *Systems Engineering and Analysis*, vol. 4 (Prentice Hall, Englewood Cliffs, 1990)
19. R.A. Brooks, Integrated systems based on behaviors. ACM Sigart Bull. **2**(4), 46–50 (1991)
20. S. Seok, A. Wang, M.Y. Chuah, D. Otten, J. Lang, S. Kim, Design principles for highly efficient quadrupeds and implementation on the mit cheetah robot, in *2013 IEEE International Conference on Robotics and Automation* (IEEE, 2013), pp. 3307–3312
21. S. Sponberg, R.J. Full, Neuromechanical response of musculo-skeletal structures in cockroaches during rapid running on rough terrain. J. Exp. Biol. **211**(3), 433–446 (2008)
22. U. Saranli, M. Buehler, D.E. Koditschek, Rhex: a simple and highly mobile hexapod robot. Int. J. Robot. Res. **20**(7), 616–631 (2001)
23. R. Blickhan, R.J. Full, Similarity in multilegged locomotion: bouncing like a monopode. J. Comp. Physiol. A Neuroethol. Sens. Neural Behav. Physiol. **173**(5), 509–517 (1993)
24. D.L. Jindrich, R.J. Full, Dynamic stabilization of rapid hexapedal locomotion. J. Exp. Biol. **205**(18), 2803–2823 (2002)
25. T.M. Kubow, R.J. Full, The role of the mechanical system in control: a hypothesis of self-stabilization in hexapedal runners. Philos. Trans. R. Soc. Lond. B Biol. Sci. **354**(1385), 849–861 (1999)
26. J. Schmitt, P. Holmes, Mechanical models for insect locomotion: Dynamics and stability in the horizontal plane i. Theory. Biol. Cybern. **83**(6), 501–515 (2000)
27. P. Holmes, R.J. Full, D. Koditschek, J. Guckenheimer, The dynamics of legged locomotion: models, analyses, and challenges. Siam Rev. **48**(2), 207–304 (2006)
28. M.H. Raibert, *Legged Robots that Balance* (MIT Press, Cambridge, 1986)
29. M.H. Raibert, Running with symmetry. Int. J. Rob. Res. **5**(4), 3–19 (1987)
30. M. Binnard, Boadicea: a small pneumatic walking robot. Master of Science Thesis, Artificial Intelligence Laboratory MIT (1995)
31. J.G. Cham, S.A. Bailey, J.E. Clark, R.J. Full, M.R. Cutkosky, Fast and robust: hexapedal robots via shape deposition manufacturing. Int. J. Robot. Res. **21**(10–11), 869–882 (2002)
32. R.D. Quinn, G.M. Nelson, R.J. Bachmann, D.A. Kingsley, J.T. Offi, T.J. Allen, R.E. Ritzmann, Parallel complementary strategies for implementing biological principles into mobile robots. Int. J. Robot. Res. **22**(3–4), 169–186 (2003)
33. S. Kim, J.E. Clark, M.R. Cutkosky, isprawl: design and tuning for high-speed autonomous open-loop running. Int. J. Robot. Res. **25**(9), 903–912 (2006)
34. D.A. Kingsley, R.D. Quinn, R.E. Ritzmann, A cockroach inspired robot with artificial muscles, in *2006 IEEE/RSJ International Conference on Intelligent Robots and Systems* (IEEE, 2006), pp. 1837–1842
35. D.E. Koditschek, R.J. Full, M. Buehler, Mechanical aspects of legged locomotion control. Arthropod Struct. Dev. **33**(3), 251–272 (2004)
36. J.G. Cham, J. Karpick, M.R. Cutkosky, Stride period adaptation for a biomimetic running hexapod. Int. J. Robot. Res. **23**(2), 141–153 (2004)
37. J.G. Cham, M.R. Cutkosky, Dynamic stability of open-loop hopping. J. Dyn. Syst. Meas. Control **129**(3), 275–284 (2007)
38. X. Xu, W. Cheng, D. Dudek, M.R. Cutkosky, R.J. Full, M. Hatanaka, Material modeling for shape deposition manufacturing of biomimetic components, in *International Design Engineering Technical Conferences and Computers and Information in Engineering Conference*, ASME IDETC, vol. 35135, pp. 205–214 (September 2000)
39. S.A. Bailey, J.G. Cham, M.R. Cutkosky, R.J. Full, Biomimetic robotic mechanisms via shape deposition manufacturing, in *Robotics Research* (Springer, Berlin, 2000), pp. 403–410
40. K.J. Waldron, R.B. McGhee, The adaptive suspension vehicle. IEEE Control Syst. Mag. **6**(6), 7–12 (1986)
41. S.-M. Song, K.J. Waldron, *Machines That Walk: The Adaptive Suspension Vehicle* (MIT Press, Cambridge, 1989)
42. M.H. Raibert, I.E. Sutherland, Machines that walk. Sci. Am. **248**(1), 44–53 (1983)
43. A. Aristotle, *History of Animals* (Harvard University Press, Cambridge, 1991)

44. K. Autumn, Y.A. Liang, S.T. Hsieh, W. Zesch, W.P. Chan, T.W. Kenny, R. Fearing, R.J. Full, Adhesive force of a single gecko foot-hair. Nature **405**(6787), 681–685 (2000)
45. K. Autumn, M. Sitti, Y.A. Liang, A.M. Peattie, W.R. Hansen, S. Sponberg, T.W. Kenny, R. Fearing, J.N. Israelachvili, R.J. Full, Evidence for van der Waals adhesion in Gecko Setae. Proc. Natl. Acad. Sci. **99**(19), 12252–12256 (2002)
46. L.F. Boesel, C. Greiner, E. Arzt, A. Del Campo, Gecko-inspired surfaces: a path to strong and reversible dry adhesives. Adv. Mater. **22**(19), 2125–2137 (2010)
47. M. Kamperman, E. Kroner, A. del Campo, R.M. McMeeking, E. Arzt, Functional adhesive surfaces with "gecko" effect: the concept of contact splitting. Adv. Eng. Mater. **12**(5), 335–348 (2010)
48. D. Sameoto, C. Menon, Recent advances in the fabrication and adhesion testing of biomimetic dry adhesives. Smart Mater. Struct. **19**(10), 103001 (2010)
49. M.K. Kwak, C. Pang, H.-E. Jeong, H.-N. Kim, H. Yoon, H.-S. Jung, K.-Y. Suh, Towards the next level of bioinspired dry adhesives: new designs and applications. Adv. Funct. Mater. **21**(19), 3606–3616 (2011)
50. M. Zhou, N. Pesika, H. Zeng, Y. Tian, J. Israelachvili, Recent advances in gecko adhesion and friction mechanisms and development of gecko-inspired dry adhesive surfaces. Friction **1**(2), 114–129 (2013)
51. P.M. Favi, S. Yi, S.C. Lenaghan, L. Xia, M. Zhang, Inspiration from the natural world: from bio-adhesives to bio-inspired adhesives. J. Adhes. Sci. Technol. **28**(3–4), 290–319 (2014)
52. D. Sameoto, Manufacturing approaches and applications for bioinspired dry adhesives, in *Bio-Inspired Structured Adhesives* (Springer, Berlin, 2017), pp. 221–244
53. K. Autumn, J. Puthoff, Properties, principles, and parameters of the gecko adhesive system, in *Biological Adhesives*, ed. by A. Smith (Springer, Berlin, 2016), pp. 245–280
54. D. Santos, M. Spenko, A. Parness, S. Kim, M. Cutkosky, Directional adhesion for climbing: theoretical and practical considerations. J. Adhes. Sci. Technol. **21**(12–13), 1317–1341 (2007)
55. W.R. Hansen, K. Autumn, Evidence for self-cleaning in gecko setae. Proc. Natl. Acad. Sci. **102**(2), 385–389 (2005)
56. E.W. Hawkes, H. Jiang, D.L. Christensen, A.K. Han, M.R. Cutkosky, Grasping without squeezing: design and modeling of shear-activated grippers. IEEE Trans. Robot. **34**(2), 303–316 (2017)
57. H. Jiang, E.W. Hawkes, C. Fuller, M.A. Estrada, S.A. Suresh, N. Abcouwer, A.K. Han, S. Wang, C.J. Ploch, A. Parness et al., A robotic device using gecko-inspired adhesives can grasp and manipulate large objects in microgravity. Sci. Robot. **2**(7) (2017)
58. K. Autumn, A. Dittmore, D. Santos, M. Spenko, M. Cutkosky, Frictional adhesion: a new angle on gecko attachment. J. Exp. Biol. **209**(18), 3569–3579 (2006)
59. E.W. Hawkes, E.V. Eason, A.T. Asbeck, M.R. Cutkosky, The gecko's toe: scaling directional adhesives for climbing applications. IEEE/ASME Trans. Mechatron. **18**(2), 518–526 (2012)
60. K. Autumn, S.T. Hsieh, D.M. Dudek, J. Chen, C. Chitaphan, R.J. Full, Dynamics of geckos running vertically. J. Exp. Biol. **209**(2), 260–272 (2006)
61. M.R. Cutkosky, Climbing with adhesion: from bioinspiration to biounderstanding. Interface Focus **5**(4), 20150015 (2015)
62. E.W. Hawkes, E.V. Eason, D.L. Christensen, M.R. Cutkosky, Human climbing with efficiently scaled gecko-inspired dry adhesives. J. R. Soc. Interface **12**(102), 20140675 (2015)
63. R. Merz, F.B. Prinz, K. Ramaswami, M. Terk, L. Weiss, *Shape Deposition Manufacturing* (Engineering Design Research Center, Carnegie Mellon University, Pittsburgh, 1994)
64. M. Binnard, M.R. Cutkosky, Design by composition for layered manufacturing. J. Mech. Des. **122**(1), 91–101 (2000)
65. X. Xu, W. Cheng, D.M. Dudek, M. Hatanaka, M.R. Cutkosky, R.J. Full, Material modeling for shape deposition manufacturing of biomimetic components, in *IDETC* (ASME, September 2000)
66. A. Parness, D. Soto, N. Esparza, N. Gravish, M. Wilkinson, K. Autumn, M. Cutkosky, A microfabricated wedge-shaped adhesive array displaying gecko-like dynamic adhesion, directionality and long lifetime. J. R. Soc. Interface **6**(41), 1223–1232 (2009)

67. Y. Xia, G.M. Whitesides, Soft lithography. Annu. Rev. Mater. Sci. **28**(1), 153–184 (1998)
68. D. Rus, M.T. Tolley, Design, fabrication and control of soft robots. Nature **52F1**(7553), 467 (2015)
69. K. Jayaram, N.T. Jafferis, N. Doshi, B. Goldberg, R.J. Wood, Concomitant sensing and actuation for piezoelectric microrobots. Smart Mater. Struct. **27**(6), 065028 (2018)
70. R.J. Wood, S. Avadhanula, R. Sahai, E. Steltz, R.S. Fearing, Microrobot design using fiber reinforced composites. J. Mech. Des. **130**(5), 052304 (2008)
71. E. Hawkes, B. An, N.M. Benbernou, H. Tanaka, S. Kim, E.D. Demaine, D. Rus, R.J. Wood, Programmable matter by folding. Proc. Natl. Acad. Sci. **107**(28), 12441–12445 (2010)
72. D.W. Haldane, C.S. Casarez, J.T. Karras, J. Lee, C. Li, A.O. Pullin, E.W. Schaler, D. Yun, H. Ota, A. Javey et al., Integrated manufacture of exoskeletons and sensing structures for folded millirobots. J. Mech. Robot. **7**(2), 021011 (2015)
73. D.M. Aukes, B. Goldberg, M.R. Cutkosky, R.J. Wood, An analytic framework for developing inherently-manufacturable pop-up laminate devices. Smart Mater. Struct. **23**(9), 094013 (2014)
74. S. Felton, M. Tolley, E. Demaine, D. Rus, R. Wood, A method for building self-folding machines. Science **345**(6197), 644–646 (2014)
75. P.S. Sreetharan, J.P. Whitney, M.D. Strauss, R.J. Wood, Monolithic fabrication of millimeter-scale machines. J. Micromech. Microeng. **22**(5), 055027 (2012)
76. B. Bhushan, M. Caspers, An overview of additive manufacturing (3d printing) for microfabrication. Microsyst. Technol. **23**(4), 1117–1124 (2017)
77. A.J. Lopes, E. MacDonald, R.B. Wicker, Integrating stereolithography and direct print technologies for 3d structural electronics fabrication. Rapid Prototyp. J. **18**(2), 129–143 (2012)
78. V.A. Webster, F.R. Young, J.M. Patel, G.N. Scariano, O. Akkus, U.A. Gurkan, H.J. Chiel, R.D. Quinn, 3D-printed biohybrid robots powered by neuromuscular tissue circuits from Aplysia californica, in *Conference on Biomimetic and Biohybrid Systems* (Springer, Berlin, 2017), pp. 475–486
79. R. Mestre, T. Patiño, S. Sánchez, Biohybrid robotics: from the nanoscale to the macroscale. Wiley Interdiscip. Rev. Nanomed. Nanobiotechnol. e01703 (2021)
80. N.W. Bartlett, M.T. Tolley, J.T.B. Overvelde, J.C. Weaver, B. Mosadegh, K. Bertoldi, G.M. Whitesides, R.J. Wood, A 3d-printed, functionally graded soft robot powered by combustion. Science **349**(6244), 161–165 (2015)
81. M.R. Cutkosky, S. Kim, Design and fabrication of multi-material structures for bioinspired robots. Philos. Trans. R. Soc. Lond. A: Math. Phys. Eng. Sci. **367**(1894), 1799–1813 (2009)
82. S.A. Suresh, D.L. Christensen, E.W. Hawkes, M. Cutkosky, Surface and shape deposition manufacturing for the fabrication of a curved surface gripper. J. Mech. Robot. **7** (2015)
83. R.R. Ma, J.T. Belter, A.M. Dollar, Hybrid deposition manufacturing: design strategies for multimaterial mechanisms via three-dimensional printing and material deposition. J. Mech. Robot. **7**(2), 021002 (2015)
84. M. Wehner, R.L. Truby, D.J. Fitzgerald, B. Mosadegh, G.M. Whitesides, J.A. Lewis, R.J. Wood, An integrated design and fabrication strategy for entirely soft, autonomous robots. Nature **536**, 451–455 (2016)
85. N. Gravish, M. Wilkinson, S. Sponberg, A. Parness, N. Esparza, D. Soto, T. Yamaguchi, M. Broide, M. Cutkosky, C. Creton et al., Rate-dependent frictional adhesion in natural and synthetic gecko setae. J. R. Soc. Interface **7**(43), 259–269 (2010)
86. O. Tricinci, E.V. Eason, C. Filippeschi, A. Mondini, B. Mazzolai, N.M. Pugno, M.R. Cutkosky, F. Greco, V. Mattoli, Approximating gecko setae via direct laser lithography. Smart Mater. Struct. **27**(7), 075009 (2018)
87. S.-J. Park, M. Gazzola, K.S. Park, S. Park, V. Di Santo, E.L. Blevins, J.U. Lind, P.H. Campbell, S. Dauth, A.K. Capulli et al., Phototactic guidance of a tissue-engineered soft-robotic ray. Science **353**(6295), 158–162 (2016)
88. H. Sato, M.M. Maharbiz, Recent developments in the remote radio control of insect flight. Front. Neurosci. **4**(119) (2010)

89. Y. Mai, A. Eisenberg, Self-assembly of block copolymers. Chem. Soc. Rev. **41**(18), 5969–5985 (2012)
90. Y. Zhang, F. Zhang, Z. Yan, Q. Ma, X. Li, Y. Huang, J.A. Rogers, Printing, folding and assembly methods for forming 3d mesostructures in advanced materials. Nat. Rev. Mater. **2**, 17019 (2017)
91. S.A. Suresh, D.L. Christensen, E.W. Hawkes, M. Cutkosky, Surface and shape deposition manufacturing for the fabrication of a curved surface gripper. J. Mech. Robot. **7**(2) (2015)
92. M.J. Spenko, G.C. Haynes, J.A. Saunders, M.R. Cutkosky, A.A. Rizzi, R.J. Full, D.E. Koditschek, Biologically inspired climbing with a hexapedal robot. J. Field Robot. **25**(4–5), 223–242 (2008)
93. S. Kim, M. Spenko, S. Trujillo, B. Heyneman, D. Santos, M.R. Cutkosky, Smooth vertical surface climbing with directional adhesion. IEEE Trans. Robot. **24**(1), 65–74 (2008)
94. S. Lee, D.A. McAdams, E. Morris, Categorizing biological information based on function-morphology for bioinspired conceptual design. Artif. Intell. Eng. Des. Anal. Manuf. **31**(3), 359 (2017)
95. P.-E. Fayemi, K. Wanieck, C. Zollfrank, N. Maranzana, A. Aoussat, Biomimetics: process, tools and practice. Bioinspir. Biomim. **12**(1), 011002 (2017)
96. C.F. Salgueiredo, A. Hatchuel, Beyond analogy: a model of bioinspiration for creative design. Artif. Intell. Eng. Des. Anal. Manuf. **30**(2), 159 (2016)
97. D. Labonte, C.J. Clemente, A. Dittrich, C.-Y. Kuo, A.J. Crosby, D.J. Irschick, W. Federle, Extreme positive allometry of animal adhesive pads and the size limits of adhesion-based climbing. Proc. Natl. Acad. Sci. **113**(5), 1297–1302 (2016)
98. T. Carleton, L. Leifer, Stanford's me310 course as an evolution of engineering design, in *Proceedings of the 19th CIRP Design Conference–Competitive Design* (Cranfield University Press, Bedford, 2009)
99. D. Stenholm, D. Moore, L. Leifer, D. Bergsjö et al., Fail early, fail often: exploring Stanford's ME310 course as a basis to improve innovation outpost efficacy, in *DS 92: Proceedings of the DESIGN 2018 15th International Design Conference* (2018), pp. 2505–2516
100. C. Kriesi, J. Blindheim, Ø. Bjelland, M. Steinert, Creating dynamic requirements through iteratively prototyping critical functionalities. Proc. CIRP **50**, 790–795 (2016)
101. C. Comer, P. Grobstein, Tactually elicited prey acquisition behavior in the frog, Rana pipiens, and a comparison with visually elicited behavior. J. Comp. Physiol. **142**(2), 141–150 (1981)
102. M.R. Cutkosky, Reach, grasp, and manipulate, in *Living Machines: A Handbook of Research in Biomimetics and Biohybrid Systems*, ed. by T.J. Prescott, N. Lepora, P.F.M.J. Verschure (Oxford University Press, Oxford, 2018)

Wheeled Robots

Marilena Vendittelli

Abstract Robotic vehicles on wheels represent the most effective robotic solution, in terms of cost and realization complexity, to achieve motion on flat terrain. The local mobility of wheeled robots, however, varies in accordance to the mechanical design of wheels, their type and their arrangement. This chapter aims at analysing how the design of wheels and their arrangement affects the mobility of vehicles and the associated modeling, planning and control problems. First, we review the main wheel types and a classification of vehicles based on their mobility, as determined by the arrangement of wheels and the associated kinematic constraints. Then, kinematic models of wheeled vehicles, having a single chassis or articulated bodies, are presented and analyzed. Particular attention is given to those aspects that determine the challenges of motion planning and control problems for wheeled robots and that have led to the development of new, specific planning and control methodologies with respect to fixed base manipulators.

1 Chapter Overview

This chapter starts by describing the constraints associated to wheel design and then illustrates how the combination of design, spatial arrangement and actuation of the wheels determine different kinds of vehicles mobility captured by mathematical models. Based on these models, the relevant motion planning and control problems related to the presence of wheels are analyzed, along with specific solutions proposed in the literature. Models and problems considered in the chapter focus on kinematics, being the constraints that determine the wheeled robots mobility of kinematic nature.

The organization of the chapter is as follows. Section 2 describes the main types of wheels and the constraints on their local mobility. Section 3 discusses how

M. Vendittelli (✉)
Sapienza University of Rome, Rome, Italy
e-mail: marilena.vendittelli@uniroma1.it

© Springer Nature Switzerland AG 2025
B. Siciliano (ed.), *Robotics Goes MOOC*,
https://doi.org/10.1007/978-3-319-75823-7_7

the mobility of wheeled robots changes with the type of wheels, their actuation, number and spatial arrangement. In Sect. 4 the kinematic model of single and multi-vehicle systems is derived, while structural properties, like, e.g., accessibility, controllability, flatness and feedback linearizability, of these vehicles are discussed in Sect. 5. The structural properties studied in Sect. 5 are reflected in the motion planning and control techniques described in Sect. 6.

2 Wheeled Mobility

This section first recalls the rolling without slipping constraint and the consequence on local mobility of a single rolling disk, the archetype of traditional wheels. Other types of wheels providing omnidirectional mobility are then considered, like Mecanum wheels and spherical wheels. Despite the less constrained mobility, omnidirectional wheels are in general mechanically more complex and less robust than traditional wheels. Omnidirectional mobility can also be achieved by proper combination of the basic wheel types reviewed in the following sections [1].

2.1 Standard Wheels

The kinematics of a disk rolling on a plane (see Fig. 1) is useful to illustrate the constraints determined by *pure rolling* (i.e., without longitudinal nor sideway slipping) of a traditional wheel on flat ground. Assuming that the disk rotates with angular speed $\dot{\psi}$ about its center, and denoting with r the wheel radius, the disk center has null velocity v_n in the direction normal to the disk sagittal plane, i.e., there is no sideway slip. This is referred to as *no-slip condition* in the following. The translational velocity v_t is, hence, directed along the sagittal plane and it is proportional to the wheel rotation speed: $v_t = r\,\dot{\psi}$. This corresponds to the absence of longitudinal slipping and is referred to as *rolling condition* in what follows. The consequence of the pure rolling constraint on the mobility of vehicles with more than one wheel is illustrated in Sect. 3.2.

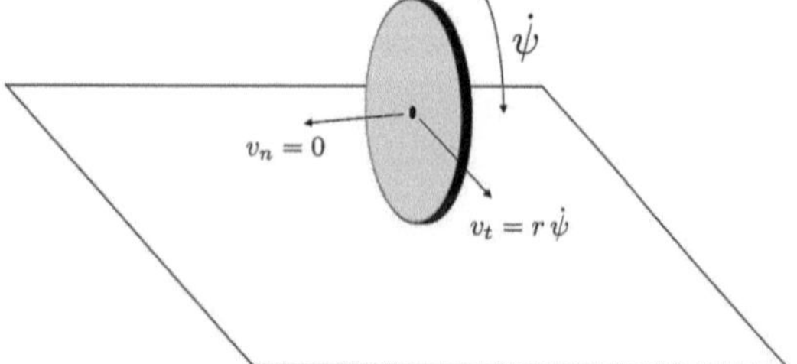

Fig. 1 Disk rolling on a plane: the velocity of the disk center lies on the disk sagittal plane (no-slip condition) and is proportional to the wheel rotation speed (rolling condition)

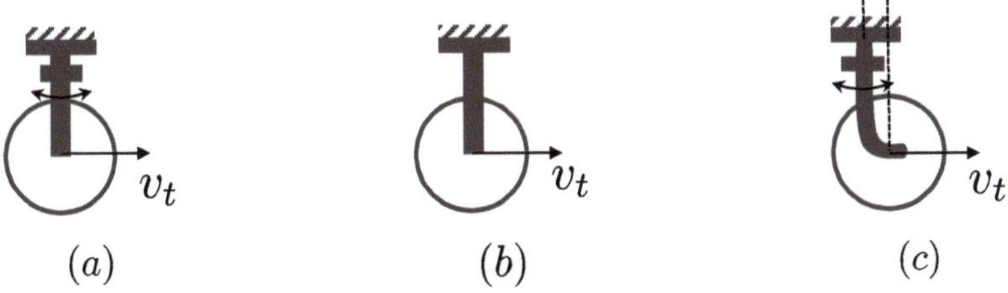

Fig. 2 Centered steerable (**a**), fixed (**b**), or off-centered (**c**) (either steerable or not) wheel

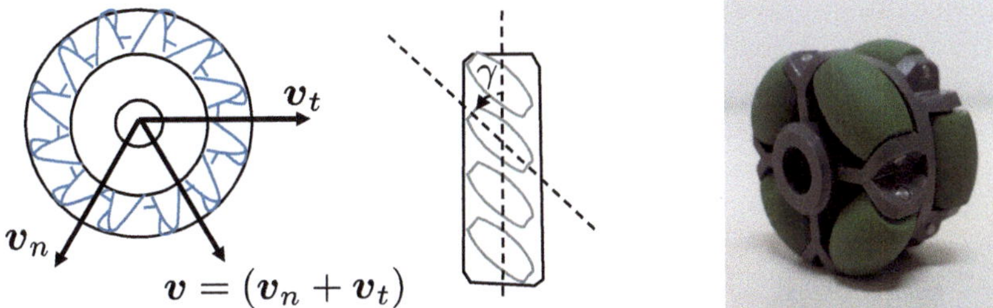

Fig. 3 Mecanum/Swedish wheel with roller axes forming an angle $\gamma = 45°$ (left), or $\gamma = 0°$ (right) with the wheel plane

Standard wheels for mobile vehicles can have different designs. Here we describe the types that are most frequently used in robotic vehicles. Figure 2 reports the sketch of centered wheels the orientation of which can be actively changed (Fig. 2a, steering wheel) or is fixed (Fig. 2b). An offset d can also be present between the wheel center and the vertical axis of steering, as shown in Fig. 2c. In all cases, the velocity $v_t = r\,\dot\psi$ can be either determined by actuators (active wheels) or the wheel can be passive. The normal velocity is $v_n = 0$ due to the no-slip constraint.

2.2 Mecanum Wheels

Mecanum or Swedish wheels[1] are characterized by the presence of passive free rollers on the wheel rim, as shown in Fig. 3. If the rollers axles are all parallel to each other and not aligned with the wheel axis, the constraint on sideslip is eliminated. The velocity of the wheel center $v = v_n + v_t$ is then the sum of the two components v_n e v_t, respectively normal and parallel to the wheel plane.

With reference to Fig. 4, denote by v_s the sliding speed of the roller in contact with the ground, by γ the angle between the rollers axles and the wheel plane, and

---

[1] Universal wheel or omniwheel are terms also used when the rollers' axes are orthogonal to the wheel axis, although the terminology is not standard.

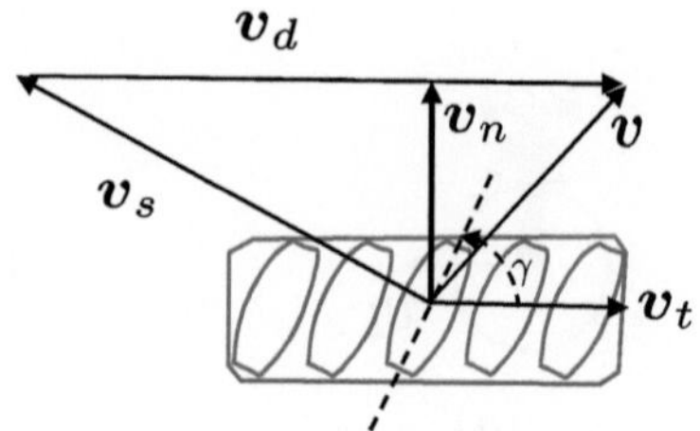

Fig. 4 Wheel velocity v results from wheel driving velocity v_d and sideslip velocity v_s due to the passive rotation of the roller in contact with the ground

by v_d the velocity due to the driving angular speed of the wheel. The normal speed is given by $v_n = v_s \cos \gamma$, while in the wheel plane $v_t = v_d - v_s \sin \gamma$.

These wheels allow to achieve vehicles omnidirectional mobility, a desirable property when moving in cluttered environments. However, with respect to conventional wheels, they present several drawbacks like, e.g., a more complex mechanical design, the load capacity limited by the diameter of the rollers rather than by the diameter of the wheels, the vertical vibrations induced by the discontinuous contact of the rollers with the ground, as discussed in [2].

2.3 Spherical Wheels

Spherical wheels can move in any direction on flat ground. The pure rolling constraint implies that the velocity of the sphere center is $v = \omega \times ||r||$, with ω and r respectively the sphere angular velocity and radius.

In most cases spherical wheels are used as passive casters. There are, however, examples of both statically stable vehicles equipped with spherical wheels [3, 4] and of robots balancing on one active spherical wheel [5–7] and even projects of future autonomous vehicles relying on spherical wheels mobility [8].

The mechanism driving the spherical wheel can take different forms, such as a ring equipped with a set of freely spinning rollers [4], actuated rollers squeezing a urethane covered ball [5], spherical induction motor [9], or the universal wheels equipped with rollers [3, 6], as illustrated in Fig. 5. The mechanism may limit the

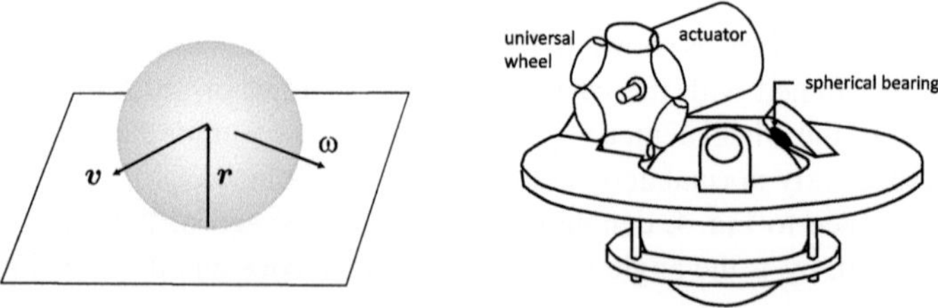

Fig. 5 A ball rolling on a plane (left) is subject to the pure rolling constraint $v = \omega \times r$. An example of actuation for a spherical wheel (reproduced from [3])

attractiveness of spherical wheels due to its complexity and bulkiness [3, 4, 6], to roller-ball excessive friction problems and elasticity in the transmission belt [5], or the limited driving torque [9].

3 From Wheels to Vehicles

This section discusses the constraints restricting the local mobility of wheeled vehicles determined by the design and spatial arrangement of wheels. The constraints are defined in coordinates on a topological space: the *configuration space* of the vehicles, introduced in Sect. 3.1. The number and nature of constraints determine the mobility of the vehicle, as described in Sect. 3.2.

3.1 Configuration Space of Wheeled Robots

The configuration q of a single rigid body is described by a set of independent parameters, called *generalized coordinates*, uniquely defining the body position and orientation (*pose*) in space. In case of articulated systems, like manipulators, the number of variables needed is given by the sum of independent variables describing the pose of each body minus the number of geometric constraints between them. The set of all configurations that a robotic system can assume is called configuration space C. A robotic system can then be considered as a point moving in C.

In the case of wheeled robots, in addition to the generalized coordinates describing the pose of the vehicle (like, e.g., the leftmost vehicle in Fig. 6), there may be present variables describing the orientation of steerable wheels as the angle ϕ for the vehicle in the middle of Fig. 6. Moreover, also the rolling angle ψ of each wheel and the orientation β of off-centered wheels (see Fig. 7) could be included

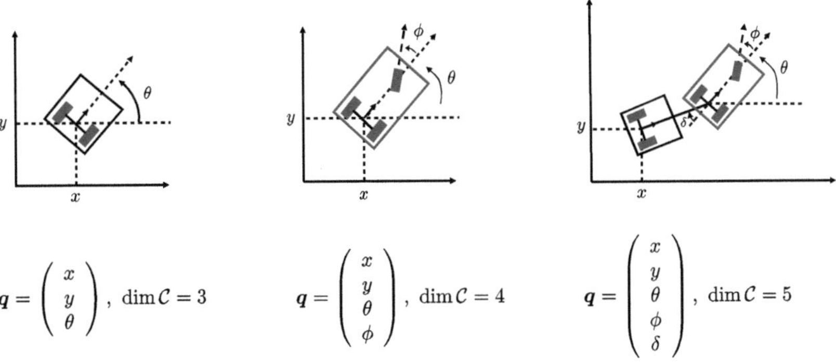

$$q = \begin{pmatrix} x \\ y \\ \theta \end{pmatrix}, \ \dim C = 3 \qquad q = \begin{pmatrix} x \\ y \\ \theta \\ \phi \end{pmatrix}, \ \dim C = 4 \qquad q = \begin{pmatrix} x \\ y \\ \theta \\ \phi \\ \delta \end{pmatrix}, \ \dim C = 5$$

Fig. 6 Examples of vehicles configuration spaces

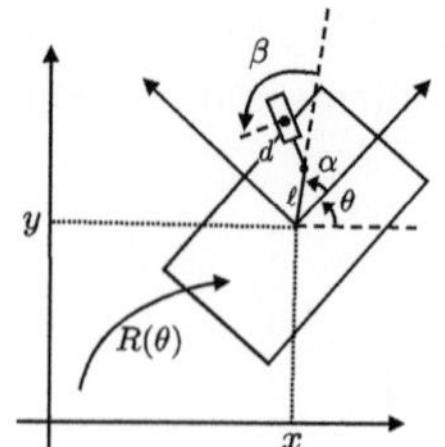
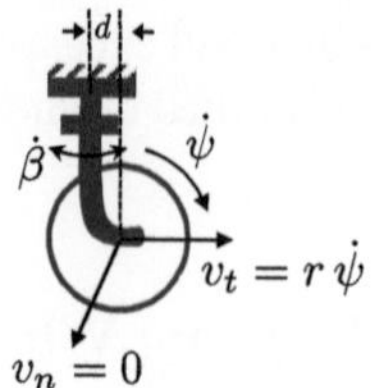

Fig. 7 General wheel type with illustration of the variables and the parameters involved in the general expression of the pure rolling constraint

in the configuration vector. These internal variables are needed in the definition of a dynamic model of the system that explicitly takes into account the actuation of the wheels. Nonetheless, as shown in [10], it is possible to obtain a feedback equivalent dynamic model of the vehicle from the corresponding kinematic one by adding one integrator on each input. Therefore, the developments that follow focus on kinematic models only, as they capture the constraints inherent to wheeled locomotion. A further simplification, often used in the literature for planning and control design purposes, is to neglect the internal state variables ψ and β associated to each wheel and concentrate on the control of the robot pose.

3.2 From Wheels to Vehicles Constraints

To determine the constraints on the mobility of wheeled robots, consider the vehicle in Fig. 7. The vector of generalized coordinates $q = (x, y, \theta)^T$ describes its configuration. In particular, the position of the vehicle body frame is given by (x, y), while θ represents the vehicle heading with respect to the fixed frame. Considering the most general case of an off-centered steerable wheel, note that its center in the body frame is defined by the constant parameters ℓ, α, d and the steering angle β, while its rolling angle is ψ.

The rolling condition (i.e., no slipping in the sagittal plane) $v_t = r\,\dot{\psi}$ and the no-slip condition $v_n = 0$ can be easily expressed (see Chap. 24 of [11]) in terms of the vehicle generalized velocity $\dot{q}$ and steering angle rate $\dot{\beta}$:

$$\left(\sin(\alpha + \beta) \quad -\cos(\alpha + \beta) \quad -\ell \cos \beta \right) R(\theta)^T \dot{q} = r\,\dot{\psi} \tag{1}$$

$$\left(\cos(\alpha + \beta) \quad \sin(\alpha + \beta) \quad d + \ell \sin \beta \right) R(\theta)^T \dot{q} + d\,\dot{\beta} = 0 \tag{2}$$

with

$$R(\theta)^T = \begin{pmatrix} \cos\theta & \sin\theta & 0 \\ -\sin\theta & \cos\theta & 0 \\ 0 & 0 & 1 \end{pmatrix}.$$

Note that, Eqs. (1) and (2), with $d \neq 0$, do not restrict the local mobility of the vehicle since for any value of $\dot{q}$ there always exist $\dot{\psi}$ and $\dot{\beta}$ solutions of (1) and (2). When $d = 0$, $\dot{q}$ is instead required to belong to the null space of $\left(\cos(\alpha + \beta)\ \sin(\alpha + \beta)\ d + \ell \sin \beta \right) R(\theta)^T$, thus constraining the vehicle local mobility. This constraint on the generalized velocity of the vehicle is not integrable to a geometric constraint and it is called *nonholonomic*. The constraints expressed by Eqs. (1) and (2) in case of centered wheels, either steerable or fixed, is obtained by setting $d = 0$ in (1) and (2). For fixed wheels β is set equal to a constant.

To describe the velocity of a Swedish/Mecanum wheel (see Fig. 3), in addition to setting $d = 0$, β equal to a constant value, it is necessary to characterize the direction of the null component of the velocity at the contact point of the wheel through the angle γ between the axle of the rollers and the wheel plane illustrated in Fig. 4. The kinematic constraint in this case is:

$$\left(\sin(\alpha + \beta + \gamma)\ -\cos(\alpha + \beta + \gamma)\ -\ell \cos(\beta + \gamma) \right) R^T(\theta)\dot{q} = \cos(\gamma)\dot{\psi}r,$$

$$(3)$$

where $\dot{\psi}$ is the rotation speed of the wheel about its axis. Note that, Eq. (3) does not restrict the vehicle mobility only if $\gamma \neq \frac{\pi}{2}$, as assumed hereafter.

It is worth to finally note that the spherical wheel is only subject to the rolling constraint that does not restrict the wheel mobility.

From the analysis above, only centered ($d = 0$), standard wheels, either fixed (constant β) or steering (variable β), introduce a differential constraint $a(q)\dot{q} = 0$ on the robot mobility. For a robot with N wheels the constraints can be written in the matrix form $A(q)\dot{q} = \mathbf{0}$, where the rows of $A(q)$ project the vehicle generalized velocity $\dot{q}$ on each wheel axle.

Clearly, $\mathrm{rank}(A(q)) \leq 3$ and any admissible $\dot{q}$ can be obtained as a linear combination of vectors spanning the null space of $A(q)$. In particular, if $\mathrm{rank}(A(q)) = 3$ there exists no velocity $\dot{q} \neq 0$ satisfying the differential constraint. Hence, the motion of a robot equipped with traditional wheels is possible if and only if $\mathrm{rank}(A(q)) < 3$.

The geometric consequence is that for each cart there could be only two independent directions for all the wheel axles which, therefore, intersect at a common point called the Instantaneous Center of Rotation (ICR). This is consistent with the observation that, during a rigid body motion, all the body points rotate instantaneously around a same point, the ICR, and the velocity due to the rotational motion is orthogonal to the line joining the ICR and the considered point. In particular, since the center of centered wheels does not change with respect to the vehicle cart, also this point rotates around the ICR. Due to the no-slip constraint its velocity is contained in the plane of the wheel (no component orthogonal to the wheel plane, i.e., along the wheel axle direction which is the zero velocity direction). Hence, the rotation axis of each wheel (orthogonal to the wheel velocity) intersects the ICR.

Based on the number, type and arrangement of the wheels, vehicles can be classified in five classes [10], characterized by the mobility type $\delta_m = 3 - \text{rank}(A(q))$ and by the degree of steerability δ_s determined by the minimum number of steering wheels of the vehicle.

The pair (δ_m, δ_s) then defines the following classes:

I : $(\delta_m, \delta_s) = (3, 0)$, i.e., $\text{rank}(A(q)) = 0$ and no steering wheels. In this case, the generalized velocity of the vehicle $\dot{q} = (\dot{x}, \dot{y}, \dot{\theta})^T$ can take value in a 3-dimensional space, i.e., any planar motion is possible. Vehicles equipped with Mecanum, off-centered or spherical wheels and without any centered wheel, either fixed or steering, belong to this class.

II : $(\delta_m, \delta_s) = (2, 0)$, i.e., $\text{rank}(A(q)) = 1$ and no steering wheels. In this case, there is one fixed wheel, or several fixed wheels on the same axis.

III : $(\delta_m, \delta_s) = (2, 1)$, i.e., $\text{rank}(A(q)) = 1$ and at least one steering wheel. There are no fixed wheels in this case. If several steering wheels are present, their orientation must be coordinated to keep $\text{rank}(A(q)) = 1$.

IV : $(\delta_m, \delta_s) = (1, 1)$, i.e., one or more fixed wheels on the same axis and one or more steering wheels; if there are more than one steering wheels, their centres cannot be located on the axis of the fixed wheels and their orientation must be coordinated.

V : $(\delta_m, \delta_s) = (1, 2)$, i.e., no fixed and at least two steering wheels. If there are more than two steering wheels, then their orientation must be coordinated to keep $\text{rank}(A(q)) = 2$.

All other cases have no practical interest either because the robot cannot move or because it can only move around a fixed point as in the case of two fixed wheels with independent axles.

4 From Constraints to Feasible Motion

Equation (2), constraining local mobility of wheeled robots when $d = 0$, is used in this section to generate feasible motions for vehicles with centered standard wheels, either fixed or steerable. For a vehicle with N standard wheels the constraint can be written in the matrix form $A(q)\dot{q} = \mathbf{0}$. Hence, feasible generalised velocities belong to the null space $\mathcal{N}(A(q))$ of the constraint matrix $A(q)$. Denoting by $G(q)$ the $n \times m$ matrix whose columns form a basis of $\mathcal{N}(A(q))$, all feasible motions are then solutions of

$$\dot{q} = G(q)\,u \tag{4}$$

for some input (pseudovelocities) $u \in \mathbb{R}^m$, $m = n - k$, with n and k denoting respectively the dimension of the configuration space and the number of constraints.

Models to generate feasible motion are determined in the following sections for various classes of vehicles. They are defined on the configuration space introduced in Sect. 3.1. The elemental model is the unicycle, presented in Sect. 4.1. The models of car-like and multi-trailers robots, presented respectively in Sects. 4.2 and 4.3, are determined from the unicycle model in an iterative fashion. Different arguments are used to derive the model of the omnidirectional vehicles presented in Sect. 4.4.

4.1 Unicycle

Ideally, a unicycle is a vehicle with a single fixed wheel. With reference to Fig. 7 (left), letting $d = \ell = \alpha = 0$ and $\beta = \pi/2$, the constraints $a(q)\dot{q} = 0$ in Eq. (2) now reads

$$a(q)\dot{q} = (0,\ 1,\ 0)R(\theta)^T \dot{q} = (-\sin\theta\ \ \cos\theta\ \ 0)\dot{q} = 0,$$

with $q = (x,\ y,\ \theta)^T$ the vehicle configuration, as illustrated in Fig. 7.

The constraint on the vehicle generalized velocity generated by a single centered and fixed wheel is therefore:

$$\dot{x}\sin\theta - \dot{y}\cos\theta = 0.$$

All feasible generalized velocities can be generated through Eq. (4) with

$$G(q) = \begin{pmatrix} \cos\theta & 0 \\ \sin\theta & 0 \\ 0 & 1 \end{pmatrix} \tag{5}$$

and some input $u \in \mathbb{R}^2$.

In particular, considering as control inputs v and ω, respectively the linear and angular speed of the unicycle, the system dynamics can be expressed as

$$\begin{pmatrix} \dot{x} \\ \dot{y} \\ \dot{\theta} \end{pmatrix} = \begin{pmatrix} \cos\theta \\ \sin\theta \\ 0 \end{pmatrix} v + \begin{pmatrix} 0 \\ 0 \\ 1 \end{pmatrix} \omega. \tag{6}$$

Different hardware design of wheeled robots are described by this mobility model. Among these, the differential drive illustrated in Fig. 8. For this vehicle the actual control inputs are ω_L and ω_R, respectively the rotational speed of the left and right wheel. The relation with the unicycle model inputs is given by:

$$v = \frac{r(\omega_L + \omega_R)}{2} \qquad \omega = \frac{r(\omega_L - \omega_R)}{d}$$

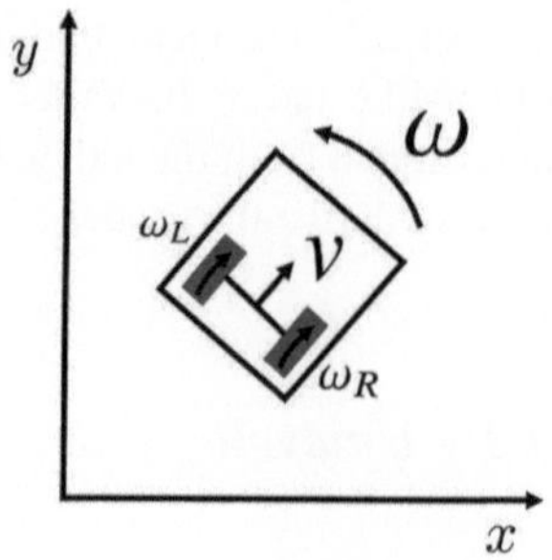

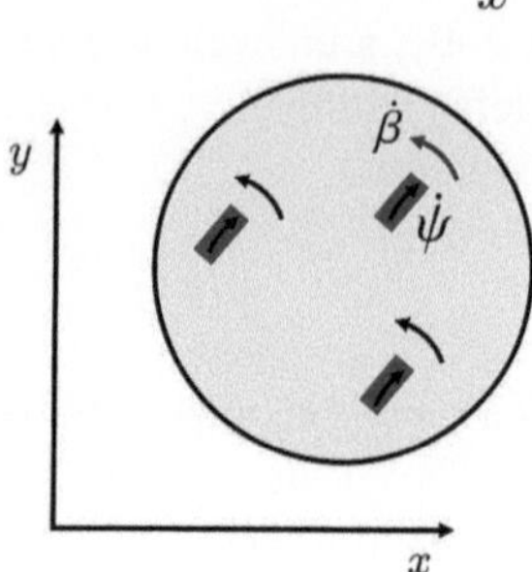

Fig. 8 Unicycle model with differential drive actuation

Fig. 9 Unicycle vehicle with synchro drive actuation

where r is the radius of the wheels and d the length of their axle. This model is readily used for the robot odometry as illustrated in [KNO2].

The synchro-drive vehicle mobility is determined by three centered steering wheels arranged as in Fig. 9. Two motors drive and steer the three wheels through transmission belts. In this case the input transformation between wheels and vehicle translational speed is $v = \dot\psi \cdot r$, where $\dot\psi$ is the (same) rolling speed of the three wheels. The steering rate $\dot\beta$ determines the direction of motion but not the orientation of the vehicle which remains fixed. A third motor is often used to change the orientation of the turret in mobile robots with synchro drive actuation. To change the orientation of the vehicle, three independently actuated steering wheels can be used. In this case, a synchronization is necessary to guarantee the existence of the ICR at all instants. This is clearly a critical feature in the control of such vehicles.

Beside representing the vehicle mobility for different actuation types, the unicycle is used as an high level model for other wheeled robots, such as car-like robots, obtained by neglecting the steering wheel orientation (an internal degree of freedom) to design collision free admissible paths (see Sect. 6.1).

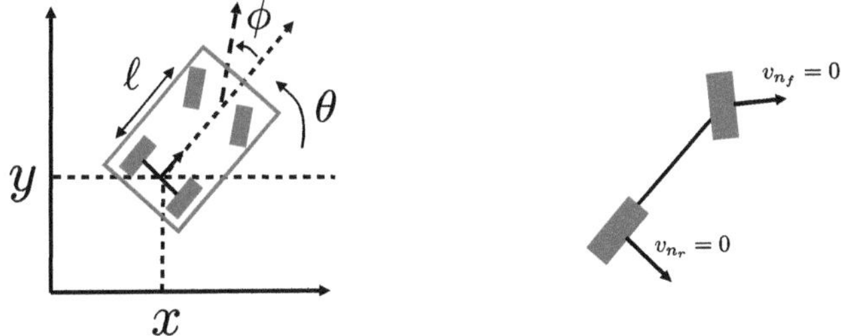

Fig. 10 Configuration variables of a car-like robot and the bicycle modeling approximation with the kinematic constraints on the front and rear wheel

4.2 Car-Like

Denote by $q = (x, y, \theta, \phi)^T$ the configuration of the vehicle in Fig. 10 (left). For modeling purposes consider the 'bicycle' model on the right obtained by collapsing the front and rear couple of wheels in a single one.[2]

The equations of the constraints for the rear and front wheel can be derived from Eq. (2) considering that in this case $\alpha = 0$ for both the rear and front wheel, while $\beta_r = \pi/2$ and $\ell_r = 0$ for the rear wheel and $\beta_f = \phi + \pi/2$ and $\ell_f = \ell$ for the front wheel:

$$A(q)\dot{q} = \begin{pmatrix} 0 & 1 & 0 & 0 \\ -\sin\phi & \cos\phi & \ell\cos\phi & 0 \end{pmatrix} \begin{pmatrix} R(\theta)^T & \mathbf{0} \\ \mathbf{0} & 1 \end{pmatrix} \dot{q} = 0.$$

The motion constraints for the car-like vehicle then are:

$$\begin{aligned} \dot{x}\sin\theta - \dot{y}\cos\theta &= 0 \\ \dot{x}\sin(\theta + \phi) - \dot{y}\cos(\theta + \phi) - \dot{\theta}\ell\cos\phi &= 0. \end{aligned} \tag{7}$$

Alternatively, the constraints for each wheel can be written as

$$\begin{aligned} \dot{x}\sin\theta - \dot{y}\cos\theta &= 0 \\ \dot{x}_f\sin(\theta + \phi) - \dot{y}_f\cos(\theta + \phi) &= 0. \end{aligned} \tag{8}$$

Taking the time derivative $(\dot{x}_f, \dot{y}_f)$ of the front wheel position

$$x_f = x + \ell\cos\theta \quad y_f = y + \sin\theta$$

and substituting it in Eq. (8), the constraints in Eq. (7) are recovered. This second method is more convenient to determine the constraints in case of multi-trailer systems and will be used in the next section.

---

[2] This is a reasonable modeling assumption if differential gears are present on the rear wheels and Ackermann steering mechanism is assumed for the front wheels.

Due to non-uniqueness of the matrix $G(q)$ choice, several options are available for generating feasible velocities through Eq. (4). A convenient approach is to make the choice so as to obtain a physical interpretation of the input u.

In particular, considering as control input the driving speed v_f and the steering rate $\omega = \dot{\phi}$ of the front wheel, the following Front-wheel Drive (FD) model is obtained:

$$\dot{q} = \begin{pmatrix} \dot{x} \\ \dot{y} \\ \dot{\theta} \\ \dot{\phi} \end{pmatrix} = \begin{pmatrix} \cos\theta\cos\phi & 0 \\ \sin\theta\cos\phi & 0 \\ 1/\ell\sin\phi & 0 \\ 0 & 1 \end{pmatrix} \begin{pmatrix} v_f \\ \omega \end{pmatrix}.$$

The Rear-wheel Drive model is obtained by choosing the speed v_r of the rear wheel in place of v_f:

$$\dot{q} = \begin{pmatrix} \dot{x} \\ \dot{y} \\ \dot{\theta} \\ \dot{\phi} \end{pmatrix} = \begin{pmatrix} \cos\theta & 0 \\ \sin\theta & 0 \\ 1/\ell\tan\phi & 0 \\ 0 & 1 \end{pmatrix} \begin{pmatrix} v_r \\ \omega \end{pmatrix}. \tag{9}$$

Note that, the Rear-Wheel Drive model has a singularity in $\phi = \pm\frac{\pi}{2}$ corresponding to the orientation of the front wheel orthogonal to the plane of the rear wheels. The model cannot represent the dynamics of the vehicle when it is in these configurations.

4.3 Multi-Trailers

Multi-trailer systems consist in a master cart pulling several trailers. Each vehicle may be equipped with standard wheels in different numbers and type of combinations. This section considers differences in the hooking system and in the number and position of steering wheels assuming Rear-wheel Drive master cart and passive trailers with fixed wheels. Centered wheels, either steering (as for the front car) or fixed, are considered because of the restriction on the local mobility, while the reason for focusing on the hooking system is that it qualifies the systems as *standard* or *general*, with different controllability properties.

Define the configuration $q = (x, y, \phi, \theta, \theta_1, \ldots \theta_N)^T$ of the car with N trailers sketched in Fig. 11. To determine the matrix of constraints, consider that the truck is subject to the car-like constraints in Eq. (7), while the constraints on the trailers are

$$\dot{x}_i \sin\theta_i - \dot{y}_i \cos\theta_i = 0, \quad i = 1, \ldots, N,$$

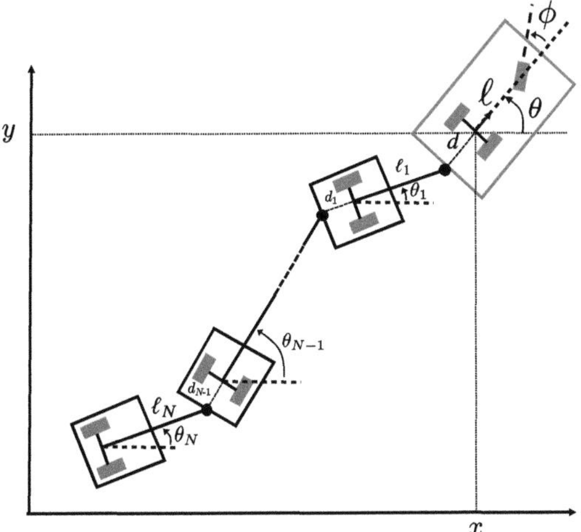

Fig. 11 A general (or non-zero hooking) car-trailer system and the generalized coordinates defining its configuration

where x_i and y_i are the Cartesian coordinates of the wheel axle midpoint of the i-th trailer and can be expressed as function of the vehicle configuration:

$$x_i = x - d\cos\theta - \sum_{\substack{1\le j\le i-1 \\ i>1}} (d_j + \ell_j)\cos\theta_j - \ell_i\cos\theta_i$$

$$y_i = y - d\sin\theta - \sum_{\substack{1\le j\le i-1 \\ i>1}} (d_j + \ell_j)\sin\theta_j - \ell_i\sin\theta_i.$$

The constraint for the i-th trailer becomes

$$\dot{x}\sin\theta_i - \dot{y}\cos\theta_i + d\cos(\theta_i - \theta)\dot{\theta}$$
$$+ \sum_{\substack{1\le j\le i-1 \\ i>1}} (d_j + \ell_j)\cos(\theta_i - \theta_j)\dot{\theta}_j + \ell_i\dot{\theta}_i, \quad i = 1, \ldots, N.$$

The constraint for the N-trailer system can then be written in matrix form

$$A(q)\dot{q} = \begin{pmatrix} A_0 & 0 & 0 & \ldots & \ldots & 0 \\ A_1 & \ell_1 & 0 & \ldots & \ldots & 0 \\ & & \vdots & & & \\ & A_i & & \ell_i & \ldots & 0 \\ & & \vdots & & & \\ & & A_N & & \ell_N & \end{pmatrix} \begin{pmatrix} \dot{x} \\ \dot{y} \\ \dot{\phi} \\ \dot{\theta} \\ \dot{\theta}_1 \\ \vdots \\ \dot{\theta}_j \\ \vdots \\ \dot{\theta}_i \\ \vdots \\ \dot{\theta}_N \end{pmatrix} = 0 \qquad (10)$$

with

$$A_0 = \begin{pmatrix} \sin\theta & -\cos\theta & 0 & 0 \\ \sin(\theta + \phi) & -\cos(\theta + \phi) & 0 & -\ell\cos\phi \end{pmatrix}$$

and

$$A_i = (\sin\theta_i \quad -\cos\theta_i \; 0 \; d\cos(\theta_i - \theta) \; \ldots \; (d_j + \ell_j)\cos(\theta_i - \theta_j)$$
$$\ldots (d_{i-1} + \ell_{i-1})\cos(\theta_i - \theta_{i-1}))$$

a $1 \times (4 + i - 1)$ matrix.

Feasible generalized velocities are obtained by determining the null space of (10). The dimension of the null space is always equal to 2 since there are $N + 2$ independent constraints and $N + 4$ generalized velocities.

Denote by G_0 a vector in the null space of the constraints associated to the front car only, by $\left(G_0^T \; G_1 \; \ldots \; G_i \; \ldots \; G_N\right)^T$ a vector in the null space of the constraints associated to the N-trailer system, and by $G = (g_1 \; g_2)$ a base of the same space,

$$g_1 = \begin{pmatrix} G_0 \\ G_1 \\ \vdots \\ G_i \\ \vdots \\ G_N \end{pmatrix} \qquad g_2 = \begin{pmatrix} 0 \\ 0 \\ 1 \\ \vdots \\ 0 \end{pmatrix}.$$

Due to the structure of the constraint (10), g_1 can be iteratively constructed as

$$G_0 = \begin{pmatrix} \cos\theta \\ \sin\theta \\ 0 \\ -\frac{1}{\ell}\tan\phi \end{pmatrix} \qquad G_1 = -\frac{1}{\ell_1}A_1 G_0 \qquad G_i = -\frac{1}{\ell_i}A_i \begin{pmatrix} G_0 \\ \vdots \\ G_{i-1} \end{pmatrix} \quad \ldots \quad .$$

$$(11)$$

4.3.1 Centered Hooking System

The case $d = d_1 = \cdots = d_{N-1} = 0$ corresponds to the zero-hooking car-trailer system, also known as *standard* car-trailer system, sketched in Fig. 12.

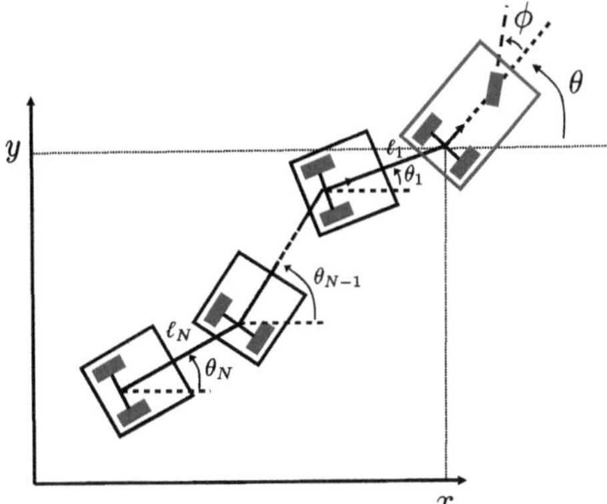

Fig. 12 A standard (or zero-hooking) car-trailer system and the generalized coordinates defining its configuration

The null space of the $N + 2$ constraints in this case can be explicitly expressed, using (11), by the two columns $\boldsymbol{g}_1$ and $\boldsymbol{g}_2$ of the matrix

$$
\boldsymbol{G(q)} =
\begin{pmatrix}
\cos\theta & 0 \\
\sin\theta & 0 \\
0 & 1 \\
-\frac{1}{\ell}\tan\phi & 0 \\
-\frac{1}{\ell_1}\sin(\theta_1 - \theta) & 0 \\
-\frac{1}{\ell_2}\cos(\theta_1 - \theta)\sin(\theta_2 - \theta_1) & 0 \\
\vdots & \vdots \\
-\frac{1}{\ell_i}\prod_{j=1}^{i-1}\cos(\theta_j - \theta_{j-1})\sin(\theta_i - \theta_{i-1}) & 0 \\
\vdots & \vdots \\
-\frac{1}{\ell_N}\prod_{j=1}^{N-1}\cos(\theta_j - \theta_{j-1})\sin(\theta_N - \theta_{N-1}) & 0
\end{pmatrix} .
\tag{12}
$$

The kinematic model of the standard N-trailer system can then be written as

$$
\dot{\boldsymbol{q}} = \boldsymbol{g}_1(\boldsymbol{q})u_1 + \boldsymbol{g}_2(\boldsymbol{q})u_2,
$$

where u_1 and u_2 are respectively the (rear wheel) driving and (front wheel) steering velocity of the pulling car. The zero-hooking system, as will be discussed in the following sections, is characterised by nice controllability properties that simplify the control and planning problems. However, the mechanical design is more expensive and complex with respect to the off-centered hooking system described in the next section.

4.3.2 Off-Centered Hooking System

Articulated vehicles with off-centered hooking are know as *general N*-trailer systems. They are obtained by simply hooking a trailer on the chassis of the previous car as shown in Fig. 11. The kinematic model is also obtained through the general expression of the constraints null space base (11), with $d \neq 0$ and $d_i \neq 0$, $i = 1, \ldots, N$. The obtained equations are quite complicated and are omitted for the sake of simplicity. The case $N = 2$ is instead explicitly described here since this model captures the salient kinematic properties, analyzed in the following sections.

The null space of the constraint matrix is spanned by the two columns of

$$
G(q) = \begin{pmatrix}
\cos\theta & 0 \\
\sin\theta & 0 \\
0 & 1 \\
-\frac{1}{\ell}\tan\phi & 0 \\
-\frac{1}{\ell_1}(\sin\phi_1 - \frac{d}{\ell}\tan\phi\cos\phi_1) & 0 \\
-\frac{1}{\ell_2}(\sin\phi_2\cos\phi_1 - \frac{d}{\ell}\tan\phi\sin\phi_2\sin\phi_1 & \\
\quad -\frac{d_1}{\ell_1}\cos\phi_2(\sin\phi_1 - \frac{d}{\ell}\tan\phi\cos\phi_1)) & 0
\end{pmatrix} \tag{13}
$$

where we have set $\phi_1 = \theta_1 - \theta$ and $\phi_2 = \theta_2 - \theta_1$.

4.4 Omnidirectional Vehicles

Vehicles whose mobility is not subject to local restrictions are called omnidirectional. Omnidirectionality can be obtained through the use of either traditional or omnidirectional wheels, as illustrated in what follows.

4.4.1 Mecanum Wheels

To build an omnidirectional robot with Mecanum wheels requires to decide the angle γ (see Fig. 4), the ratio of the roller to the hub radius, the number of rollers per hub, together with the location of the wheels with respect to the robot chassis, the number of wheels and possibly how many of them are actuated. Kinematic isotropy was introduced in [12] to enhance accuracy of both forward and inverse kinematics. Isotropy is achieved by requiring a unitary condition number of the Jacobian matrices to be inverted for mapping wheels rotation velocity to vehicle's twist, and vice versa. General conditions on relative wheels arrangement to guarantee the absence of kinematic singularities and the possibility of decoupling linear and angular robot commanded velocities are provided in [13].

For an omnidirectional robot belonging to class I, i.e., with $(\delta_m, \delta_s) = (3, 0)$, at least three Mecanum wheels are necessary. The vehicle illustrated in Fig. 13, with

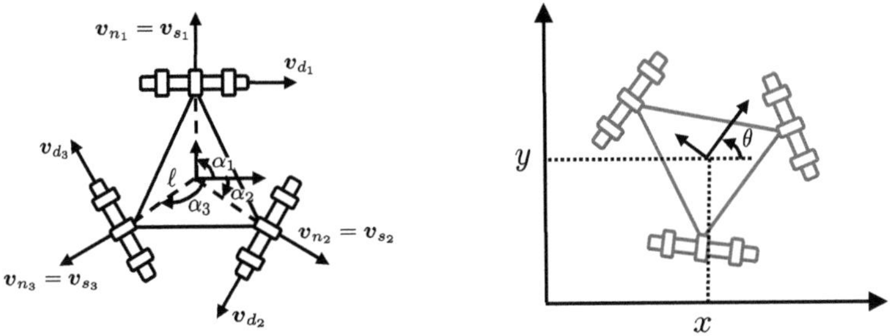

Fig. 13 Mecanum wheels with $\gamma = 0$ mounted at the vertices of an equilateral triangle (left) and the configuration variables of the vehicle (right)

three omniwheels mounted at the vertices of an equilateral triangle, satisfies the conditions for kinematic isotropy [12]: (1) the angle $\gamma = 0$ is the same for the three wheels; (2) all wheels' centers are equidistant from the centroid of the triangular platform; (3) each wheel plane has the same orientation with respect to the plane perpendicular to the floor that contains the line joining the centroid of the platform and the wheel center.

The kinematic model is easily obtained by determining the relationship between the vehicle velocity and the steering velocity given by Eq. (3) for each wheel and then by stacking them up. As an example, we consider here the case of rollers axles in the wheel plane, i.e., with the angle γ in Eq. (3) equal to zero.

According to the parameters definition in Fig. 7, Eq. (3) for each of the three wheels, with same radius r, of the vehicle in Fig. 13 reads as:

1. wheel 1: $\gamma_1 = 0, \beta_1 = 0, \alpha_1 = \frac{\pi}{2}$

$$(1 \quad 0 \quad -\ell)\, \boldsymbol{R}(\theta)^T \dot{\boldsymbol{q}} = r\dot{\psi}_1$$

2. wheel 2: $\gamma_2 = 0, \beta_2 = 0, \alpha_2 = -\frac{\pi}{6}$

$$\left(-\frac{1}{2} \quad -\sin(\pi/3) \quad -\ell\right) \boldsymbol{R}(\theta)^T \dot{\boldsymbol{q}} = r\dot{\psi}_2$$

3. wheel 3: $\gamma_3 = 0, \beta_3 = 0, \alpha_3 = -\frac{5}{6}\pi$

$$\left(-\frac{1}{2} \quad \sin(\pi/3) \quad -\ell\right) \boldsymbol{R}(\theta)^T \dot{\boldsymbol{q}} = r\dot{\psi}_3.$$

Considering $\dot{\psi}_i$, $i = 1, 2, 3$, as the input angular speed of each of the three wheels, the kinematic model analogous to the one provided in case of noholonomic vehicles is $\dot{q} = R(\theta)u$. The input $u \in \mathbb{R}^3$ is mapped to the wheels rotation velocity by

$$\frac{1}{r}\begin{pmatrix} 1 & 0 & -\ell \\ -\frac{1}{2} & -\sin(\pi/3) & -\ell \\ -\frac{1}{2} & \sin(\pi/3) & -\ell \end{pmatrix} u = \begin{pmatrix} \dot{\psi}_1 \\ \dot{\psi}_2 \\ \dot{\psi}_3 \end{pmatrix}. \tag{14}$$

4.4.2 Actuated Caster Wheels

Omnidirectional mobility can also be achieved with standard off-centered wheels. As shown in [10], the minimum number of motors necessary to obtain omnidirectional mobility is 4. Figure 14 presents the case of a three-wheeled vehicle in which two of the three off-centered wheels are steerable (i.e., both wheel rolling and steering are actuated), while the third is passive.

Considering $\dot{\beta}_1$, $\dot{\beta}_2$, $\dot{\psi}_1$, $\dot{\psi}_2$ the corresponding input velocities, as illustrated in Fig. 14, and denoting by r the radius of the wheels, Eq. (1) and (2) take the form:

1. wheel 1: $\alpha_1 = 0$

$$(\sin\beta_1 \quad -\cos\beta_1 \quad -\ell\cos\beta_1)\, R(\theta)^T \dot{q} = r\dot{\psi}_1$$

$$(\cos\beta_1 \quad \sin\beta_1 \quad d+\ell\sin\beta_1)\, R(\theta)^T \dot{q} = -d\dot{\beta}_1$$

2. wheel 2: $\gamma_2 = 0$, $\beta_2 = 0$, $\alpha_2 = -\frac{\pi}{6}$

$$(-\sin\beta_2 \quad \cos\beta_2 \quad -\ell\cos\beta_2)\, R(\theta)^T \dot{q} = r\dot{\psi}_2$$

$$(-\cos\beta_2 \quad -\sin\beta_2 \quad d+\ell\sin\beta_2)\, R(\theta)^T \dot{q} = -d\dot{\beta}_2.$$

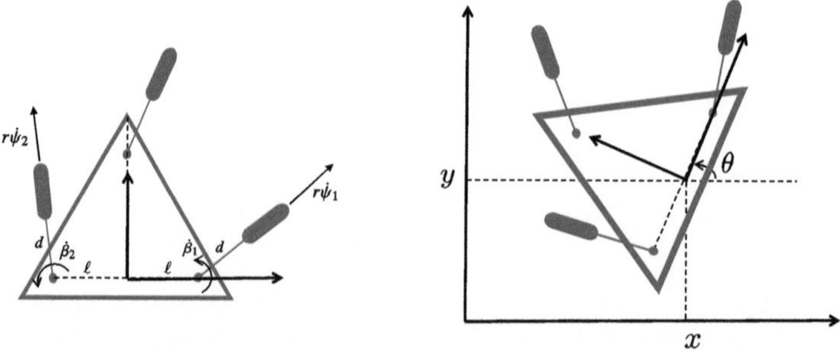

Fig. 14 An omnidirectional vehicle with at least two actuated caster wheels

Overall the equations describing the vehicle motion can be rearranged in matrix form:

$$\begin{pmatrix} \frac{1}{r} & 0 & 0 & 0 \\ 0 & \frac{1}{r} & 0 & 0 \\ 0 & 0 & -\frac{1}{d} & 0 \\ 0 & 0 & 0 & -\frac{1}{d} \end{pmatrix} \begin{pmatrix} \sin\beta_1 & -\cos\beta_1 & -\ell\cos\beta_1 \\ -\sin\beta_2 & \cos\beta_2 & -\ell\cos\beta_2 \\ \cos\beta_1 & \sin\beta_1 & d+\ell\sin\beta_1 \\ -\cos\beta_2 & -\sin\beta_2 & d+\ell\sin\beta_2 \end{pmatrix} \boldsymbol{R}(\theta)^T \dot{\boldsymbol{q}} = \begin{pmatrix} \dot{\psi}_1 \\ \dot{\psi}_2 \\ \dot{\beta}_1 \\ \dot{\beta}_2 \end{pmatrix}.$$

$$(15)$$

Also in this case the kinematic model is $\dot{\boldsymbol{q}} = \boldsymbol{R}(\theta)\boldsymbol{u}$ and the map between the model and actual input is

$$\begin{pmatrix} \frac{1}{r} & 0 & 0 & 0 \\ 0 & \frac{1}{r} & 0 & 0 \\ 0 & 0 & -\frac{1}{d} & 0 \\ 0 & 0 & 0 & -\frac{1}{d} \end{pmatrix} \begin{pmatrix} \sin\beta_1 & -\cos\beta_1 & -\ell\cos\beta_1 \\ -\sin\beta_2 & \cos\beta_2 & -\ell\cos\beta_2 \\ \cos\beta_1 & \sin\beta_1 & d+\ell\sin\beta_1 \\ -\cos\beta_2 & -\sin\beta_2 & d+\ell\sin\beta_2 \end{pmatrix} \boldsymbol{u} = \begin{pmatrix} \dot{\psi}_1 \\ \dot{\psi}_2 \\ \dot{\beta}_1 \\ \dot{\beta}_2 \end{pmatrix}. \qquad (16)$$

Note that the matrix multiplying the input $\boldsymbol{u}$ is always full rank and that there is some redundancy in the input mapping to deal with when controlling the motion of these systems.

4.4.3 Spherical Wheels

Omnidirectional vehicles with standard or mecanum wheels, of the kind illustrated in the previous section, present some drawbacks including vibrations due to discontinuous contact of the rollers with the ground, limited load capacity, complexity in the design of the control law [2]. To overcome these limitations omnimobile platforms with spherical wheels have been proposed in the literature both in statically stable vehicles, like e.g., [3, 4], and in robots dynamically balancing on a single spherical wheel, also known as *ballbots* [5, 6]. Focusing on statically stable vehicles, we report here the model of a vehicle with three spherical wheels driven by Universal wheels, proposed in [3].

The vehicle kinematic model is obtained using the same methodological approach illustrated in the previous sections. Consider the vehicle in Fig. 15, with a triangular base. At the vertices of the equilateral triangle, the spherical wheels described in Sect. 2.3 and illustrated in Fig. 5 are a schematically represented. The spheres axes parallel to the Mecanum wheels axles point toward the centre of the platform. This vehicle corresponds to the Rollmobs concept [3]. According to [12], this configuration provides the robot with isotropic properties. The configuration of the vehicle is $\boldsymbol{q} = (x,\ y,\ \theta)^T$, while $\dot{\boldsymbol{\xi}} = (\omega_1,\ \omega_2,\ \omega_3)^T$ is the vector of the angular velocities of the three Mecanum wheels, and $\dot{\boldsymbol{\psi}} = (\Omega_1,\ \Omega_2,\ \Omega_3)^T$ that of the spheres around an axis passing through the sphere centre and parallel to the driving Mecanum wheel shaft. The kinematic constraints for each wheel correspond

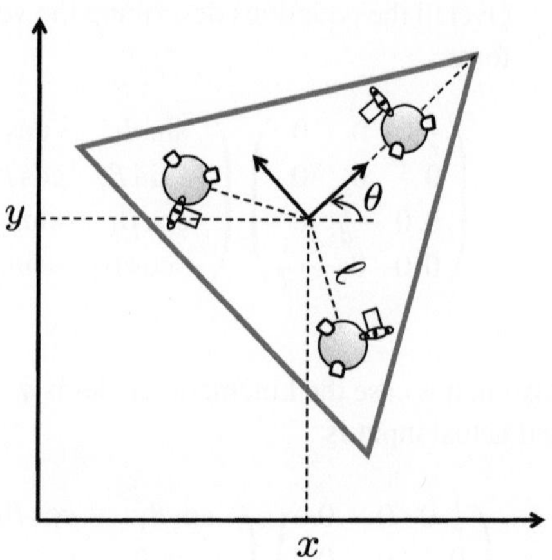

Fig. 15 Concept of the omnidirectional vehicle with spherical wheels in [3]

to a rolling with no-slip condition between the Universal wheel and the sphere in the plane of the wheel, and to a no-slip condition between the sphere and the ground. These constraints relate the angular velocities of the wheels $\dot{\boldsymbol{\xi}}$ to the angular position θ and generalized velocity $\dot{\boldsymbol{q}}$ of the vehicle. Taking ℓ as the distance between the spheres and the frame centre, r_s as sphere radius and r_w as wheel radius, the kinematic constraints for the i-th wheel are written as [3]:

- sphere–ground contact

$$(\sin\alpha_i \quad -\cos\alpha_i \quad -\ell)\, \boldsymbol{R}(\theta)^T \dot{\boldsymbol{q}} = r_s \Omega_i$$

with $\alpha_i = (i-1)\frac{2\pi}{3}$, the angle that the sphere rotation axis form with the abscissa axis of the body frame, (see Fig. 7);
- wheel–sphere contact $r_s \Omega_i = -r_w \omega_i$.

The complete kinematic model of the robot can then be written as

$$\frac{1}{r_w}\begin{pmatrix} 0 & 1 & \ell \\ -\frac{\sqrt{3}}{2} & -\frac{1}{2} & \ell \\ \frac{\sqrt{3}}{2} & -\frac{1}{2} & \ell \end{pmatrix} \boldsymbol{R}(\theta)^T \dot{\boldsymbol{q}} = \dot{\boldsymbol{\xi}}. \tag{17}$$

Also in this case one could consider the kinematic model $\dot{\boldsymbol{q}} = \boldsymbol{R}(\theta)\boldsymbol{u}$ and the map between the model and actual input can be easily deduced.

Although this type of omnidirectional vehicles present attractive properties in terms of load carrying and capacity to surmount obstacles on the ground, spherical wheels driving mechanisms still present the challenges mentioned in Sect. 2.3.

5   Structural Properties

The kinematic model of a wheeled robot provides all feasible directions of instantaneous motion and describes the relationship between the velocity input commands and the derivatives of the generalized coordinates. This model is useful to study system controllability and other structural properties useful for generating kinematically feasible trajectories and for designing control algorithms.

The structural properties of omnidirectional vehicles are quite similar to those of fully actuated manipulators. Controllability is easy to check through static feedback linearization and linear controllability test. Due to full actuation, all the trajectories in the configuration space are feasible, they are smoothly stabilizable and they admit *universal controllers* providing stabilization to arbitrary trajectories, either persistent or not persistent (like, e.g., those leading to a fixed, desired, configuration).

The above properties are not shared by vehicles whose local mobility is restricted by the no-slip constraint, i.e., vehicles equipped with at least one centered fixed or steerable wheel. In the following sections we will focus on the structural properties of this class of vehicles.

5.1   *Controllability, Linearizability, Stabilizability*

For a mobile robot with constrained local mobility, the first interesting question to answer is: given two arbitrary points q_s and q_g in the robot configuration space, does a connecting trajectory, satisfying the kinematic constraints, exist? This is a controllability problem. Controllability can have several qualifying adjectives. We recall here two properties of interest in this chapter, their respective implications and the mathematical tools available for testing controllability.

Consider the nonlinear control system

$$\dot{x} = f(x) + \sum_{j=1}^{m} g_j(x) u_j, \quad x \in \mathcal{M} \simeq \mathbb{R}^n \tag{18}$$

where $f, g_1, \ldots, g_m$ are C^∞ vector fields on $\mathcal{M}$, and the input vector $u = (u_1(t), \ldots, u_m(t))$, in the class $\mathcal{U}$ of piecewise-continuous time functions, takes values on $\mathbb{R}^m$.

System (18) is controllable if $\forall (x_1, x_2) \in \mathcal{M}, \exists T < \infty, \exists u : [0, T] \to \mathcal{U} : x(T, 0, x_1, u) = x_2$. The set of states reachable from x_0 within time $T > 0$, with trajectories contained in a neighborhood V of x_0, is denoted by

$$\mathcal{R}(x_0)_T^V = \bigcup_{\tau \leq T} \mathcal{R}(x_0)^V(x_0, \tau)$$

where

$$\mathcal{R}(x_0)^V(x_0, \tau) = \{x \in \mathcal{M} \mid x(\tau, 0, x_0, u) = x, \; \forall t \in [0, \tau], \; x(t, 0, x_0, u) \in V\}.$$

In particular, we have the following

Definition 1 System (18) is locally accessible (LA) from x_0 if $\forall V$, neighborhood of x_0, and $\forall T > 0$, $\mathcal{R}(x_0)_T^V \supset \Omega$, with Ω some non-empty open set.

Definition 2 System (18) is small-time locally controllable (STLC) from x_0 if $\forall V$, neighborhood of x_0, and $\forall T > 0$, $\mathcal{R}(x_0)_T^V \supset \Psi$, with Ψ some neighborhood of x_0.

Note that, small-time local controllability implies controllability which implies local accessibility. The converse is not true.

Denoted with $[f, g] = \partial g \cdot f - \partial f \cdot g$ the Lie bracket of vector fields f and g, and by $\bar{\Delta}$ the involutive closure of a distribution Δ, i.e., the closure of the distribution under Lie bracketing. To test the controllability of a given system we have the following

Theorem 1 (Chow's Theorem [14]) *Let $\bar{C}$ be the involutive closure of the distribution associated to the vector fields $\{f, g_1, g_2, \ldots, g_m\}$, a system is locally accessible from x_0 if and only if $\dim \bar{C}(x_0) = n$.*

Accessibility can be tested by recursively building[3] $\bar{C} = \mathrm{span}\{v \in \bigcup_{k \geq 0} C^k\}$:

$$\begin{cases} C^0 = \mathrm{span}\{f, g_1, \ldots, g_m\} \\ C^k = C^{k-1} + \mathrm{span}\{[f, v], [g_j, v], j = 1, \ldots, m : v \in C^{k-1}\}. \end{cases}$$

A system is accessible if it is locally accessible from any $x_0 \in \mathcal{M}$.

Only sufficient conditions [14] exist for small time local controllability of nonlinear systems with drift like (18).

However, since the models of wheeled vehicles determined in the previous sections have all the form

$$\dot{q} = \sum_i g_i(q)u_i \tag{19}$$

which is that of a driftless, control-affine system, LA implies STLC and vice-versa. If a system represented by the model (19) satisfies Chow's condition, then it is STLC.

The above result can be stated in terms of Lie algebra: system (19) is STLC if and only if the elements of the Lie Algebra generated by the input vector fields span the tangent space $T_{q_0}\mathcal{M}$ at each point q_0 (Lie Algebra Rank Condition, or LARC).

---

[3] Any P. Hall basis of the free algebra generated by the input vector fields can actually be used to find a basis of the tangent space as in the proposed construction.

The analysis of the Lie Algebra provides additional information. Let $L^s(x_0)$ be the vector space generated by the values at $x_0 \in \mathcal{M}$ of the brackets of $g_1 \ldots, g_m$ of length $s, s = 1, 2, \ldots$ (the g_i's are brackets of length 1). The LARC guarantees that there exists a smallest integer $r = r(x_0)$ such that $\dim L^r(x_0) = n$, called the *degree of nonholonomy* at x_0. In other words, the degree of nonholonomy at x_0 is the minimum length of the Lie brackets required to span the tangent space at x_0.

Let $n_s(x) = \dim L^s(x), s = 1, \ldots, r$. A point x_0 is *regular* if the *growth vector* $(n_1(x), \ldots n_r(x))$ is constant around x_0; otherwise x_0 is *singular*. In particular, points where r changes are singular. Regular points are an open and dense set in $\mathcal{M}$.

The growth vector accounts for the maneuverability of the system and, as will be made clear later, it is directly linked to the flatness property, widely exploited in motion planning, as discussed in the next section.

Examples of controllable systems include the unicycle, the standard n-trailer system in Fig. 12, with arbitrary N, the general two-trailer system in Fig. 11, with $N = 2$. In particular, it is easy to verify that the tangent space of the unicycle is spanned by the vector fields $\{g_1, g_2, [g_1, g_2]\}$, with the input vector fields g_1 and g_2 belonging to the distribution (5), that the growth vector is (2, 3) and the degree of nonholonomy is two. The controllability proof for the standard N-trailer system has been given in [15], where a family of vector fields in the system control Lie Algebra is determined that spans the whole configuration space. To prove the controllability of the general two-trailer system it is convenient to consider a unicycle model for the car towing the two off-hooked trailers[4] and to choose the configuration vector $q = (x_1, y_1, \theta_1, \phi_1, \phi_2)^T$, with (x_1, y_1) the coordinate of the first trailer, $\phi_1 = \theta_1 - \theta$ and $\phi_2 = \theta_2 - \theta_1$ (see Fig. 11). Setting, for simplicity, $d = d_1 = \ell_1 = \ell_2 = 1$ in Eq. (13), the kinematic model of the system can then be written as:

$$
\begin{aligned}
\dot{x}_1 &= \cos\theta_1 v_1 \\
\dot{y}_1 &= \sin\theta_1 v_1 \\
\dot{\theta}_1 &= \omega_1 \\
\dot{\phi}_1 &= -\sin\phi_1 v_1 - (1 + \cos\phi_1)\omega_1 \\
\dot{\phi}_2 &= \sin\phi_2 v_1 + (1 + \cos\phi_2)\omega_1.
\end{aligned}
\tag{20}
$$

In the above equations, the driving and steering velocities of the first trailer, respectively v_1 and ω_1, are related to v and ω, the driving and steering velocities of the car (the actual inputs), by the input transformation:

$$
\begin{aligned}
v &= v_1 \cos\phi_1 - \omega_1 \sin\phi_1 \\
\omega &= -v_1 \sin\phi_1 - \omega_1 \cos\phi_1.
\end{aligned}
$$

---

[4] This does not affect the controllability result which holds also in case of a car-like pulling the trailers.

If $\phi_1 = \pi$ or $\phi_2 = \pi$, the system is clearly not controllable (model singularity). The configuration manifold is then defined as $C = \mathbb{R}^2 \times S^1 \times (S^1 - \pi)^2$. Denote by g_1, g_2 the input vector fields of system (20), and consider the vector fields g_1, g_2, $g_3 = [g_1, g_2]$, $g_4 = [g_1, [g_1, g_2]]$, $g_5 = [g_2, [g_1, g_2]]$, $g_6 = [g_1, [g_1, [g_1, g_2]]]$. Vector fields g_1, g_2, g_3, g_4, g_5 span the tangent space of C at points such that $\phi_1 \neq \phi_2$ (regular points), while g_1, g_2, g_3, g_4, g_6 span the tangent space everywhere, including points such that $\phi_1 = \phi_2$ (singular points). Hence, the system is controllable and the degree of nonholonomy is 3 at regular points and 4 at singular points.

LARC guarantees STLC for driftless systems (19) because of the symmetric dynamics. In other words, it is required that any trajectory run backwards is also a trajectory. Kinematic models of wheeled robots which are not symmetric are quite rare, with the exception of the so called Dubins car [16], i.e., a unicycle-like vehicle that can move only forward, which is locally accessible but not small-time controllable. It is worth noting at this point that while the linear approximation of omnidirectional vehicles is controllable and their dynamics fully linearizable through static feedback, these properties do not hold for wheeled robots with restricted local mobility [10]. However, the class of vehicles that can be transformed in chained form (see Sect. 5.2), including the unicycle, can be linearized via dynamic feedback [17].

Finally, the stabilization at a point of wheeled nonholonomic vehicles cannot be obtained trough continuous, time-invariant feedback laws, due to the celebrated Brocket's theorem [18] on smooth stabilizability of nonlinear systems. In addition, there not exist universal control laws that can stabilize these systems to arbitrary trajectories, either persistent or not persistent [19].

5.2 *Differential Flatness and Chained Form Transformability*

A nonlinear control system of the form (18) is differentially flat if there exists a *flat* (or *linearizing*) output vector y, with the same dimension m of the input, such that the state and the input can be expressed algebraically in terms of y and its derivatives

$$x = x(y, \dot{y}, \ddot{y}, \ldots, y^{(r)})$$

$$u = u(y, \dot{y}, \ddot{y}, \ldots, y^{(r)}).$$

For a two-input driftless system (as the vehicles considered in this chapter), define iteratively the vector space Δ^k as $\Delta_0 = span\{g_1, g_2\}$, $\Delta_1 = span\{g_1, g_2, [g_1, g_2]\}$ and $\Delta_{i+1} = \Delta_0 + [\Delta_i, \Delta_i]$ with $[\Delta_i, \Delta_i] = span\{[f, g]$, $f \in \Delta_i$, $g \in \Delta_i\}$. A necessary and sufficient condition for flatness is that $rank(\Delta_i) = 2 + i$.

In terms of the growth vector, a two-input driftless system is flat if and only if the value of the growth vector entries increases by one, i.e., there are no "jumps" in the components of the growth vector. It can be easily verified that the standard N-trailer in Fig. 12 is flat, while the general N-trailer in Fig. 11 is not flat if $N > 1$. The flat output in the first case is the wheels axle midpoint of the last trailer, while the linearizing output for the general one-trailer system does not have an easy geometric interpretation [20]. It is worth noticing that there is no general method to compute the linearizing/flat output.

For two-input driftless systems, flatness is equivalent to chained-form transformability. The $(2, n)$ chained form, a 2-input driftless control system, is:

$$\dot{z}_1 = v_1$$

$$\dot{z}_2 = v_2$$

$$\dot{z}_3 = z_2 v_1$$

$$\dot{z}_4 = z_3 v_1$$

$$\vdots$$

$$\dot{z}_n = z_{n-1} v_1. \tag{21}$$

Denoting with

$$
g_1 = \begin{pmatrix} 1 \\ 0 \\ z_2 \\ z_3 \\ \vdots \\ z_{n-1} \end{pmatrix}, \qquad
g_2 = \begin{pmatrix} 1 \\ 0 \\ 0 \\ 0 \\ \vdots \\ 0 \end{pmatrix}
$$

the input vector fields of (21), and the repeated Lie brackets as

$$\mathrm{ad}_{g_1} g_2 = [g_1, g_2] \qquad \mathrm{ad}_{g_1}^k g_2 = [g_1, \mathrm{ad}_{g_1}^{k-1} g_2],$$

we have

$$
\mathrm{ad}_{g_1}^k g_2 = \begin{pmatrix} 0 \\ \vdots \\ (-1)^k \\ \vdots \\ 0 \end{pmatrix}
$$

where the nonzero term $(-1)^k$ is the $(k+2)$-th entry of the vector $\mathrm{ad}_{g_1}^k g_2$. It is then easily verified that system (21) is controllable, with degree of nonholonomy equal to $n - 1$, being the n vectors $\{g_1, g_2, \ldots, \mathrm{ad}_{g_1}^i g_2, \ldots, \mathrm{ad}_{g_1}^{n-2} g_2\}$ independent.

There exist necessary and sufficient conditions for transforming a control system in the form (19) into a chained form through input transformation $v = \alpha(q)u$ and coordinates change $z = T(q)$ (see [21]). For a 2-input, n-dimensional system, a constructive sufficient condition is as follows. Define the distributions

$$\Delta_0 = \mathrm{span}\{g_1, g_2, \mathrm{ad}_{g_1}g_2, \ldots, \mathrm{ad}_{g_1}^{n-2}g_2\}$$
$$\Delta_1 = \mathrm{span}\{g_2, \mathrm{ad}_{g_1}g_2, \ldots, \mathrm{ad}_{g_1}^{n-2}g_2\}$$
$$\Delta_2 = \mathrm{span}\{g_2, \mathrm{ad}_{g_1}g_2, \ldots, \mathrm{ad}_{g_1}^{n-3}g_2\}.$$

If, for some open set, one has (i) dim $\Delta_0 = n$, (ii) Δ_1, Δ_2 are involutive, (iii) there exists a scalar function $h_1(q)$ whose differential $dh_1(q)$ is such that $dh_1 \cdot \Delta_1 = 0$ and $dh_1 \cdot g_1 = 1$, then the system can be transformed in chained form. Denoting with $L_g^n f$ the Lie derivative of order n of the function f along the vector field g, the change of coordinates is given by

$$z_1 = h_1$$
$$z_2 = L_{g_1}^{n-2}h_2$$
$$\vdots$$
$$z_{n-1} = L_{g_1}h_2$$
$$z_n = h_2$$

with h_2 independent from h_1 and such that $dh_2 \cdot \Delta_2 = 0$. The input transformation is given by

$$v_1 = u_1$$
$$v_2 = (L_{g_1}^{n-1}h_2)u_1 + (L_{g_2}L_{g_1}^{n-2}h_2)u_2.$$

Noting that, given the above definition of z_1 and z_n, and the input transformation, to obtain a chained form in the z coordinates, we must have:

$$z_1 = h_1$$

$$z_2 = \frac{\dot{z}_3}{\dot{z}_1}$$

$$\vdots$$

$$z_{n-2} = \frac{\dot{z}_{n-1}}{\dot{z}_1} \tag{22}$$

$$z_{n-1} = \frac{\dot{z}_n}{\dot{z}_1}$$

$$z_n = h_2.$$

This form of the coordinate change is often used in practice.

5.2.1 Chained Form Models of Wheeled Robots

We review here the coordinates and input transformations allowing to transform in chained form the models of the most common wheeled robots.

Unicycle

A change of coordinates for the unicycle model (6) is

$$z_1 = x$$
$$z_2 = \tan \theta$$
$$z_2 = y$$

that coupled with the input transformation, including a preliminary scaling of the driving velocity by $\cos \theta$,

$$v = v_1 / \cos \theta$$
$$\omega = v_2 \cos^2 \theta$$

yields

$$\dot{z}_1 = v_1$$
$$\dot{z}_2 = v_2$$
$$\dot{z}_3 = z_2 v_1.$$

Car-Like (Rear-Wheel Drive)

A change of coordinates for the Rear-wheel Drive model of a car-like (9) is

$$z_1 = x$$
$$z_2 = \tfrac{1}{\ell} \sec^2 \theta \tan \phi$$
$$z_3 = \tan \theta$$
$$z_4 = y$$

that coupled with the input transformation, including a preliminary scaling of the driving velocity by $\cos \theta$,

$$v_r = v_1 / \cos \theta$$
$$\omega = -\frac{2}{\ell} v_1 \sin^2 \phi + \frac{1}{\ell} v_2 \cos^2 \theta \cos^2 \phi$$

yields

$$\dot{z}_1 = v_1$$
$$\dot{z}_2 = v_2$$
$$\dot{z}_3 = z_2 v_1$$
$$\dot{z}_4 = z_3 v_1.$$

Standard 2-Trailer

Generalizing the approach used for the unicycle and the car-like, denote by x and y the coordinates of the last trailer wheels' axle midpoint and express the system dynamics (for the Cartesian part only) in terms of these coordinates. It is easy to verify that the kinematics (12), takes the form:

$$\begin{cases} \dot{x} = \cos\theta_2 \cos(\theta_2 - \theta_1)\cos(\theta_1 - \theta)u_1 \\ \dot{y} = \sin\theta_2 \cos(\theta_2 - \theta_1)\cos(\theta_1 - \theta)u_1 \\ \dot{\phi} = u_2 \\ \dot{\theta} = -\frac{1}{\ell}\tan\phi\, u_1 \\ \dot{\theta}_1 = -\frac{1}{\ell_1}\sin(\theta_1 - \theta)u_1 \\ \dot{\theta}_2 = -\frac{1}{\ell_2}\sin(\theta_2 - \theta_1)cos(\theta_1 - \theta)u_1 \end{cases}$$

Operating a preliminary input transformation

$$\tilde{u}_1 = \cos\theta_2 \cos(\theta_2 - \theta_1)cos(\theta_1 - \theta)u_1$$

the system kinematics takes the form

$$\begin{cases} \dot{x} = \tilde{u}_1 \\[2ex] \dot{y} = \tan\theta_2\, \tilde{u}_1 \\[2ex] \dot{\phi} = u_2 \\[2ex] \dot{\theta} = -\frac{1}{\ell}\tan\phi\, \dfrac{\tilde{u}_1}{\cos\theta_2 \cos(\theta_2 - \theta_1)cos(\theta_1 - \theta)} \\[2ex] \dot{\theta}_1 = -\frac{1}{\ell_1}\dfrac{1}{\cos\theta_2 \cos(\theta_2 - \theta_1)}\tan(\theta_1 - \theta)\, \tilde{u}_1 \\[2ex] \dot{\theta}_2 = -\frac{1}{\ell_2}\dfrac{1}{\cos\theta_2}\tan(\theta_2 - \theta_1)\, \tilde{u}_1 \end{cases}$$

The following change of coordinates (excessively long expressions are omitted):

$$\begin{cases} z_1 = x \\[2ex] z_2 = \dot{z}_3/\tilde{u}_1 \\[2ex] z_3 = \dot{z}_4/\tilde{u}_1 \\[2ex] z_4 = \dot{z}_5/\tilde{u}_1 = -\dfrac{1}{\ell_2}\dfrac{\tan(\theta_2-\theta_1)}{\cos^3\theta_2}. \\[2ex] z_5 = \dot{z}_6/\tilde{u}_1 = \tan\theta_2 \\[2ex] z_6 = y \end{cases}$$

and the input transformation

$$\begin{cases} v_1 = \tilde{u}_1 = \cos\theta_2\cos(\theta_2-\theta_1)cos(\theta_1-\theta)u_1 \\ v_2 = \dot{z}_2 \end{cases}$$

put the system in chained form for every point of the configuration space except those corresponding to configurations such that the angle between bodies are equal to $\frac{\pi}{2}$ (singularity of the transformation). This result was generalized in [22] for the standard N-trailer system.

6 Motion Planning and Control of Wheeled Robots

The motion planning problem for wheeled robots consists in finding a path in their configuration space that does not collide with obstacles in the environment. If the instantaneous velocity of the vehicle is not constrained (like, e.g., for omnidirectional robots), standard methods solving the classical piano mover problem can be applied [23, 24]. Some of these methods are reviewed in [KNO3] for fixed-base manipulators. For vehicles subject to non-integrable kinematic constraints (i.e., nonholonomic constraints), not all the paths in the configuration space are feasible for the system, hence geometric motion planning algorithms cannot (directly) provide a solution. The same constraints, when dealing with kinematic models, make feedback control of wheeled vehicles a difficult problem since Brockett's theorem [18] prevents the application of classical control methods developed for fully actuated fixed-base serial manipulators and applicable to omnidirectional robots.

For these reasons, the following sections will review the main planning and control problems related to wheeled vehicles equipped with at least one centered fixed or steerable wheel restricting their local mobility, i.e., nonholonomic vehicles.

6.1   Noholonomic Motion Planning

The problem of motion planning for nonholonomic vehicles consists in finding, if it exists, a *feasible* (i.e., satisfying the kinematic constraints) path joining a start to a goal point, respectively q_s and q_g, in the vehicle configuration space C that does not collide with obstacles in the environment. Defined C_{free} is the configuration space manifold in which the robot does not overlaps or touches the obstacles, the seeked path is a continuous mapping $\sigma : \tau \in [0, 1] \to \sigma(\tau) \in C_{\text{free}}$, such that $\sigma(0) = q_s$ and $\sigma(1) = q_g$, with τ some normalized path parameter.

The following sections report the main results of the insightful analysis of this problem provided in [25]. Although the discussion makes reference to the generation of collision-free feasible *paths*, the same steering methods can be used to generate trajectories, i.e., paths with an associated timing law specifying how the configuration of the vehicle evolves, along the path, over time. In any case, in view of their kinematic feasibility, appropriate velocity inputs exist allowing to follow these configuration space paths (see, e.g., [26] for a complete discussion on the topic). This property can be exploited to design timing laws associated to the paths that can comply with velocity bounds and continuity requirements. Clearly, time optimality properties can be lost in this case.

Prior to solve a motion planning problem it is necessary to check if it is decidable, i.e., if there exists an exact algorithm that decides in finite time wether a solution exists (the *decision problem*); decidability proofs are not necessarily constructive. To solve the *complete problem*, a motion planning algorithm must be designed that returns the solution path or a failure if the path does not exist.

In [25] it has been proved that the decision problem is solved for any symmetric small-time controllable system and that the existence of a feasible collision-free path between two configurations is equivalent to the existence of *any* collision-free path between the same configurations.

Exact solutions for the complete problem are available only for some special classes of small-time controllable systems. Unicycle, car-like, standard N-trailer, and general one-trailer vehicles are small-time locally controllable systems for which exact solutions to the complete problem exist. For the general N-trailer, with $N > 1$, an exact solution to the complete problem is not available. Nonetheless, a globally convergent algorithm iteratively steering general nonholonomic systems to the desired configuration has been provided in [27]. The method applies to the general two-trailer system. The convergence to the goal configuration remains, however, asymptotic.

There are no general results for systems that are not small-time controllable like, e.g., the Dubins car. The decision problem for this system has a solution only for a point-shaped vehicle [28]. The complete problem has an approximate solution [29], while exact solutions exist for environments with obstacles represented by generalized polygons whose boundaries are admissible Dubins paths [30, 31] (*moderate* obstacles). The decision problem is open for Dubins vehicles with polygonal shape.

A General Approach to Nonholonomic Motion Planning

To solve the complete problem of nonholonomic motion planning it is possible to formulate a general planning paradigm:

- (*i*) plan a collision-free path without considering the velocity constraints (*geometric path*);
- (*ii*) approximate the *geometric path* by a finite sequence of *feasible* and collision-free paths joining via-points belonging to the geometric path.

This two-step approach relies on the decidability result for symmetric, small-time controllable systems and on the equivalent existence of kinematically feasible and unconstrained paths.

The peculiar difficulty of nonholonomic motion planning is captured by the second step. In particular, to turn this general paradigm into a planning algorithm it is necessary to device *steering methods* to generate feasible paths between any two given configurations without considering the presence of obstacles. The via-points to be joined with these *local paths* can be generated iteratively, and according to different strategies, starting from an initial guess, until a sequence of feasible, collision-free paths is generated.

A sufficient condition for the convergence of step (*ii*) is that a steering method must generate paths in configuration space such that, for any configuration x and any neighborhood U of x, there exists a neighborhood V of x such that for any configuration $\tilde{x} \in V$, the path generated by the steering method to connect x to $\tilde{x}$ is contained in U. In other words, for the convergence of step (*ii*) it is sufficient that the paths generated by the steering method are contained in a neighborhood of decreasing radius as the two configurations to be connected get closer. Equivalent conditions for achieving this behaviour are available in literature [32–34].

It is worth noticing that exact steering methods are known only for special classes of wheeled vehicles. An algorithm based on the above approach is, therefore, not applicable *as such* to any nonholonomic system. Its structure, however, allows highlighting the difficulties in steering general nonholonomic systems even without considering obstacles and despite their possible controllability.

Alternative approaches to motion planning for general nonholonomic vehicles are represented by numerical methods [35–37]. These methods, out of the scope of this chapter, lead to approximate solutions and usually produce wandering paths. Moreover, a careful implementation is appropriate to limit the computational cost.

The following section describes the main methods, either exact or approximate, to steer a nonholonomic system between two given configurations.

6.2 Steering Nonholonomic Vehicles

A *steering method* solves the motion planning problem without taking into account the geometric constraints on the robot configurations due to the presence of obstacles. There is currently no algorithm providing an exact solution to the steering problem for *any* nonholonomic system, i.e., any wheeled vehicle with standard wheels. Special classes of vehicles admitting exact solutions include nilpotent and flat systems.

6.2.1   Steering Flat or Nilpotent Systems

Flatness is a particularly convenient property to plan the motion of nonholonomic systems when the flat output has a simple geometrical relationship with the vehicle configuration. For example, for the car-like it is the midpoint of the rear wheels axle. In this cases, any interpolation method can be used to join the initial and goal point in the Cartesian plane with the appropriate boundary conditions. Due to the system flatness, algebraic transformations provide the (feasible) trajectories of the other system variables and of the control inputs.

Although the method could be extended to any flat system, in some cases it would not provide geometrical intuition on the system evolution or it does not reflect the local controllability of the system. This is a critical point when local properties of the system trajectories are required to stay clear from obstacles. An example is provided by the standard N-trailer system which has been shown to be flat irrespective of the number N of the trailers [20]. The flat output is the middle point of the last trailer wheels axle. A flatness-based steering method has been proposed for this system in [38] which does not allow cusp points (corresponding to motion inversion) along the Cartesian path. However, in tasks like parking, or for close approximation of collision-free unfeasible paths, cusps might be required. This limitation is removed in [39] for a car with one trailer, thus fully exploiting its small-time controllability. The approach is then extended to the standard N-trailer system in [40].

For driftless two-input systems, flatness is equivalent to transformability in chained form. For the $(2, n)$ chained form (21), the flat outputs are z_1 and z_n. By definition (see Eq. (22)), all the other variables and the inputs can be computed from these variables and their derivatives. Although two-input chained systems are relevant models of the most popular vehicles, 3-input nonholonomic systems have also been studied in [41], with the fire-truck vehicle as an example of practical relevance.[5] Because of their special structures, chained forms can be easily steered (i.e., the defining equations can be integrated in closed form) using sinusoidal, polynomial, and piecewise constant inputs [42–44].

It is worth noticing that chained forms belong to the wider class of nilpotent systems [45]. A system is nilpotent if the Lie brackets of the input vector fields of length greater than some value (the order of nilpotency) vanish. It is easy to verify, for example, that system (21) has degree of nilpotency equal to $n - 2$. Nilpotency was first used by Lafferiere and Sussmann [46] to compute the input to exactly steer nonholonomic nilpotent systems from a starting to a desired goal configuration, given a holonomic trajectory joining the same start and goal configurations.

---

[5] The fire-truck system is a car-like robot with one trailer with steerable wheels.

6.2.2 Steering Based on Optimal Paths

Early planning methods for car-like robots tried to use primitives of minimum length to link configurations pairs (see for instance [47, 48]). Analytic solutions, however, exist only for unicycle-like systems. Examples include the *Reeds and Shepp* and *Dubins* car, respectively corresponding to the unicycle model (6) with $|v| \equiv 1$ and $v \equiv 1$. A review of the rich literature results for these vehicles is described in the following. Shortest paths for differential-drive robots have been determined in [49]. Analogous studies for omnidirectional robots include, e.g., [50]. For general systems numerical methods (like, e.g., [51]) remain the only available option.

Shortest Paths for Reeds and Shepp and Dubins Car

In the absence of obstacles, shortest paths for the Reeds and Shepp car are contained in a sufficient family of 48 shortest paths [52] and are made up of a finite sequence of at most five elementary pieces which are either straight line segments or arcs of a circle of minimum radius. Sussmann and Tang reduced the sufficient family to 46 canonical paths using optimal control principles [53].

Using these results Souères and Laumond computed a synthesis of the shortest paths, i.e., a partition of $\mathbb{R}^2 \times S^1$ into cells reachable by only one type of shortest path (among the 46 of the sufficient family) [54]. They also computed the exact shape of the shortest path metric, i.e., the shape of the domain in $\mathbb{R}^2 \times S^1$ reachable by paths with bounded length [55] from a given configuration assumed, without loss of generality, to be the origin of the configuration space. The projection of those domains on $\mathbb{R}^2$ provides the accessibility domain in the plane [56]. A result of this work, of particular interest for determining the distance of noholonomic vehicles to workspace obstacles, is the synthesis of the shortest path from an initial configuration to a point in the plane.

Shortest paths for the Dubins car are finite sequences of at most three pieces consisting of straight line segments or arcs of a circle of minimum radius [16]. Accessibility domains have been studied in [57–59]. A comprehensive overview on Reeds and Shepp and Dubins optimal paths is [60].

Nonholonomic Metric

The cost of the shortest paths induces a special metric in the configuration space which is called nonholonomic [61], or singular [62], or Carnot-Carathéodory [63], or sub-Riemannian [64] metric. By definition, any optimal path of length ℓ is contained in the nonholonomic ball with radius ℓ centered at the starting configuration. The nonholonomic metric induces the same topology as the "natural" metric, i.e., the one-norm distance $d_{CS}(\boldsymbol{q}_1, \boldsymbol{q}_2) = \sum_{i=1}^{n} |q_{i,1} - q_{i,2}|$. As a consequence, for any nonholonomic ball $B_{nh}(\boldsymbol{q}, \ell)$ with radius ℓ, there exist two strictly positive real numbers ϵ and η such that $B(\boldsymbol{q}, \eta) \subset B_{nh}(\boldsymbol{q}, \ell) \subset B(\boldsymbol{q}, \epsilon)$. Due to the continuity of the nonholonomic distance, contracting steering methods based on shortest paths can be obtained to solve the complete motion planning problem.

The exact shape of the nonholonomic balls is known only for the Reeds and Shepp car. For the Dubins car, balls associated to shortest paths of a given length are also known but the shortest paths do not define a distance because they are not symmetric. The general approximated shape of the nonholonomic balls for regular systems has been studied, e.g., in [65]. An estimate of the nonholonomic ball for the general two-trailer system (13), i.e., a system with singularities, is obtained through the nonhomogeneous approximation of the system [66].

The radius of the nonholonomic balls can be used as the appropriate distance to the closest obstacle for wheeled vehicles subject to nonholonomic constraints in a motion planning scheme [67]. In [68] this distance (pseudo-distance in case of the Dubins car) has been determined for unicycle-like vehicles neglecting the car footprint. This hypothesis has been removed in [69] and [70] extending the result to vehicles with polygonal shape.

Steering Non-flat, Non-nilpotent Systems

The method in [46] developed for nilpotent systems can also be applied to general non-nilpotent systems considering a properly chosen nilpotency order k. The generated path in this case does not converge to the goal. However, since the steering method is contracting it can be used in an iterated way. The error in reaching the goal configuration can be made arbitrarily small by increasing the number of iterations. Not surprisingly, the approach produces solutions with an high number of elementary pieces. This number can be reduced by using special basis of the free Lie algebra associated to the input vector field [71].

A different approach consists in explicitly considering nilpotent approximations [72, 73] of systems that cannot be transformed in a nilpotent canonical form. Nilpotent approximations allow to transform any controllable system in a triangular nilpotent one, preserving the structural properties of the original system, like controllability, degree of nonholonomy and growth vector. This makes nilpotent approximations a useful tool for studying controllability conditions for systems with drift [74, 75], stabilizability properties for non-smoothly stabilizable systems [76, 77], complexity of the motion planning problem [78] through the estimation of the sub-Riemannian distance provided by the Ball-Box Theorem [72].

The approximation procedure requires first to express the original dynamics in a privileged coordinate system centered at the approximation point, defined on the basis of the control Lie algebra. Then, the transformed vector fields are expanded in Taylor series; by truncating the expansion at a proper order, one obtains a nilpotent system which is polynomial, homogeneous and triangular. As shown in [79], homogeneity may be essential for preserving controllability and stabilizability; for this reason, homogeneous nilpotent approximations have been used in the above applications.

There is however a situation in which homogeneous approximations exhibit a drawback: in the neighborhood of singular points (where the growth vector changes), both the privileged coordinate system and the truncation order change. Hence, in the presence of singularities, homogeneous approximations vary discontinuously with the approximation point. Another consequence is that the sub-Riemannian distance estimate is not uniform around singularities: when approaching a singular point through a sequence of regular points, the region of validity for such estimate tends to zero [80].

The above problem is particularly critical when nilpotent approximations are used to design approximate steering laws to be applied iteratively in a feedback fashion. Within this scheme, continuity is essential for estimating the steering error due to the approximation in order to prove stability. In [66] it was shown that continuity in the presence of singularities may be achieved by giving up the homogeneity property. For 5-dimensional 2-input driftless systems with generic singularities, *nonhomogeneous* nilpotent approximations were built preserving structural properties and continuously varying with the approximation point over a finite number of subsets covering the singular locus. In this way, it is possible to associate a continuous approximation procedure to each point. As a byproduct, a uniform estimate of the sub-Riemannian distance is also obtained.

The car pulling two off-hooked trailers falls in this category. By applying the nonhomogeneous nilpotent approximation procedure [66] it is possible to compute an approximation that is valid at every point of the configuration manifold which allows also to consider the appropriate estimation of the distance between any pair of configurations. The obtained approximation can be used to devise steering methods that account for the small-time controllability of the system [81].

6.3 Feedback Control of Nonholonomic Vehicles

The basic feedback control problems for nonholonomic wheeled mobile robots are typically discussed in the literature adopting the unicycle vehicle as a target system. This is due to the fact that the model of most of the single-body wheeled mobile robots can be reduced to that of a unicycle and that most of the techniques developed for the unicycle can be extended to chained form transformable systems.

The basic motion tasks addressed in the literature are: posture stabilization and tracking of feasible Cartesian trajectories. Because of the difficulty implied by Brockett's theorem, the two problems have classically been approached separately. In fact, a universal controller has been shown to not exist [19].

Posture stabilization solves the same point-to-point motion problem as steering but uses feedback. Because it is a non-square (for the unicycle, state has dimension three, while the inputs available for control are two) regulation problem, the celebrated Brockett's Theorem [18], applied to driftless systems, implies that a state-discontinuous or time-varying feedback is necessary. The literature offers two main approaches to posture stabilization: discontinuous, time invariant, control laws [82] and time-varying, either continuous [83] or not continuous at the desired configuration [84, 85], stabilizers. Although solving the same problem, posture stabilization algorithms are not considered an efficient alternative to steering methods in a motion planning paradigm due to the complexity in controlling the transient behaviour.

The trajectory tracking problem has a continuous feedback solution since the error dynamics is square. Beside being an easier problem to solve, the developed methods are of practical interest for tracking feasible trajectories that have been planned to avoid obstacles in the environment. A solution to the problem can, in fact, only exist if the trajectory to be tracked is feasible for the system, i.e., it must satisfy the nonholonomic constraints. Planning feasible Cartesian trajectories is an easy task using the flatness property. Considering the example of a unicycle, the flat output is the midpoint of the wheels axle. Given a trajectory for this point, there is a unique associated state trajectory and input time evolution. The control action is then composed of two terms: a feedforward command and an error feedback. The feedforward is the input ideally necessary to follow the assigned trajectory in the absence of errors and, for the unicycle, it can be determined using the equations of motion (6) and the derivative of the flat output, up to the second order. The state error is provided by the desired Cartesian trajectory and the evolution of its tangent providing the vehicle reference orientation. The main approaches to the problem of Cartesian trajectory tracking for the unicycle are based on approximate linearization of the error dynamics around the reference trajectory, on dedicated nonlinear control design, on dynamic feedback linearization (see Chapter 49 of [11]). The feedforward part is either based on the desired inputs [83, 86] determined as mentioned above, or directly on the flat outputs and their derivatives, like e.g., in approaches based on dynamic feedback linearization [17].

Most of the methods developed for the unicycle can be extended to higher order systems (e.g., car-like and multi-trailers). In particular, for the systems transformable in chained form it is possible to device methods that, in principle, are valid for state spaces of arbitrarily high dimensions. Textbooks like, e.g., [11, 26, 87] provide an excellent overview and specific literature pointers.

It is worth noticing that in many practical situations it is of interest pursuing a less demanding objective than the asymptotic stabilization of the origin of some error dynamics, such as the stabilization of a small set containing the origin of the error dynamics. It is then possible to design feedback controllers providing *practical* stabilization of arbitrary reference trajectories, including fixed postures and non-feasible trajectories using the transverse function approach [88]. In principle, this approach allows to uniformly bound the tracking error independently of the reference trajectory, that could also be unfeasible with respect to the nonholonomic constraint. It can therefore be adopted for tracking collision-free trajectories in a motion planning scheme unaware of the constraint or, more in general, a priori unknown trajectories.

Note, in conclusion, that the above mentioned feedback control methods of nonholonomic vehicles can rely either on proprioceptive information, like e.g., odometry measurements [KNO2], or can integrate information from exteroceptive sensors, like e.g., in SLAM methods [KNO8]. Alternatively, feedback control can consist in servoing the vehicle motion to the achievement of a task defined in the sensor space, like e.g., in visual servoing schemes [KNO5] widely used in mobile robotics [89–91].

7 Problems

7.1 Problem 1

A bi-steerable vehicle has dependent front and rear steering wheels. Denoting by φ the front wheel steering angle, prove that if the function relating the rear to the front wheel steering angle is $f(\varphi) = -k\varphi$, with $k \neq 1$, the system is flat.

7.2 Problem 2

Check controllability of the standard N-trailer system.

7.3 Problem 3

Determine the trajectories of the car with two off-hooked trailers corresponding to singular configurations.

7.4 Problem 4

Prove that a car pulling a Hoff-hooked trailer followed by a zero-hooking second trailer is not flat.

7.5 Problem 5

Determine the kinematic model of a general one-trailer system in which the joint linking the car to the trailer is actuated with velocity input $\dot{\phi}$.

7.6   Problem 6

Propose a contracting steering method for the standard one-trailer system.

Cross-References

This chapter contains cross-references to content published in other chapters belonging to the four "Robotic Goes MOOC" books: KNO2, KNO3, KNO5, KNO8. The full list of chapters abbreviations is available in the Preface.

Acknowledgments Jean-Paul Laumond read these notes and helped me find a synthesis of the vast literature on nonholonomic motion planning. His contributions were invaluable, and his memory continues to inspire.

References

1. H. Taheri, C.X. Zhao, Omnidirectional mobile robots, mechanisms and navigation approaches. Mech. Mach. Theory **153**, 103958 (2020) [Online]. Available: https://www.sciencedirect.com/science/article/pii/S0094114X20301798
2. L. Ferriere, B. Raucent, G. Campion, Design of omnimobile robot wheels, in *IEEE Int. Conf. on Robotics and Automation*, vol. 4 (1996), pp. 3664–3670
3. L. Ferrière, G. Campion, B. Raucent, ROLLMOBS, a new drive system for omnimobile robots. Robotica **19**(1), 1–9 (2001)
4. M. West, H. Asada, Design of ball wheel mechanisms for omnidirectional vehicles with full mobility and invariant kinematics. ASME J. Mech. Des. **119**(2), 153–161 (1997)
5. U. Nagarajan, G. Kantor, R. Hollis, The ballbot: An omnidirectional balancing mobile robot. Int. J. Robot. Res. **33**(6), 917–930 (2014)
6. L. Hertig, D. Schindler, M. Bloesch, C.D. Remy, R. Siegwart, Unified state estimation for a ballbot, in *IEEE Int. Conf. on Robotics and Automation*, May (2013), pp. 2471–2476
7. M. Kumagai, T. Ochiai, Development of a robot balancing on a ball, in *2008 International Conference on Control, Automation and Systems*, Oct (2008), pp. 433–438
8. GoodYear, Goodyear eagle 360 urban (2017) [Online]. Available: https://www.youtube.com/watch?v=KAdw09M-F-g
9. A. Bhatia, M. Kumagai, R. Hollis, Six-stator spherical induction motor for balancing mobile robots, in *IEEE Int. Conf. on Robotics and Automation* (2015), pp. 226–231
10. G. Campion, G. Bastin, B. Dandrea-Novel, Structural properties and classification of kinematic and dynamic models of wheeled mobile robots. IEEE Trans. Robot. Autom. **12**(1), 47–62 (1996)
11. B. Siciliano, O. Khatib (eds.), *Springer Handbook of Robotics* (Springer International Publishing, 2016)
12. S.K. Saha, J. Angeles, J. Darcovich, The design of kinematically isotropic rolling robots with omnidirectional wheels. Mech. Mach. Theory **30**(8), 1127–1137 (1995)
13. G. Indiveri, Swedish wheeled omnidirectional mobile robots: Kinematics analysis and control. IEEE Trans. Robot. **25**(1), 164–171 (2009)
14. H.J. Sussmann, A general theorem on local controllability. SIAM J. Control Optim. **25**(1), 158–194 (1987)

15. J.P. Laumond, Controllability of a multibody mobile robot. IEEE Trans. Robot. Autom. **9**(6), 755–763 (1993)
16. L.E. Dubins, On curves of minimal length with a constraint on average curvature and with prescribed initial and terminal positions and tangents. Am. J. Math. **79**, 497–516 (1957)
17. G. Oriolo, A. De Luca, M. Vendittelli, WMR control via dynamic feedback linearization: design, implementation, and experimental validation. IEEE Trans. Control Syst. Technol. **10**(6), 835–852 (2002)
18. R.W. Brockett, *Asymptotic Stability and Feedback Stabilization* (Birkhäuser Boston, 1983)
19. D.A. Lizárraga, Obstructions to the existence of universal stabilizers for smooth control systems. Math. Control Signals Syst. **16**, 255–277 (2004)
20. P. Rouchon, J.L.M. Fliess, P. Martin, Flatness and motion planning: the car with n trailers, in *European Control Conference* (1993), pp. 1518–1522
21. R. Murray, Nilpotent bases for a class on nonintegrable distributions with applications to trajectory generation for nonholonomic systems. Math. Control Signal Syst. **7**, 58–75 (1994)
22. O.J. Sørdalen, Conversion of a car with n trailers into a chained form, in *IEEE Int. Conf. on Robotics and Automation* (1993), pp. 382–387
23. J.C. Latombe, *Robot Motion Planning* (Kluwer Academic Publishers, 1991)
24. S.L.L. Valle, *Planning Algorithms* (Cambridge University Press, 2006)
25. J.P. Laumond, S. Sekhavat, F. Lamiraux, *Guidelines in Nonholonomic Motion Planning for Mobile Robots* (Springer Berlin Heidelberg, Berlin, 1998), pp. 1–53 [Online]. Available: https://www.di.ens.fr/jean-paul.laumond/promotion/chap1.pdf
26. B. Siciliano, L. Sciavicco, L. Villani, G. Oriolo, *Robotics* (Springer London, 2009)
27. Y. Chitour, F. Jean, R. Long, A global steering method for nonholonomic systems. J. Differential Equations **254**(4), 1903–1956 (2013) [Online]. Available: https://www.sciencedirect.com/science/article/pii/S0022039612004433
28. S. Fortune, G. Wilfong, Planning constrained motions, in *ACM STOC* (1988), pp. 445–459
29. P. Jacobs, J. Canny, Planning smooth paths for mobile robots, in *IEEE Int. Conf. on Robotics and Automation* (1989)
30. P.K. Agarwal, P. Raghavan, H. Tamaki, Motion planning for a steering-constrained robot through moderate obstacles, in *ACM Symp. on Computational Geometry* (1995)
31. J.D. Boissonnat, S. Lazard, A polynomial time algorithm for computing a shortest path of bounded curvature amidst moderate obstacle, in *ACM Symp. on Computational Geometry* (1996)
32. S. Sekhavat, J. Laumond, Topological property for collision-free nonholonomic motion planning: the case of sinusoidal inputs for chained form systems. IEEE Trans. Robot. Autom. **14**(5), 671–680 (1998)
33. G. Oriolo, M. Vendittelli, A framework for the stabilization of general nonholonomic systems with an application to the plate-ball mechanism. IEEE Trans. Robot. **21**(2), 162–175 (2005)
34. G.W. Haynes, H. Hermes, Nonlinear controllability via lie theory. SIAM J. Control **8**, 450–460 (1970)
35. J. Barraquand, J.C., Latombe, Nonholonomic multibody mobile robots: Controllability and motion planning in the presence of obstacles. Algorithmica **10**(4), 121–155 (1993)
36. A.W. Divelbiss, J.T. Wen, A path space approach to nonholonomic motion planning in the presence of obstacles. IEEE Trans. Robot. Autom. **13**(3), 443–451 (1997)
37. S.M. LaValle, J.J. Kuffner, Randomized kinodynamic planning. Int. J. Robot. Res. **20**(5), 378–400 (2001)
38. M. Fliess, J. Lévine, P. Martin, P. Rouchon, Flatness and defect of non-linear systems: introductory theory and examples. Int. J. Control **61**, 1327–1361 (1995)
39. S. Sekhavat, F. Lamiraux, J.P. Laumond, G. Bauzil, A. Ferrand, Motion planning and control for Hilare pulling a trailer: experimental issues, in *IEEE/RSJ Int. Conf. on Intelligent Robots and Systems* (1997)
40. F. Lamiraux, J. Laumond, Flatness and small-time controllability of multibody mobile robots: application to motion planning. IEEE Trans. Autom. Control **45**(10), 1878–1881 (2000)

41. L. Bushnell, D. Tilbury, S. Sastry, Steering three-input nonholonomic systems: the fire-truck example. Int. J. Robot. Res. **14**(4), 366–381 (1995)
42. D. Tilbury, R. Murray, S. Sastry, Trajectory generation for the n-trailer problem using Goursat normal form. IEEE Trans. Autom. Control **40**(5), 802–819 (1995)
43. S. Sekhavat, Planification de mouvements sans collisions pour systèmes non holonomes, Ph.D. dissertation, INPT, LAAS-CNRS, Toulouse, 1996
44. S. Monaco, D. Normand-Cyrot, An introduction to motion planning under multirate digital control, in *IEEE Conference on Decision and Control* (1992), pp. 1780–1785
45. R. Murray, S. Sastry, Nonholonomic motion planning: steering using sinusoids. IEEE Trans. Autom. Control **38**(5), 700–716 (1993)
46. Z. Li, J. Canny (eds.), *A Differential Geometric Approach to Motion Planning*, vol. 192. The Kluwer International Series in Engineering and Computer Science (1992)
47. J. Laumond, P. Jacobs, M. Taïx, R. Murray, A motion planner for nonholonomic mobile robot. IEEE Trans. Robot. Autom. **10**, 577 (1994)
48. J. Latombe, A fast path planner for a car-like indoor mobile robot, in *Ninth National Conf. on Artificial Intelligence, AAAI* (1991), pp. 659–665
49. H. Chitsaz, S.M. LaValle, D.J. Balkcom, M.T. Mason, Minimum wheel-rotation paths for differential-drive mobile robots. Int. J. Robot. Res. **28**(1), 66–80 (2009)
50. D.J. Balkcom, P.A. Kavathekar, M.T. Mason, *The Minimum-Time Trajectories for an Omni-Directional Vehicle* (Springer Berlin Heidelberg, Berlin, 2008), pp. 343–358 [Online]. Available: https://doi.org/10.1007/978-3-540-68405-3_22
51. C. Fernandes, L. Gurvits, Z. Li, A variational approach to optimal nonholonomic motion planning, in *IEEE Int. Conf. on Robotics and Automation* (1991), pp. 680–685
52. J.A. Reeds, R.A. Shepp, Optimal paths for a car that goes both forward and backwards. Pacific J. Math. **145**, 367 (1990)
53. H.J. Sussmann, W. Tang, Shortest paths for the Reeds and Shepp car: a worked out example of the use of geometric techniques in nonlinear optimal control, Report SYCON-91-10, Rutgers Univ., 1991
54. P. Souères, J.P. Laumond, Shortest path synthesis for a car-like robot, in *Proc. European Control Conference* (1993)
55. J.P. Laumond, P. Souères, Metric induced by the shortest paths for a car-like mobile robot, in *Proc. IEEE Int. Conf. on Intelligent Robots and Systems* (1993)
56. P. Souères, J. Fourquet, J.P. Laumond, Set of reachable positions for a car. IEEE Trans. Autom. Control **41**, 1626–1630 (1994)
57. E.J. Cockayne, G.W.C. Hall, Plane motion of a particle subject to curvature constraints. SIAM J. Control **13**, 197 (1975)
58. X.N. Bui, P. Souères, J.D. Boissonnat, J.P. Laumond, Shortest path synthesis for Dubins non-holonomic robots, in *IEEE Int. Conf. on Robotics and Automation* (1994)
59. J.D. Boissonnat, X.N. Bui, Accessibility region for a car that only move forwards along optimal paths, Research Report INRIA 2181, Sophia-Antipolis, 1994
60. P. Souères, J.D. Boissonnat, Optimal trajectories for nonholonomic robots, in *Robot Motion Planning and Control*, ed. by J. Laumond, vol. 229 (Springer, 1998)
61. A. Vershik, V.Y. Gershkovich, Nonholonomic problems and the theory of distributions. Acta Appl. Math. **12**, 181–209 (1988)
62. R.W. Brockett, *Control Theory and Singular Riemannian Geometry* (Springer New York, New York, NY, 1982), pp. 11–27
63. J. Mitchell, On Carnot-Carathéodory metrics. J. Differential Geom. **21**, 35–45 (1985)
64. R. Strichartz, Sub-riemannian geometry. J. Differential Geom. **24**, 221–263 (1986)
65. A. Bellaïche, F. Jean, J. Risler, Geometry of nonholonomic systems, in *Robot Motion Planning and Control*, ed. by J. Laumond, vol. 229, Lecture Notes in Control and Information Sciences (Springer, 1998)
66. M. Vendittelli, G. Oriolo, F. Jean, J.P. Laumond, Nonhomogeneous nilpotent approximations for nonholonomic systems with singularities. IEEE Trans. Autom. Control **49**(2), 261–266 (2004)

67. M. Khatib, H. Jaouni, R. Chatila, J.P. Laumond, Dynamic path modification for car-like nonholonomic mobile robots, in *International Conference on Robotics and Automation*, vol. 4 (1997), pp. 2920–2925
68. M. Vendittelli, J.P. Laumond, C. Nissoux, Obstacle distance for car-like robots. IEEE Trans. Robot. Autom. **15**(4), 678–691 (1999)
69. P. Robuffo Giordano, M. Vendittelli, J. Laumond, P. Souères, Nonholonomic distance to polygonal obstacles for a car-like robot of polygonal shape. IEEE Trans. Robot. **22**(5), 1040–1047 (2006)
70. P. Robuffo Giordano, M. Vendittelli, Shortest paths to obstacles for a polygonal dubins car. IEEE Trans. Robot. **25**(5), 1184–1191 (2009)
71. G. Jacob, Lyndon discretization and exact motion planning, in *European Control Conference* (1991), pp. 1507–1512
72. A. Bellaïche, *The Tangent Space in Sub-Riemannian Geometry*, ed. by A. Bellaïche, E.J.-J. Risler, Sub-Riemannian Geometry (Birkhäuser, 1996)
73. H. Hermes, Nilpotent approximations of control systems and distributions. SIAM J. Control Optim. **24**, 731–736 (1986)
74. R.M. Bianchini, G. Stefani, Controllability along a trajectory: A variational approach. SIAM J. Control Optim. **31**, 900–927 (1993)
75. H.J. Sussmann, A general theorem on local controllability. SIAM J. Control Optim. **25**, 158–194 (1987)
76. H. Hermes, Smooth homogeneous asymptotically stabilizing feedback controls. ESAIM Control Optim. Calc. Var. **2**, 13–32 (1997)
77. M. Kawski, Homogeneous stabilizing feedback laws. C-TAT **6**(4), 497–516 (1990)
78. F. Jean, Complexity of nonholonomic motion planning. Int. J. Control **74**(8), 776–782 (2001)
79. H. Hermes, *Vector Field Approximations; Flow Homogeneity*, ed. by J. Wiener, J.K. Hale, Ordinary and Delay Differential Equations (Longman, 1991)
80. F. Jean, Uniform estimation of sub-Riemannian balls. J. Dyn. Control Syst. **7**(4), 473–500 (2001)
81. A. De Luca, G. Oriolo, M. Vendittelli, *Control of Wheeled Mobile Robots: An Experimental Overview* (Springer Berlin Heidelberg, Berlin, 2001), pp. 181–226
82. M. Aicardi, G. Casalino, A. Bicchi, A. Balestrino, Closed loop steering of unicycle like vehicles via Lyapunov techniques. IEEE Robot. Autom. Mag. **2**(1), 27–35 (1995)
83. C. Samson, Time-varying feedback stabilization of car-like wheeled mobile robots. Int. J. Robot. Res. **12**(1), 55–64 (1993)
84. R. M'Closkey, R. Murray, Exponential stabilization of driftless nonlinear control systems using homogeneous feedback. IEEE Trans. Autom. Control **42**(5), 614–628 (1997)
85. P. Morin, C. Samson, Control of nonlinear chained systems: from the Routh-Hurwitz stability criterion to time-varying exponential stabilizers. IEEE Trans. Autom. Control **45**(1), 141–146 (2000)
86. C. Samson, Path following and time-varying feedback stabilization of a wheeled mobile robot, in *Int. Conf. Autom. Robotics Comput. Vis.* (1992)
87. J.P. Laumond (ed.), *Robot Motion Planning and Control*, vol. 229, Lectures Notes in Control and Information Sciences (Springer, 1998)
88. P. Morin, C. Samson, Control of nonholonomic mobile robots based on the transverse function approach. IEEE Trans. Robot. **25**(5), 1058–1073 (2009)
89. M. Ferro, A. Paolillo, A. Cherubini, M. Vendittelli, Vision-based navigation of omnidirectional mobile robots. IEEE Robot. Autom. Lett. **4**(3), 2691–2698 (2019)
90. A. Paolillo, P. Gergondet, A. Cherubini, M. Vendittelli, A. Kheddar, Autonomous car driving by a humanoid robot. J. Field Robot. **35**(2), 169–186 (2018)
91. A. Cherubini, F. Chaumette, G. Oriolo, An image-based visual servoing scheme for following paths with nonholonomic mobile robots, in *IEEE Int. Conf. on Control, Automation, Robotics and Vision* (2008), pp. 108–113

Humanoids

Kensuke Harada

Abstract This section of the book describes the mechanical design of humanoid robots which are human-like robots with human-like features such as a head, two arms, and two legs. To design such humanoid robot, we have to realize required functions with compact design. In this section, some mechanical components and the full-body design examples are presented. After discussing the historical perspectives in Sect. 1, I briefly describe the design of each mechanical component, i.e., the leg/foot, multi-fingered hand, and face mechanisms. Lastly, I briefly describe an example full-body design.

1 Introduction

One of the ultimate goals of robotics research is the realization of a robot that looks and moves like a human. To achieve this goal, we need to design the mechanisms of individual parts, such as the hands, fingers, feet, and head of the robot, such that the specific requirements of each part are met. For example, the hand must be capable of generating a gripping force that allows grasping objects used in daily life; furthermore, the foot must be able to support the weight of the robot and enable human-like walking. These individual elements need to be integrated into a full-body robot design that best mimics the proportions of a human body.

To date, there have been several attempts to design full-body humanoid robots that include such individual elements. As pioneers of humanoid robotics Kato et al. developed the humanoid robot WABOT-1 [1] in 1973; it had two arms, two legs, and a head. Furthermore, it could communicate with humans by engaging in simple Japanese language conversation using an artificial mouth, recognize an object with artificial eyes, move by bipedal walking, and grasp an object with both hands that were equipped with tactile sensors (Fig. 1).

K. Harada (✉)
Graduate School of Engineering Science, Osaka University, Toyonaka, Japan
c-mail: harada@sys.es.osaka u.ac.jp

© Springer Nature Switzerland AG 2025
B. Siciliano (ed.), *Robotics Goes MOOC*,
https://doi.org/10.1007/978-3-319-75823-7_8

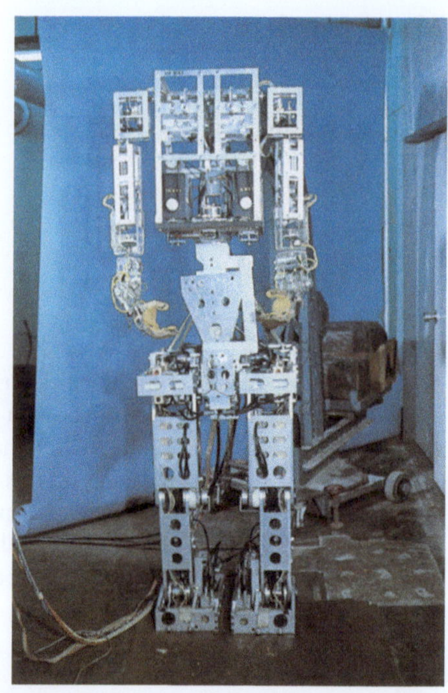

Fig. 1 Overview of WABOT-1 [1] (Image courtesy of Humanoid Robotics Institute, Waseda University)

Fig. 2 Somersault by the humanoid robot Atlas [2]

Since then, several humanoid robots have been continuously developed. More recently, a hydraulically driven humanoid robot, Atlas [2], was developed by a company named Boston Dynamics. They have not published many original papers, and therefore the technical details are not easily accessible. However, they have released a video footage showing Atlas traversing various uneven terrains, such as grass, jumping up a step, and performing a somersault (Fig. 2). This type of movement can be achieved by designing the legs and other parts such that they are extremely lightweight, highly rigid, and compact. Additionally, the actuators must be high output and have a quick response speed.

In this chapter, we provide a general overview of the mechanical design of humanoid robots. To design such humanoid robot, we have to realize required functions with compact design. We begin by briefly discussing various leg mechanism design examples such as the light-weight leg design of the humanoid robot HRP-2 and HRP-3 [3, 4], compliant leg-joint mechanism of the humanoid robot COMAN [5], and a toe-joint mechanism [6]. Then, we briefly discuss various examples of multi-fingered hand designs such as a finger link mechanism actuated by electric motors [7], an under-actuated finger mechanism with a tendon system [8], and an under-actuated link mechanism [9]. We also present an example of the human-like face and head mechanism for the humanoid robot HRP-4C [10]. Lastly, we discuss the full-body design of the humanoid robot HRP-4C [10].

2 Foot/Leg Mechanism

In this section, we introduce three foot/leg mechanism designs for biped humanoid robots.

2.1 *High-Stiffness Leg Design*

The leg mechanism is a crucial component in the design of humanoid robots. Specifically, non-optimized elasticity in the legs causes the torso to be unstable, resulting in structural oscillation when the robot walks. Additionally, increasing the weight of the legs makes the robot unstable due to the inertial effect of the swing leg. Various factors related to its mechanical design, such as the kinematic structure, mass distribution, actuators, drive mechanisms, and sensor layout, continue to hinder the development of biped robots with fast and human-like walking capabilities.

To date, there have been several attempts to optimize the humanoid robot leg design; some examples include ASIMO [11], HRP-2 [3], HRP-3 [4], HUBO [12], Atlas [2], Valkyrie [13]. Here, as an example, we especially discuss HRP-2 [3] and HRP-3 [4], biped humanoid robots developed in 2002 and 2008, respectively in collaboration with the National Institute of Advanced Industrial Science and Technology (AIST) and Kawada Industries, Japan (Fig. 3). At a weighting approximately 60 kg, HRP-2 stands 154cm tall and has 30 actuated DOFs, two and six of which constitute the pelvis and leg, respectively (Fig. 4).

One of unique configurations of these robots is that the hip joint has a cantilever-type structure. The cantilever-type leg structure enables to have less collision between two upper-limbs and to have a wide area of landing point of the swinging leg contributing to appropriately shape the support polygon during the double support phase. HRP-2 is able to pass a very narrow path without making collision between two upper-limbs.

The range of motion of each joint is designed such that the robot can take a seat, sit on the floor with the Japanese manner, walk straight with 40 cm steps, and climb up and down stairs with 20 cm height. Using physics simulation of these five motions, the joint movable range of the leg is determined.

(a)　　　　　　　　　　　　　　(b)

Fig. 3 Outlook of HRP series humanoid robot [3, 4]. (**a**) HRP-2. (**b**) HRP-3

The leg was designed to balance structural stiffness and actuator performance with the lightness of the mechanical parts. The leg structure was designed to simultaneously meet the high stiffness of link structure and the thin-walled cross section with high second moment of area. To meet such requirements, the leg links of HRP-2 and HRP-3 are manufactured by investment casting.

2.2 Compliant Leg-Joint Mechanism

Next, we explain the passive compliant mechanism introduced to the leg joints. Implementing passive compliance introduced in the leg joints is beneficial in many ways. First, it reduces the impact force generated when the foot is in contact with the floor. Second, the gait will be energy-efficient because the energy can be stored

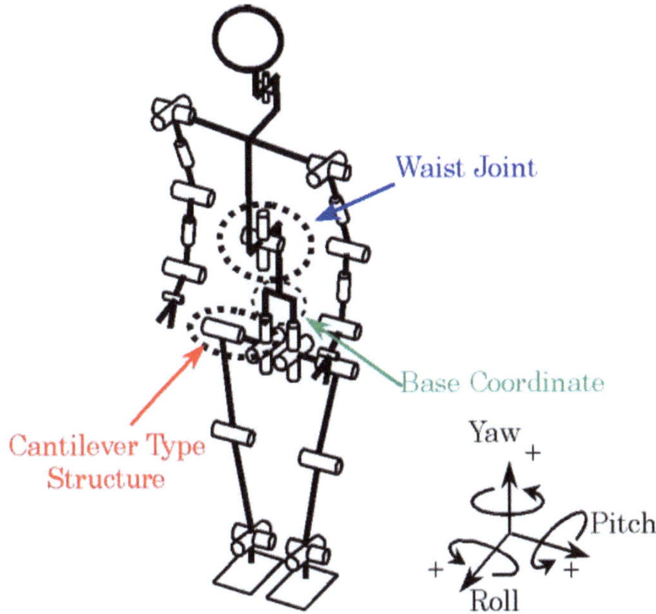

Fig. 4 Joint configuration of HRP-2 [3]

in the mechanical spring. Third, the robot will be safe against the contact with a human. As an example, we introduce the leg design of the humanoid robot COMAN (i.e., the COmpliant huMANoid). An overview of the COMAN is shown in Fig. 5; COMAN has a mass of 34 kg, a 3-DoF waist, and two 6-DoF legs. Additionally, all of the joints have active (torque-controlled) compliance properties, but a series of elastic actuation (SEA)-based passive compliance system was added to some DoFs, including the flexion/extension DoFs of the legs. Figure 6 shows the compact design of the SEA unit installed in the COMAN, which includes a DC motor, encoder, gear reduction, torque sensor, spring, and spring deflection encoder. Figure 7 shows the SEA units actually installed in the COMAN. By appropriately tuning the leg compliance, it was proven that the robot can maintain the stability while standing on two legs.

2.3 Foot Mechanism

The foot of a human plays an essential role in maintaining a stable gait. In particular, a human's foot comprises three arch structures and a toe-joint mechanism. The arch structures include the anterior transverse arch, the lateral longitudinal arch, and the medial longitudinal arch. When a human walks, the arch structures work as a shock absorber that enhances the foot's adaptability to the ground. In contrast,

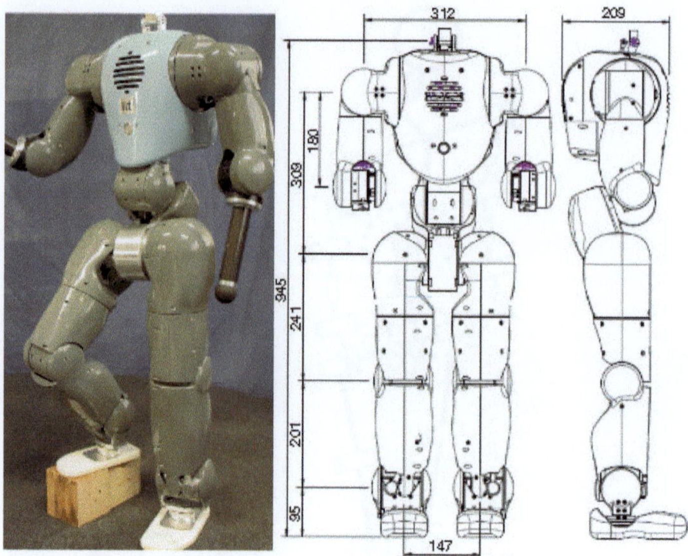

Fig. 5 Overview of COMAN [5]

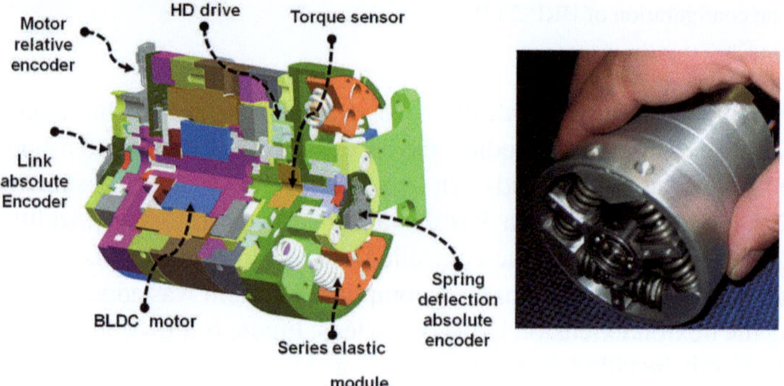

Fig. 6 SEA unit [5]

the toe joint is purposed to flex upward to push the foot off of the ground. This contributes to the fast and stable biped gait [14]. However, when a foot is pushed off of the ground, the toe joint may have to support the robot's weight. To avoid this problem, the toe-thenar mechanism, which involves a parallel four-bar linkage, was proposed as shown in Fig. 8. When Joint 1 is actuated, Links A and C roll over the ground, maintaining contact. This mechanism ensures that the thenar and toe are simultaneously in contact with the ground during the heel-off phase of the biped gait. Figure 9 shows the toe-thenar mechanism installed in the humanoid robot UT-μ2/magnum.

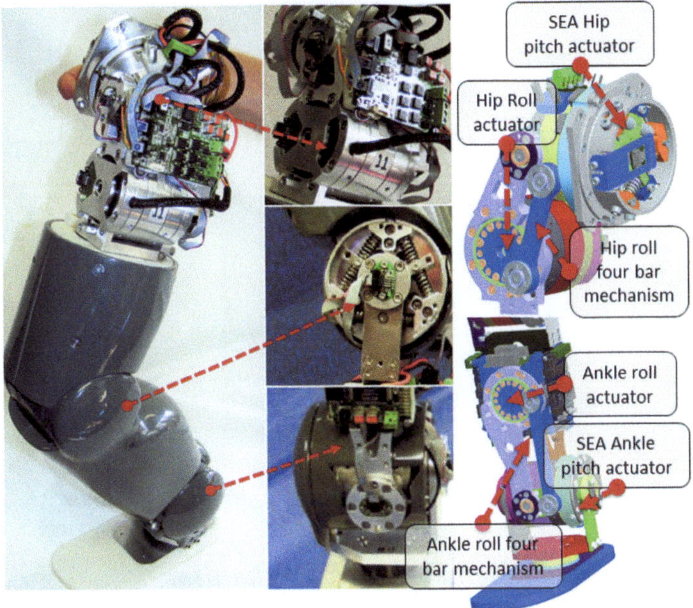

Fig. 7 SEA unit installed in COMAN [5]

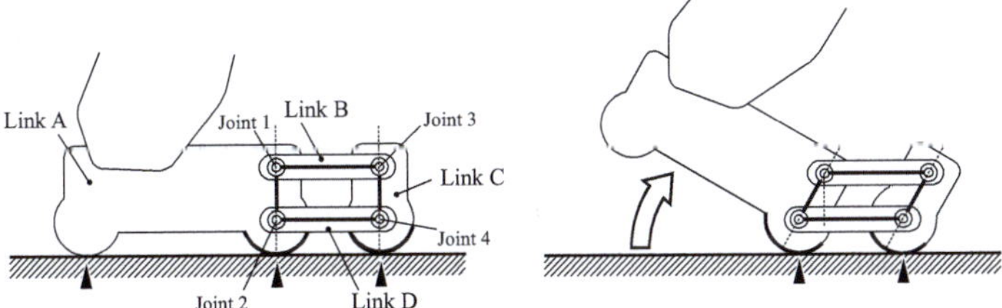

Fig. 8 Toe-thenar mechanism with parallel four-bar linkage [6]

3 Hand

In this section, we introduce several examples of robotic hand mechanisms. There are variety of power sources for multi-fingered hands, i.e., hydraulic pressure, pneumatic pressure, and electric motor sources. Additionally, many mechanisms, such as serial-link mechanism, parallel-link mechanism, and tendon-driven mechanism, have been proposed to transmit the power to each finger link. Furthermore, the relationship between the number of the DoF and actuators can be used to determine whether a hand is over-actuated or under-actuated. This section primary focuses on electric motor-actuated link mechanisms without using the tendons, the tendon-driven mechanism actuated by electric motors, and the under-actuation link mechanism.

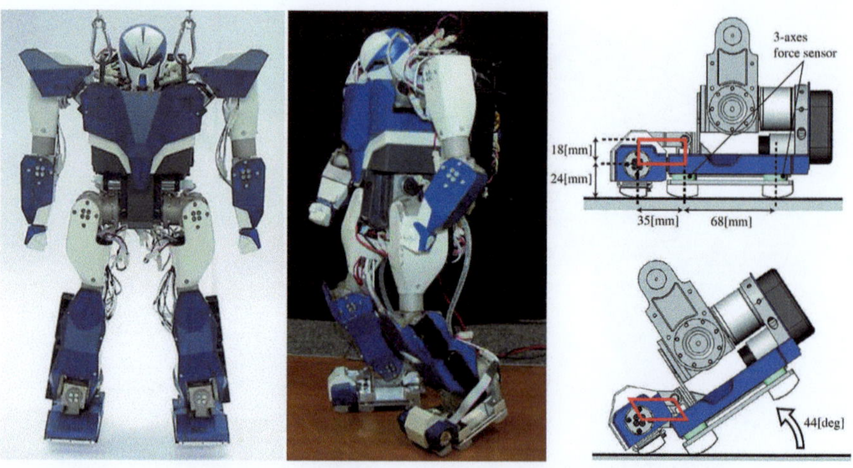

Fig. 9 Toe-thenar mechanism installed in UT-μ2/magnum humanoid robot [6]

3.1 Electric Motor-Actuated Link Mechanism

We begin by focusing on a life-sized humanoid robot hand with a tendonless electric motor-actuated link mechanism. Figure 10 shows a multi-fingered hand, HDH-1, attached to a life-sized humanoid robot. The hand has four fingers with 17 joints, i.e., 13 active joints and four linked joints.

As shown in Fig. 11, each finger has a modularized structure with a joint configuration in which only the thumb has one additional joint, which is called the carpometacarpal (CM) joint. Each finger module has four joints, i.e., three active joints and one linked joints. A distal interphalangeal (DIP) joint works together with a proximal interphalangeal (PIP) joint. The pitch axis of the metacarpophalangeal (MP) joint enables adduction and abduction, whereas other axes enable flexion and extension. Each active joint is independently driven by its own electrical servomotor via a transmission system with a harmonic drive gear.

The MP joint was designed to have two sub-joints, one of which is the MP roll joint between the metacarpal and the base links; the other sub-joint is the MP pitch joint between the proximal and metacarpal links. The MP roll joint is driven by the servomotor installed in the base link. Additionally, a parallel crank system is employed between the harmonic drive gear's output shaft and metacarpal link. The MP pitch joint is driven by the servomotor installed in the proximal link.

The PIP and DIP joints were designed to be mechanically coupled and driven by one actuator. A double-rocker mechanism was adopted to couple these joints.

With this mechanism, we can realize approximately 7.8 N of maximum force at the fingertip with an outstretched finger and approximately 17 N maximum force with a bent finger.

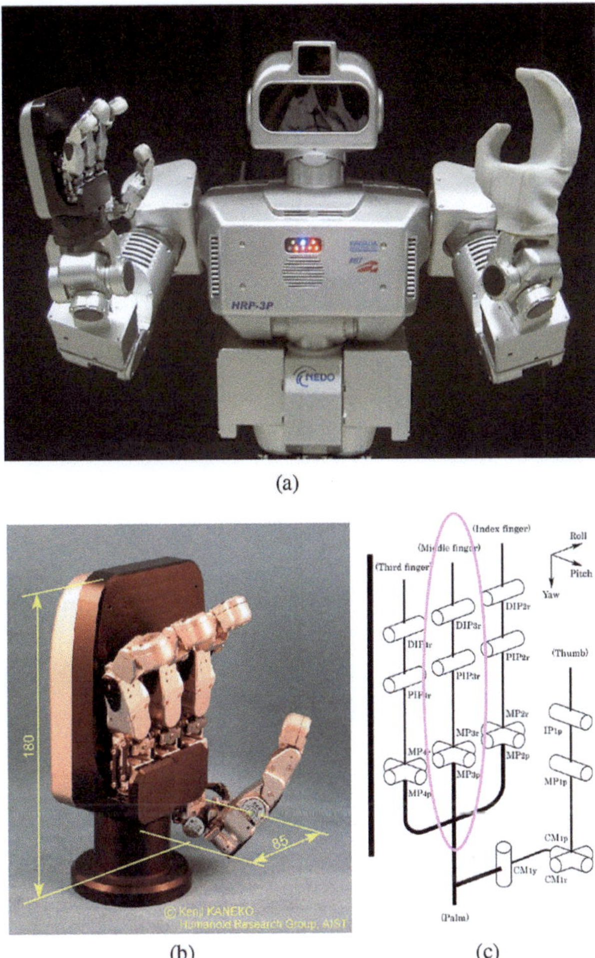

(b)   (c)

Fig. 10 Multi-fingered hand HDH-1 [7]. (**a**) Multi-fingered hand mounted on the humanoid robot HRP-3P. (**b**) Overview. (**c**) Joint configuration

3.2 *Tendon-Driven Under-Actuation*

Humans move their fingers by contracting the muscles attached to the bones. Thus, an intuitive approach to design a multi-fingered robotic hand entails the use of tendons to drive finger movement. The Utah/MIT hand [15], Stanford/JPL hand [16], and ETL hand [17] are the earliest examples of modern tendon-driven robotic multi-fingered hands. The Shadow hand [18], DLR hand-arm system [19] and Pisa/IIT hand [20] are more recent examples of the use of tendon-driven system. In this section, we especially focus on the under-actuated tendon-driven multi-fingered hand designs in which the number of DoFs is larger than the number of actuators.

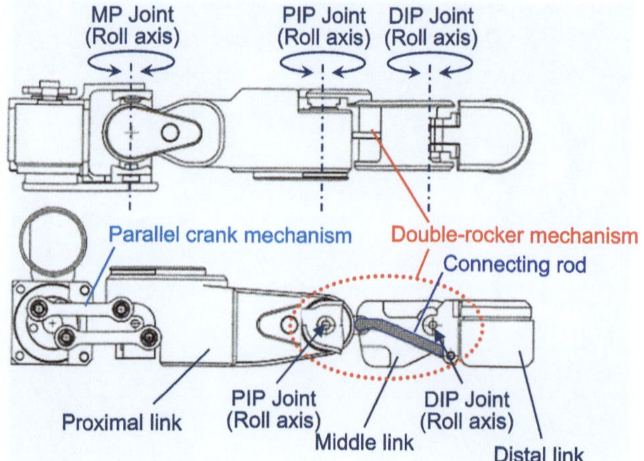

Fig. 11 HDH-1 finger mechanism [7]

Fig. 12 Joint placement
DoFs of the under-actuated
tendon-driven hand [8]

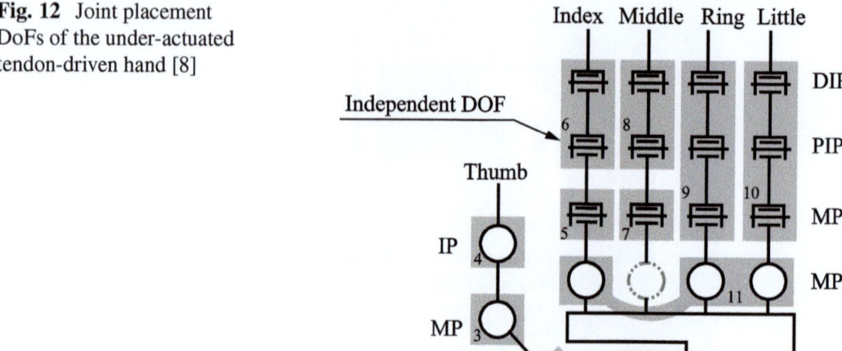

Generally, the under-actuated hand design is expected to ensure the mechanical adaptability required for object grasping.

Kaminaga et al. [8] developed an under-actuated hand with five fingers, 20 joints, and 12 tendons (Fig. 12). The index and middle fingers are equipped with two active flextion DoFs, and the DIP and PIP joints are coupled with a spring. Alternatively, the ring and little fingers have one active DoF. The abduction joints are driven by an under-actuated mechanism. Additionally, the thumb was set opposite to the middle finger.

The tendon configurations for the 2-DoF fingers and 1-DoF fingers are shown in Figs. 13 and 14, respectively. Pulley layers were installed in the 3D-printed finger phalanges to control finger slack.

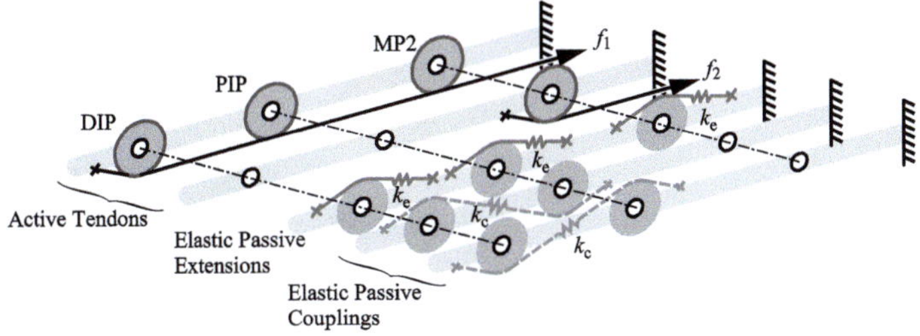

Fig. 13 Tendon routing and spring framework for 2-DoF fingers [8]

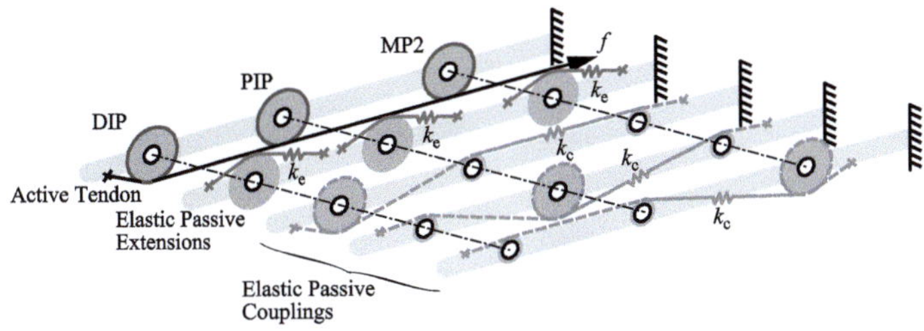

Fig. 14 Tendon routing and spring framework for 1-DoF fingers [8]

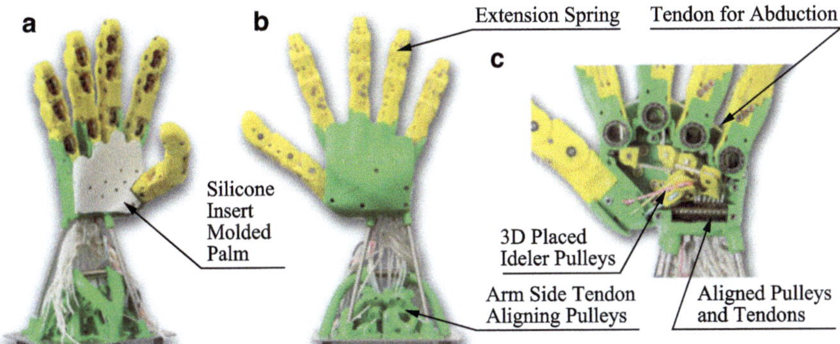

Fig. 15 Views of an under-actuated hand. (**a**) Palm view (anterior), (**b**) Backside view (posterior), and (**c**) Tendon aligning pulley structure (posterior) [8]

Different views of the hand are shown in Fig. 15a, b. The wires are aligned to enable connections between the 14 axes of a small linear cluster electro-hydrostatic actuator. Figure 15c provides a close-up view of the tendon-aligning pulleys. Figure 16 shows under-actuated hand performing a power grasping. Specifically, Fig. 16b shows the posture when the hand is closed without an object. Figure 16c shows the posture when the hand is grasping a cylindrical object. In Fig. 16b, c, the tension was the same for tendons, but the hand was able to adapt to the object because of under-actuation.

Fig. 16 Power grasping with
under-actuated hand. (**a**)
Open hand, (**b**) Closing hand,
and (**c**) Cylindrical-object
power grasping [8]

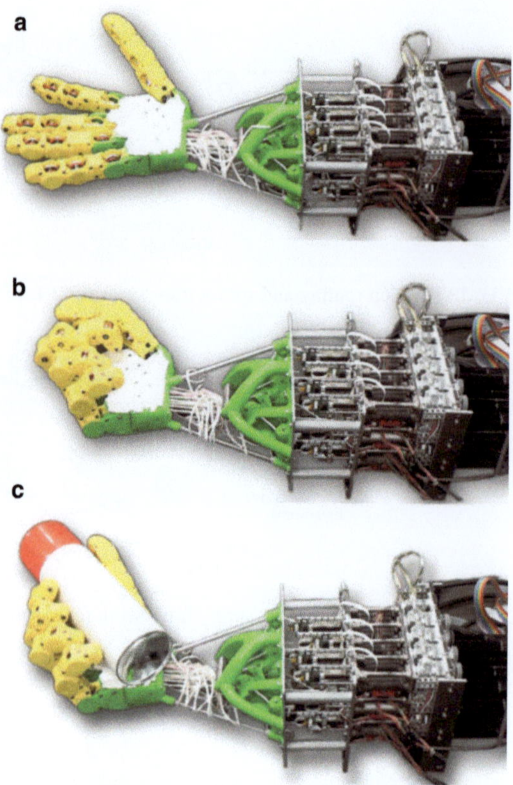

3.3 *Link Mechanism-Based Under-Actuation*

Next, we discuss the incorporation of mechanical linkages into the design of under-actuated robotic hands; in this design, the mechanical linkages serve to ensure the mechanical adaptation to the grasped object. While the hand introduced in this subsection is not human-like, underactuation with link mechanism is one of the key elements to realize compact human-like multi-fingered hand. The general principle of a linkage-based under-actuated finger is illustrated in Fig. 17 and represented here as a simple two-phalanx five-bar under-actuated finger. At rest, the two phalanges are maintained in an aligned configuration by an extension spring and a mechanical limit (as can be seen in a). The actuator, shown by the arrows in the figure, is used to close the finger on an object. When contact is established between the object and distal phalanx (as observed in b), the actuator force extends the spring, which separates the mechanical limit (as can be seen in c), and contact is then established between the object and distal phalanx (as shown in d). At this stage, the actuator force is distributed between the two phalanges, and the adaptive mechanism alone has allowed the finger to adapt to the shape of the object.

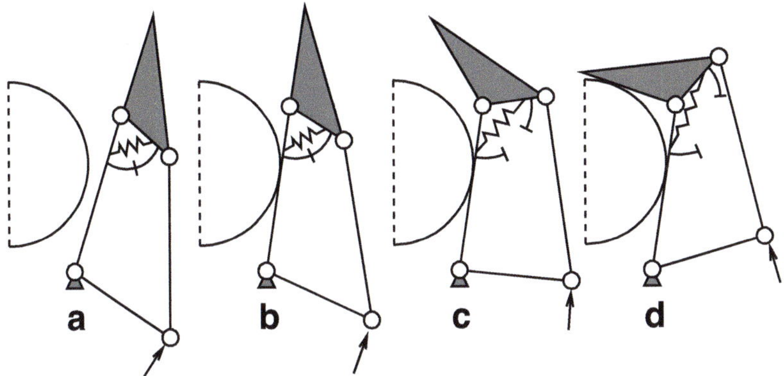

Fig. 17 Grasping sequence for a two-phalanx under-actuated finger with a five-bar linkage system [9]

Fig. 18 Grasping sequence for a two-phalanx under-actuated finger with a five-bar linkage system [9]

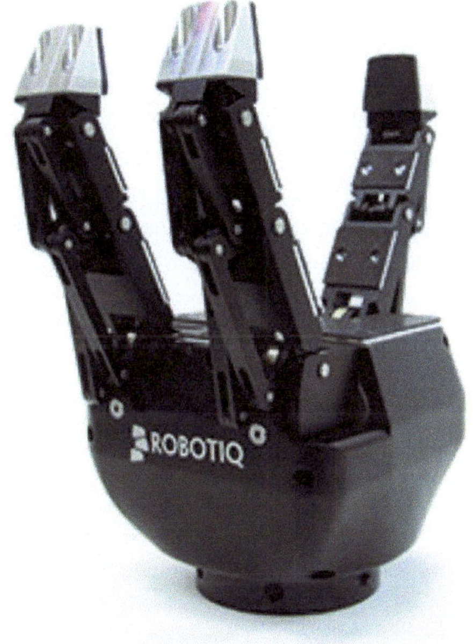

The design of compact linkage-based fingers is difficult because linkages tend to require more space than pulleys and tendons. Figure 18 shows a commercialized robotic hand (Robotiq hand S) that has 10 DoFs and two degrees of actuation. One actuator serves to close the fingers, and the other actuator rotates the finger around the base.

4 Face and Head Mechanism

The human face plays several important roles in our daily life. Most importantly, humans look at faces to recognize each other. Additionally, by observing the subtle changes in the facial expressions of a person with whom they are speaking, humans can infer how the person is feeling and adjust the topic of conversation according to the person's feelings. There are approximately 30 different muscles in a human's face; these muscles can be partitioned into two broad categories: the facial muscles that are responsible for facial expressions, and those that are responsible for chewing food. Facial muscles are thin, superficial plate muscles that are attached to the underside of the skin of the head, and cover the entire face. A person can make approximately 100 different facial expressions by utilizing the facial expression muscles. It is a great challenge to realize such a function of the human face with a robot.

This section introduces and discusses an example of mechanical face design. An overview of the face mechanism of the humanoid robot HRP-4C [21] is shown in Fig. 19. Silicone rubber was used to mimic the skin of the face, as it can be deformed by an internalized mechanism to create a variety of facial expressions. To best reproduce the head size of an average young female while minimizing weight, the creators of HRP-4C limited the numbers of DoFs in the head. Table 1 lists the numbers of DoFs applied to create facial expression.

Figure 19 also shows the assignment of the face-driving actuators. Each pair of eyebrows, eyelids, as well as the eyeball pan and tilt are mechanically linked and actuated by a single servomotor with a gear head. Although this design allowed the developers to reduce the number of actuators, it prevent the robot from performing asymmetric face motions, such as a wink. In contrast, the mouth motion is controlled by three motors for opening and closing of the mouth, and upper and lower lip movement, respectively. This design was necessary to realize sufficient lip motion for talking and singing. Regarding the head mechanism, servomotors with planetary gear heads regulate the head movement. Unlike leg actuators, which require backlash-free harmonic drive gears for balance control, gear backlash is not a significant concern for face actuation.

Table 1 Head joints of HRP-4C [10]

Face part	Joint name	DoF
Eyebrows	EYEBROW_P	1
Eyelids	EYELID_P	1
Eyeballs pan	EYE_Y	1
Eyeballs tilt	EYE_P	1
Mouth	MOUTH_P	1
Upper lip	UPPERLIP_P	1
Lower lip	LOWERLIP_P	1
Cheek	CHEEK_P	1
Total		8

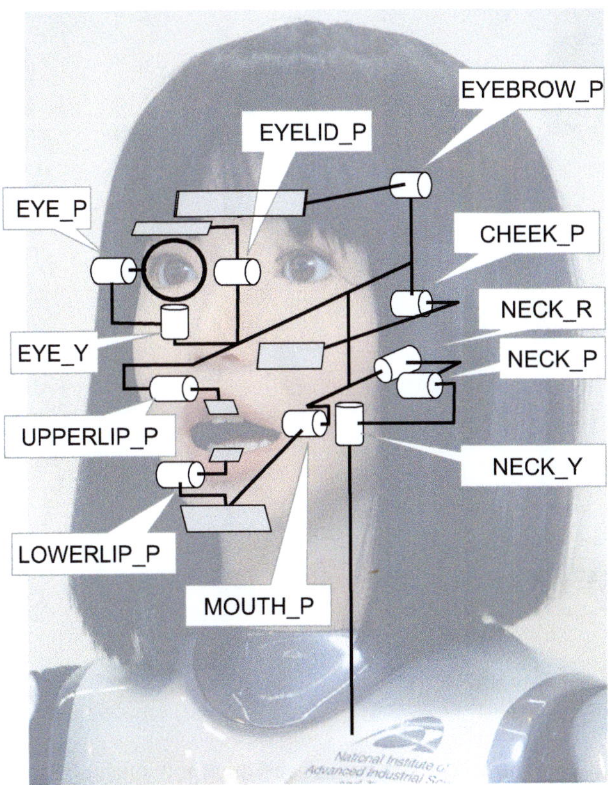

Fig. 19 Head of the humanoid robot HRP-4C [10]

5 Full-Body Design

Next, we will introduce a full-body humanoid robot design example and briefly discuss its leg, foot, arm, hand and head mechanisms. Here, we present the design of the humanoid HRP-4C, the body dimensions of which were designed to match the average young Japanese female. It can not only walk and move like a human, but it can interact with humans using speech recognition, speech synthesis, and gestures. Such robots can be employed in the entertainment industry; specific examples include exhibitions and fashion shows. This type of robot can also be used as a human simulator to evaluate devices purposed for human use (Fig. 20).

The anthropometric database for the Japanese population was used to determine the target shape and dimensions of the HRP-4C. The average data on young females (i.e., aged 19–29) were applied in consideration of the HRP-4C's potential for entertainment-related applications, such as incorporation into fashion shows [22]. Figure 21 shows select dimensions provided by the database. To reproduce the graceful movement of women, the developers asked a professional walking model

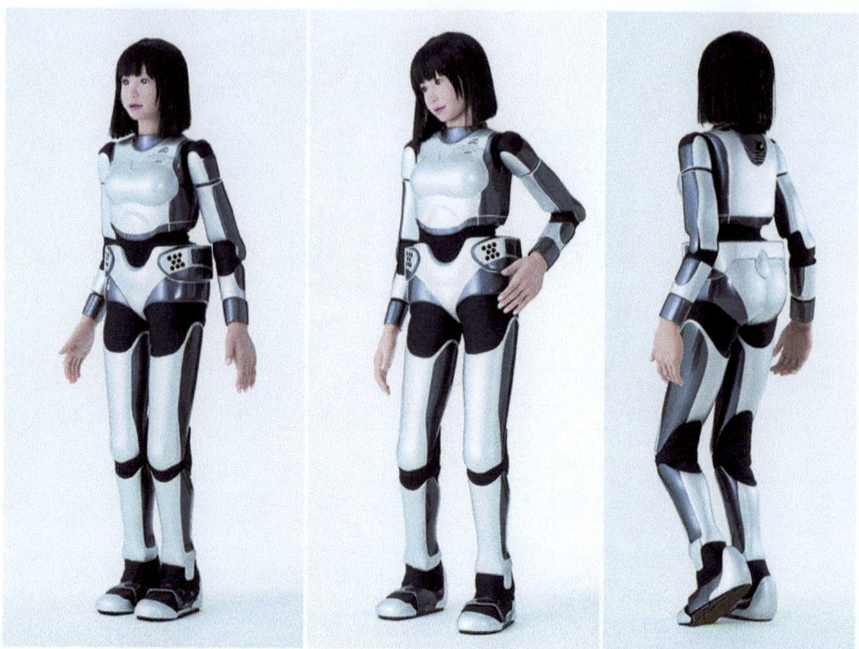

Fig. 20 Humanoid robot HRP-4C [10]

to perform walking, turning, chair-sitting, and other movement-related tasks. These data were used to evaluate the different joint configurations proposed for the HRP-4C framework. The joint angles were calculated to reproduce the captured motion, and the necessary movable range was subsequently estimated. The developers also estimated the motor power required for biped walking to determine the appropriate leg joint configuration.

Figure 20 shows the images of the HRP-4C, which has a height of 1.6 m, mass of 48 kg, and 44 DoF including the toe joints. The principal dimensions and specifications are outlined in Table 2. Figure 22 shows the joint configurations.

Wrist and neck roll axes were included to optimize human-like behavior (Fig. 22a). Regarding the overall frame of the robot, the forearm and thigh links were angled to better mimic the human body (Fig. 22b), and a standard hip joint structure was used to realize a natural waist line (Fig. 22c). Additionally, compact motor drivers were employed to ensure that the robot had a slender, human-like body frame. Lastly, active toe joints were added to enable wide-stride biped walking (Fig. 22d).

The hands of HRP-4C consist of silicone rubber skin with an internalized mechanism. The developers minimized the number of DoFs in their design to mimic movements that can be used in dance performances. Specifically, the first actuator rotates the thumb, and the second actuator flexes the remaining four fingers.

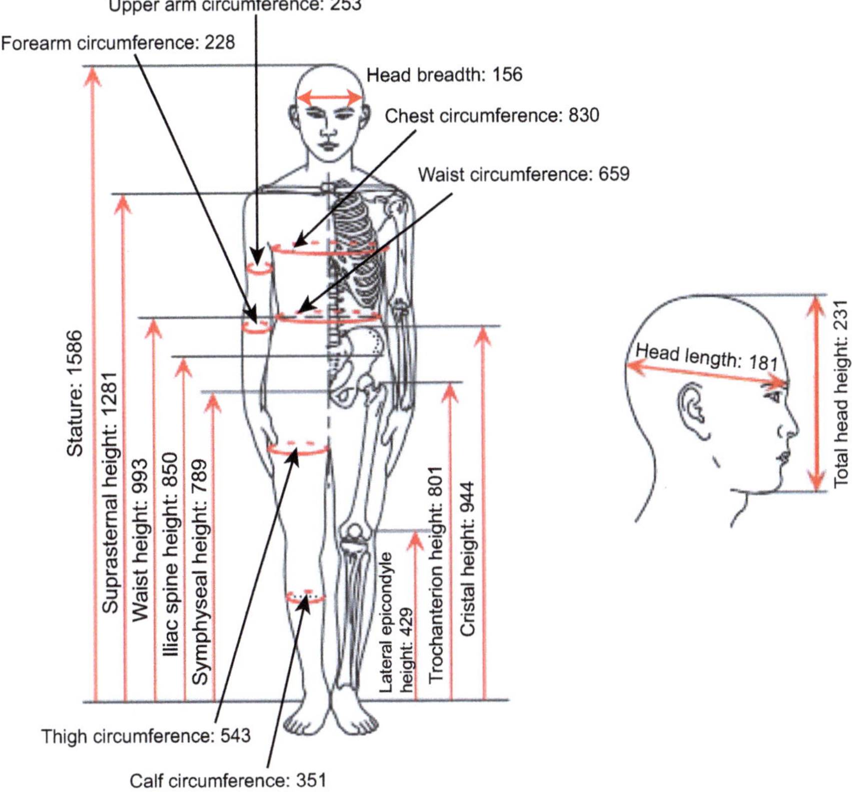

Fig. 21 Averaged human dimensions [22]

Table 2 Dimension of HRP-4C [10]

Height	1600 mm
Weight	48 kg (with batteries)
Total DoF	44 DoF
Face	8 DoF
Neck	3 DoF
Arm	6 DoF × 2
Hand	2 DoF × 2
Waist	3 DoF
Leg	7 DoF ×2
CPU	Intel Pentium M 1.6 GHz
Sensors	
Head	CCD camera
Body	Inertial measurement unit (IMU)
Sole	6-Axis force sensor × 2
Batteries	NiMH 48 V

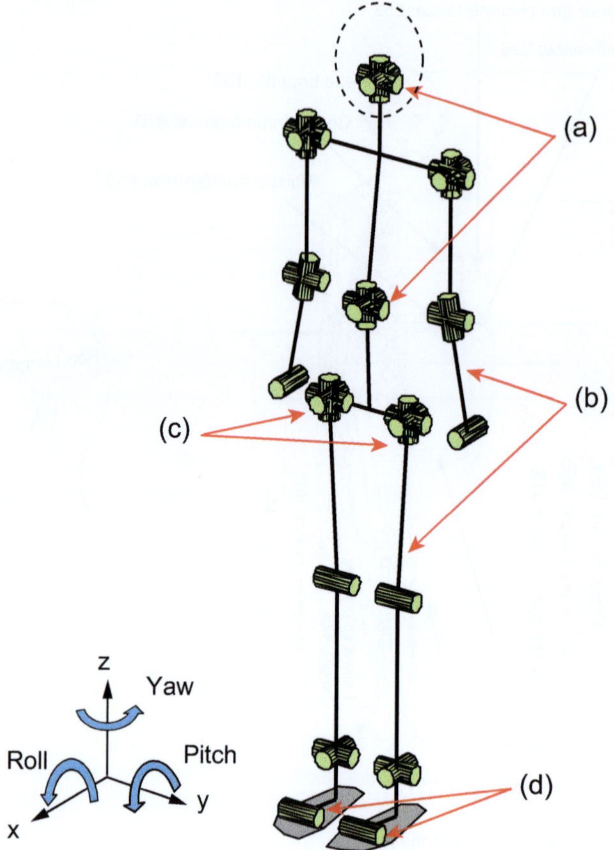

Fig. 22 Joint configurations of HRP-4C (Face and hand joints have been omitted) [10]

6 Conclusions

In this section of the book, we briefly discussed the mechanical design of humanoid robots. We first introduced three foot/leg mechanism examples. Then, we presented multi-fingered hand examples, and provided an overview of the human-like head/face mechanism. Lastly, we briefly described a full-body humanoid robot design.

Discussions

1. We have introduced the cantilever-type leg structure. Discuss the pros and cons of such leg structure.

2. We have introduced several actuation mechanism of multi-fingered hands. Discuss the pros and cons of each actuation mechanism.
3. You can find the motion of HRP-4C at AIST YouTube channel. Do you think that it is like a human's one. If not, discuss how to improve its joint structure.

Cross-References

This chapter contains cross-references to content published in other chapters belonging to the four "Robotic Goes MOOC" books: DES1, DES2, DES3, DES6. The full list of chapters abbreviations is available in the Preface.

References

1. I. Kato, Development of a biped robot WABOT-1. J. Biomech. Soc. Jpn. **2**, 173–174 (1973)
2. Boston Dynamics YouTube Channel. https://www.youtube.com/user/BostonDynamics/videos
3. K. Kaneko, F. Kanehiro, S. Kajita, H. Hirukawa, T. Kawasaki, M. Hirata, K. Akachi, T. Isozumi, Humanoid robot HRP-2, in *Proceedings of IEEE International Conference on Robotics & Automation* (2004), pp. 1083–1090
4. K. Kaneko, K. Harada, F. Kanehiro, G. Miyamori, K. Akachi, Humanoid robot HRP-3, in *IEEE/RSJ Int. Conf. on Intelligent Robots and Systems* (2008)
5. N.G. Tsagarakis, G.M. Cerda, D.G. Caldwell, Compliant leg mechanism of Coman, in *Humanoid Robotics: A Reference*, ed. by A. Goswami, P. Vadakkepat (2019), pp. 407–434
6. K. Yamamoto, Human-like toe joint mechanism, in *Humanoid Robotics: A Reference*, ed. by A. Goswami, P. Vadakkepat (2019), pp. 435–456
7. K. Kaneko, K. Harada, F. Kanehiro, Development of multi-fingered hand for life-size humanoid robots, in *Proceedings of IEEE International Conference on Robotics and Automation* (2007), pp. 913–920
8. H. Kaminaga, Wire driven multi-fingered hand, in *Humanoid Robotics: A Reference*, ed. by A. Goswami, P. Vadakkepat (2019), pp. 457–479
9. C. Gosselin, Underactuation with link mechanisms, in *Humanoid Robotics: A Reference*, ed. by A. Goswami, P. Vadakkepat (2019), pp. 523–533
10. S. Kajita, Mechanism drsign of human-like HRP-4C, in *Humanoid Robotics: A Reference*, ed. by A. Goswami, P. Vadakkepat (2019), pp. 597–613
11. S. Shigemi, ASIMO and humanoid robot research at Honda, in *Humanoid Robotics: A Reference*, ed. by A. Goswami, P. Vadakkepat (2019), pp. 55–90
12. J.-W. Heo, J. Lee, I.H. Lee, J. Lim, H.-H. Oh, History of Hubo: Korean humanoid robot, in *Humanoid Robotics: A Reference*, ed. by A. Goswami, P. Vadakkepat (2019), pp. 117–130
13. J. Yamokoski, N. Radford, *Humanoid Robotics: A Reference*, ed. by A. Goswami, P. Vadakkepat (2019), pp. 201–214
14. T. Saida, H. Ohta, Y. Yokokohji, Function analysis of human-like mechanical foot, using mechanically constrained shoes, in *Proc. of IEEE/RSJ Int. Conf. on Intelligent Robots and Systems* (2004)
15. S. Jacobsen, E. Iversen, D. Knutti, R. Johnson, K. Biggers, Design of the Utah/M.I.T. dextrous hand, in *Proc. of IEEE Int. Conf. on Robotics and Automation* (1986)
16. C. Loucks, V. Johnson, P. Boissiere, G. Starr, J. Steele, Modeling and control of the stanford/JPL hand, in *Proc. of IEEE Int. Conf. on Robotics and Automation* (1987)

17. T. Okada, Computer control of multijoined finger system, in *Proc. of the 6th Int. Joint Conf. on Artificial intelligence* (1979), pp. 693–695
18. Shadow. https://www.shadowrobot.com/products/dexterous-hand
19. M. Grebenstein, M. Chalon, M.A. Roa, C. Borst, DLR multi-fingered hands, in *Humanoid Robotics: A Reference*, ed. by A. Goswami, P. Vadakkepat (2019), pp. 481–522
20. M.G. Catalano, G. Grioli, E. Farnioli, A. Serio, C. Piazza, A. Bicchi, Adaptive synergies for the design and control of the Pisa/IIT SoftHand. Int. J. Robot. Res. **33**(5) (2014), pp. 768–782
21. S. Kajita, K. Kaneko, F. Kanehiro, K. Harada, M. Morisawa, S. Nakaoka, K. Miura, K. Fujiwara, E.S. Neo, I. Hara, I. Hara, K. Yokoi, H. Hirukawa, Cybernetic human HRP-4C: a humanoid robot with human-like proportions, in *Robotics Research, The 14th International Symposium ISRR*, ed. by C. Pradalier, R. Siegward, C. Hirzinger (Springer, Berlin/Heidelberg, 2011), pp. 301–314
22. M. Kouchi, M. Mochimaru, H. Iwasawa, S. Mitani, Anthropometric database for Japanese population 1997–98. Japanese Industrial Standards Center (AIST, MITI) (2000). http://riodb.ibase.aist.go.jp/dhbodydb

Correction to: Robotics Goes MOOC

Bruno Siciliano

**Correction to:
B. Siciliano (ed.), *Robotics Goes MOOC*,
https://doi.org/10.1007/978-3-319-75823-7**

Due to an unfortunate error, cross-references to content published in other chapters belonging to the four "Robotic Goes MOOC" books appearing in each chapter were added as links redirecting to a fake page. The wording of these sections has been updated accordingly, the links have been deleted, and the appropriate references are given in a table in the FM giving an overview of the four books, chapter abbreviations, and full chapter titles.

The updated version of this book can be found at
https://doi.org/10.1007/978-3-319-75823-7

MIX
Papier aus verantwortungsvollen Quellen
Paper from responsible sources
FSC® C105338

If you have any concerns about our products,
you can contact us on
ProductSafety@springernature.com

In case Publisher is established outside the EU,
the EU authorized representative is:
Springer Nature Customer Service Center GmbH
Europaplatz 3, 69115 Heidelberg, Germany

Printed by Libri Plureos GmbH
in Hamburg, Germany